Applications of Algebraic Geometry to Coding Theory, Physics and Computation

NATO Science Series

A Series presenting the results of scientific meetings supported under the NATO Science Programme.

The Series is published by IOS Press, Amsterdam, and Kluwer Academic Publishers in conjunction with the NATO Scientific Affairs Division

Sub-Series

I. Life and Behavioural Sciences	IOS Press
II. Mathematics, Physics and Chemistry	Kluwer Academic Publishers
III. Computer and Systems Science	IOS Press
IV. Earth and Environmental Sciences	Kluwer Academic Publishers

The NATO Science Series continues the series of books published formerly as the NATO ASI Series.

The NATO Science Programme offers support for collaboration in civil science between scientists of countries of the Euro-Atlantic Partnership Council. The types of scientific meeting generally supported are "Advanced Study Institutes" and "Advanced Research Workshops", and the NATO Science Series collects together the results of these meetings. The meetings are co-organized bij scientists from NATO countries and scientists from NATO's Partner countries – countries of the CIS and Central and Eastern Europe.

Advanced Study Institutes are high-level tutorial courses offering in-depth study of latest advances in a field.
Advanced Research Workshops are expert meetings aimed at critical assessment of a field, and identification of directions for future action.

As a consequence of the restructuring of the NATO Science Programme in 1999, the NATO Science Series was re-organized to the four sub-series noted above. Please consult the following web sites for information on previous volumes published in the Series.

http://www.nato.int/science
http://www.wkap.nl
http://www.iospress.nl
http://www.wtv-books.de/nato-pco.htm

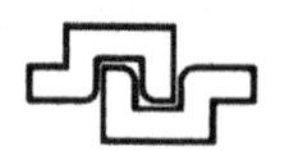

Series II: Mathematics, Physics and Chemistry – Vol. 36

Applications of Algebraic Geometry to Coding Theory, Physics and Computation

edited by

Ciro Ciliberto
Dipartimento di Matematica,
Università di Roma "Tor Vergata",
Roma, Italy

Friedrich Hirzebruch
Max Planck Institut für Mathematik,
Bonn, Germany

Rick Miranda
Department of Mathematics,
Colorado State University,
Fort Collins, CO, U.S.A.

and

Mina Teicher
Department of Mathematics and Computer Science,
Bar-Ilan University,
Ramat Gan, Israel

Kluwer Academic Publishers

Dordrecht / Boston / London

Published in cooperation with NATO Scientific Affairs Division

Proceedings of the NATO Advanced Research Workshop on
Applications of Algebraic Geometry to Coding Theory, Physics and Computation
Eilat, Israel
25th February to 1 March 2001

A C.I.P. Catalogue record for this book is available from the Library of Congress.

ISBN 1-4020-0004-9 (HB)
ISBN 1-4020-0005-7 (PB)

Published by Kluwer Academic Publishers,
P.O. Box 17, 3300 AA Dordrecht, The Netherlands.

Sold and distributed in North, Central and South America
by Kluwer Academic Publishers,
101 Philip Drive, Norwell, MA 02061, U.S.A.

In all other countries, sold and distributed
by Kluwer Academic Publishers,
P.O. Box 322, 3300 AH Dordrecht, The Netherlands.

Printed on acid-free paper

Printed in the Netherlands.

CONTENTS

PREFACE

A NATO Advanced Research Workshop on *Applications of Algebraic Geometry to Coding Theory, Physics and Computation*, was held at the Eilat Princess Hotel in Eilat, Israel, between February 25 and March 1, 2001. The purpose of the Workshop was to bring together researchers interested in significant applications of algebraic geometry, as well as in fundamental research, for a week of presentations of recent contributions and proposed new problems, as well as intensive discussions of possible collaborative efforts.

Application areas discussed explicitly during lectures of the workshop included computational algebra, singularity theory algorithms, unified field theory, numerical solutions of polynomial systems, coding theory, communication networks, and computer vision. A demonstration of the computer algebra system SINGULAR was kindly presented by Prof. G.-M. Greuel, who is one of its main developers.

Lectures on more fundamental aspects of algebraic geometry included expositions related to counting points on varieties over finite fields, Mori theory, real and pseudo-holomorphic curves, linear systems, curves on abelian varieties and abelian varieties with everywhere good reduction, elliptic genera, fundamental groups of complements of cuspidal curves, configuration spaces, and bitangent reconstruction of plane curves. Many of the lecturers have submitted written accounts of their presentations to this volume, and other related papers by several participants have also been included.

The variety of application areas, and the array of algebro-geometric techniques being utilized more and more frequently, emerged as a main theme of the Workshop. This was especially brought out at a Round Table Discussion held on the second day, at which many of the participants expressed a perception of renewed enthusiasm for algebraic geometry contributions (both from mathematicians and physicists, engineers, and medical practitioners). In addition, it was recognized that there are still many areas of science and technology which could benefit from an infusion of mathematical ideas and techniques, and that, as working mathematicians interested in applications, it is our responsibility to make an effort to bridge this gap whenever feasible. We believe that the articles in the book reflect this in large measure.

The presence of some interesting surveys makes this volume not simply a collection of papers presented at the Workshop, but also an up-to-date report on the present situation of important research topics in algebraic geometry and its applications.

The organizers and participants especially thank the financial sponsors of the Workshop: the NATO Scientific Affairs Division, the Ministry of Science of Israel, the Emmy Noether Research Institute of Mathematics at Bar-Ilan University and EAGER (the European Union Research Network in Algebraic Geometry).

The editors want to thank drs. A. Calabri, B. Kuniavskiĭ and F. Tovena for their substantial help in the organization of the workshop. Special thanks go to A. Calabri for having attended to most of the editing process of the present volume.

The Editors:
Ciro Ciliberto
Friedrich Hirzebruch
Rick Miranda
Mina Teicher

LIST OF PARTICIPANTS

NATO Advanced Research Workshop
Eilat, Israel, February 25–March 1, 2001

Herbert ABELS
abels@mathematik.uni-bielefeld.de
University of Bielefeld, Germany

Meirav AMRAM
meirav@macs.biu.ac.il
Bar-Ilan University, Israel

Marco ANDREATTA
andreatt@science.unitn.it
University of Trento, Italy

Enrico ARBARELLO
ea@mat.uniroma1.it
Univ. of Rome "La Sapienza", Italy

Tzachi BEN-ITZHAK
benitzi@macs.biu.ac.il
Bar-Ilan University, Israel

Igor I. BURBAN
burban@mathematik.uni-kl.de
University of Kaiserslautern, Germany

Alberto CALABRI
calabri@mat.uniroma2.it
Univ. of Rome "Tor Vergata", Italy

Ciro CILIBERTO
cilibert@mat.uniroma2.it
Univ. of Rome "Tor Vergata", Italy

Maurizio CORNALBA
cornalba@dimat.unipv.it
University of Pavia, Italy

Wolfram DECKER
decker@math.uni-sb.de
University of Saarlandes, Germany

Ron DONAGI
donagi@math.upenn.edu
University of Pennsylvania, U.S.A.

David GARBER
garber@macs.biu.ac.il
Bar-Ilan University, Israel

Francis O. A. GARDEYN
fgardeyn@cage.rug.ac.be
Ghent University, Belgium

Gerard van der GEER
geer@wins.uva.nl
Univ. of Amsterdam, The Netherlands

Brent GORDON
bgordon@nsf.gov
National Science Foundation, U.S.A.

Gert-Martin GREUEL
greuel@mathematik.uni-kl.de
University of Kaiserslautern, Germany

Phillip A. GRIFFITHS
pg@ias.edu
Institute for Advanced Study, U.S.A.

Klaus HULEK
hulek@math.uni-hannover.de
University of Hannover, Germany

Vasilii A. ISKOVSKIKH
iskovsk@mi.ras.ru
Steklov Institute, Moscow, Russia

Elham IZADI
izadi@math.uga.edu
University of Georgia, U.S.A.

Jeremy KAMINSKI
jeremy@cs.huji.ac.il
Bar-Ilan University, Israel

Vassil KANEV
kanev@math.unipa.it
University of Palermo, Italy

Serguei A. KOUDRIAVTSEV
koudr@mi.ras.ru
Steklov Institute, Moscow, Russia

Victor KULIKOV
kulikov@mi.ras.ru
Steklov Institute, Moscow, Russia

Boris KUNIAVSKIĬ
kunyav@macs.biu.ac.il
Bar-Ilan University, Israel

Adrian LANGER
langer@maths.warwick.ac.uk
University of Warwick, U.K.

Anatoly LIBGOBER
libgober@math.uic.edu
Univ. of Illinois at Chicago, U.S.A.

Vladimir LIN
vlin@techunix.technion.ac.il
Technion - Israel Inst. of Technology

Ron LIVNÉ
rlivne@math.huji.ac.il
The Hebrew Univ., Jerusalem, Israel

Massimiliano MELLA
mll@dns.unife.it
University of Ferrara, Italy

Rick MIRANDA
miranda@math.colostate.edu
Colorado State University, U.S.A.

Ferruccio ORECCHIA
orecchia@unina.it
Univ. of Naples “Federico II”, Italy

Claudio PEDRINI
pedrini@dima.unige.it
University of Genova, Italy

Thomas PETERNELL
Thomas.Peternell@uni-bayreuth.de
University of Bayreuth, Germany

Inna SCHERBAK
scherbak@post.tau.ac.il
Tel Aviv University, Israel

René SCHOOF
schoof@mat.uniroma2.it
Univ. of Rome "Tor Vergata", Italy

Edoardo SERNESI
sernesi@mat.uniroma3.it
University of Rome 3, Italy

Eugenii SHUSTIN
shustin@post.tau.ac.il
Tel Aviv University, Israel

Andrew J. SOMMESE
sommese@nd.edu
University of Notre Dame, U.S.A.

Balázs SZENDRŐI
balazs@maths.warwick.ac.uk
University of Warwick, U.K.

Peter TANNENBAUM
peter_tannenbaum@csufresno.edu
California State University, U.S.A.

Mina TEICHER
teicher@macs.biu.ac.il
Bar-Ilan University, Israel

Francesca TOVENA
tovena@mat.uniroma2.it
Univ. of Rome "Tor Vergata", Italy

Mohammed ULUDAG
uludam@macs.biu.ac.il
Bar-Ilan University, Israel

Jonathan M. WAHL
jmwahl@unc.edu
University of North Carolina, U.S.A.

Bronek WAJNRYB
wajnryb@techunix.technion.ac.il
Technion – Israel Inst. of Technology

Shira ZUR
zur@macs.biu.ac.il
Bar-Ilan University, Israel

LIST OF CONTRIBUTORS

Igor I. BURBAN
burban@mathematik.uni-kl.de
Fachbereich Mathematik
Universität Kaiserslautern
Erwin-Schrödinger-Straße
D-67663 Kaiserslautern, Germany

Alberto CALABRI
calabri@mat.uniroma2.it
Dipartimento di Matematica
Università di Roma "Tor Vergata"
Via della Ricerca Scientifica
I-00133 Roma, Italy

Luca CHIANTINI
chiantini@unisi.it
Dipartimento di Matematica
Università di Siena
Via de Capitano 15
I-53100 Siena, Italy

Ciro CILIBERTO
cilibert@mat.uniroma2.it
Dipartimento di Matematica
Università di Roma "Tor Vergata"
Via della Ricerca Scientifica
I-00133 Roma, Italy

Francesca CIOFFI
cioffifr@unina.it
Dip. di Matematica e Applicazioni
Università di Napoli "Federico II"
Complesso di Monte S. Angelo
Via Cintia
I-80126 Napoli, Italy

Wolfram DECKER
decker@math.uni-sb.de
Fachrichtung Mathematik
Universität des Saarlandes
66041 Saarbrücken
GERMANY

Yuriy DROZD
yuriy@drozd.org
Dept. of Mechanics and Mathematics
Kyiv Taras Shevchenko University
Volodimirska 64
252033 Kyiv, Ukraine

Michael FRYERS
fryers@math.uni-hannover.de
Institut für Mathematik
Universität Hannover
D-30167 Hannover, Germany

Gerard van der GEER
geer@science.uva.nl
Dept. of Math. and Computer Science
University of Amsterdam
Plantage Muidergracht 24
1018 TV Amsterdam, The Netherlands

Gert-Martin GREUEL
greuel@mathematik.uni-kl.de
Fachbereich Mathematik
Universität Kaiserslautern
Erwin-Schrödinger-Straße
D-67663 Kaiserslautern, Germany

Klaus HULEK
hulek@math.uni-hannover.de
Institut für Mathematik
Universität Hannover
D-30167 Hannover, Germany

Elham IZADI
izadi@math.uga.edu
Department of Mathematics
University of Georgia
Athens, GA 30602-7403
U.S.A.

Jeremy Y. KAMINSKI
kaminsj@macs.biu.ac.il
Dept. of Math. and Computer Science
Bar-Ilan University
Ramat Gan 52900, Israel

Anatoly LIBGOBER
libgober@math.uic.edu
Department of Mathematics
University of Illinois at Chicago
851 South Morgan Street
Chicago, IL 60607-7045, U.S.A.

Ron LIVNÉ
rlivne@math.huji.ac.il
Mathematics Institute
The Hebrew University of Jerusalem
Givat-Ram, Jerusalem 91904, Israel

Christoph LOSSEN
lossen@mathematik.uni-kl.de
Fachbereich Mathematik
Universität Kaiserslautern
Erwin-Schrödinger-Straße
D-67663 Kaiserslautern, Germany

Rick MIRANDA
miranda@math.colostate.edu
Department of Mathematics
Colorado State University
Fort Collins, CO 80523-1874, U.S.A.

Ferruccio ORECCHIA
orecchia@unina.it
Dip. di Matematica e Applicazioni
Università di Napoli "Federico II"
Complesso di Monte S. Angelo
Via Cintia
I-80126 Napoli, Italy

Thomas PETERNELL
Thomas.Peternell@uni-bayreuth.de
Mathematisches Institut
Universität Bayreuth
D-95440 Bayreuth, Germany

Renè SCHOOF
schoof@mat.uniroma2.it
Dipartimento di Matematica
Università di Roma "Tor Vergata"
Via della Ricerca Scientifica
I-00133 Roma, Italy

Mathias SCHULZE
mschulze@mathematik.uni-kl.de
Fachbereich Mathematik
Universität Kaiserslautern
Erwin-Schrödinger-Straße
D-67663 Kaiserslautern, Germany

Andrew J. SOMMESE
sommese@nd.edu
http://www.nd.edu/~sommese
Department of Mathematics
University of Notre Dame
Notre Dame, IN 46556, U.S.A.

Jeroen SPANDAW
spandaw@math.uni-hannover.de
Institut für Mathematik
Universität Hannover
D-30167 Hannover, Germany

Balázs SZENDRŐI
balazs@maths.warwick.ac.uk
Mathematics Institute
University of Warwick
Coventry CV4 7AL, United Kingdom

Mina TEICHER
teicher@macs.biu.ac.il
Dept. of Math. and Computer Science
Bar-Ilan University
Ramat Gan 52900, Israel

Jan VERSCHELDE
jan@math.uic.edu
http://www.math.uic.edu/~jan
Department of Mathematics
University of Illinois at Chicago
851 South Morgan (M/C 249)
Chicago, IL 60607-7045, U.S.A.

Charles W. WAMPLER
Charles.W.Wampler@gm.com
General Motors Research Laboratories
Mail Code 480-106-359
30500 Mound Road
Warren, MI 48090-9055, U.S.A.

VECTOR BUNDLES ON SINGULAR PROJECTIVE CURVES

I. BURBAN*
University of Kaiserslautern, Germany

YU. DROZD†
Kiev Taras Shevchenko University, Ukraine

G.-M. GREUEL‡
University of Kaiserslautern, Germany

Abstract. In this survey article we report on recent results known for vector bundles on singular projective curves (see (Drozd and Greuel; Drozd, Greuel and Kashuba; Yudin). We recall the description of vector bundles on tame and finite configurations of projective lines using the combinatories of matrix problems. We also show that this combinatorics allows us to compute the cohomology groups of a vector bundle, the dual bundle of a vector bundle, the tensor product of two vector bundles, the dimension of the homomorphism spaces between two vector bundles, and finally to classify simple vector bundles.

Key words: tame and wild representation type, matrix problems, vector bundles on curves.

Mathematics Subject Classification (2000): 14H60, 16G60, 16G50.

1. Introduction

Let X be a projective curve over an algebraically closed field k. For any two coherent sheaves (in particular vector bundles) $\mathcal{E}$ and $\mathcal{F}$ we have

$$\dim_k(\mathrm{Hom}(\mathcal{E}, \mathcal{F})) < \infty.$$

This implies that in the category of vector bundles the generalized Krull-Schmidt theorem holds:

$$\mathcal{E} \cong \bigoplus_{i=1}^{s} \mathcal{F}_i^{m_i},$$

* Partially supported by the DFG project "Globale Methoden in der komplexen Geometrie" and CRDF Grant UM2-2094.

† Partially supported by CRDF Grant UM2-2094.

‡ Partially supported by the DFG project "Globale Methoden in der komplexen Geometrie".

C. Ciliberto et al. (eds.),
Applications of Algebraic Geometry to Coding Theory, Physics and Computation, 1–15.

where the vector bundles $\mathcal{F}_i$ are indecomposable and m_i, $\mathcal{F}_i$ are uniquely determined.
Our aim is to describe all indecomposable vector bundles on X.

What is known about the classification of indecomposable vector bundles on smooth projective curves?

1. Let $X = \mathbb{P}^1_k$. Then indecomposable vector bundles are just the line bundles $\mathcal{O}_{\mathbb{P}^1}(n), n \in \mathbb{Z}$ (Grothendieck).
2. Let X be an elliptic curve. The indecomposable vector bundles are described by two discrete parameters r, d, rank and degree and one continuous parameter (point of the curve X), see (Atiyah).
3. It is well-known that with the growth of the genus g of the curve the moduli spaces of vector bundles become bigger and bigger. For smooth curves of genus $g \geq 2$ it was shown (Drozd and Greuel; Scharlau) that the classification problem of vector bundles is wild. "Wild" means

   a) "geometrically": we have n-parameter families of indecomposable non-isomorphic vector bundles for arbitrary large n;
   b) "algebraically": for every finite-dimensional k-algebra Λ there is an exact functor $(\Lambda\text{-mod}) \longrightarrow \mathrm{VB}_X$ from the category of Λ-modules to the category of vector bundles on X mapping non-isomorphic objects to non-isomorphic and indecomposable to indecomposable ones

   Note that the implication (b) $\Longrightarrow$ (a) is easy, while the equivalence of (a) and (b) appeared to be hard, see (Drozd and Greuel).

   Moreover, in (Drozd and Greuel) the following trichotomy was proved:

1. The category VB_X is finite (indecomposable objects are described by discrete parameters) if X is a configuration of projective lines of type A_n (in this case indecomposable vector bundles are just line bundles)

Figure 1.

2. VB_X is tame (intuitively this means that indecomposable objects are parametrized by one continuous parameter and several discrete parameters, see (Drozd and Greuel) for a precise definition) if

   a) X is an elliptic curve

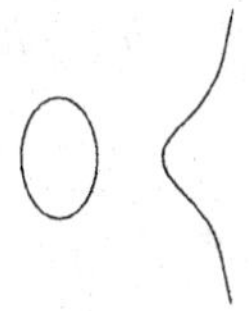

Figure 2.

b) X is a rational curve with one simple node

Figure 3.

c) X is a configuration of projective lines of type $\tilde{A}_n$

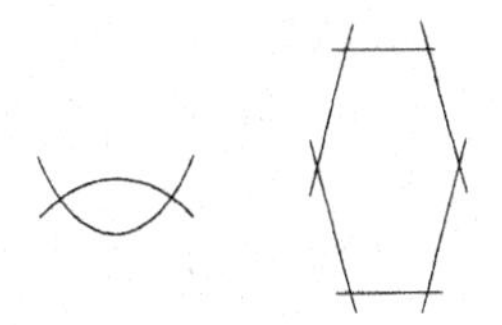

Figure 4.

3. VB_X is wild in all other cases.

2. Category of triples

Let X be a singular curve, $\tilde{X} \xrightarrow{\pi} X$ its normalization, $\tilde{\mathcal{O}} = \pi_*(\mathcal{O}_{\tilde{X}})$. Since π is affine, the categories $\mathrm{Coh}_{\tilde{X}}$ and $\mathrm{Coh}(\tilde{\mathcal{O}}\text{–mod})$ are equivalent, see (Hartshorne).

Let $\mathcal{J} = \mathrm{Ann}_{\mathcal{O}}(\tilde{\mathcal{O}}/\mathcal{O})$ be the conductor. It is the biggest common ideal sheaf of the sheaves of the rings $\mathcal{O}$ and $\tilde{\mathcal{O}}$ such that $\mathcal{J}\mathcal{O} = \mathcal{J}\tilde{\mathcal{O}}$. The usual way to deal with vector bundles on a singular curve is to lift them up to the normalization, and then work on the smooth curve. Surely we lose some information, since non-isomorphic vector bundles can have isomorphic inverse images. To avoid this problem we introduce the following definition:

DEFINITION 2.1. The category of triples, *denoted by* T_X, *is defined as follows*

1. objects are the triples $(\tilde{\mathcal{F}}, \mathcal{M}, i)$, *where*
$\tilde{\mathcal{F}}$ is a locally free $\tilde{\mathcal{O}}$-module,
$\mathcal{M}$ is a locally free $\mathcal{O}/\mathcal{J}$-module and

$i: \mathcal{M} \longrightarrow \tilde{\mathcal{F}} \otimes_{\tilde{\mathcal{O}}} \tilde{\mathcal{O}}/\mathcal{J}$ is an inclusion of $\mathcal{O}/\mathcal{J}$-modules, which induces an isomorphism
$\tilde{i}: \mathcal{M} \otimes_{\mathcal{O}} \tilde{\mathcal{O}}/\mathcal{J} \longrightarrow \tilde{\mathcal{F}}/\mathcal{J}\tilde{\mathcal{F}}$.

2. *morphisms $(\tilde{\mathcal{F}}_1, \mathcal{M}_1, i_1) \xrightarrow{(\Phi,\varphi)} (\tilde{\mathcal{F}}_2, \mathcal{M}_1, i_2)$ are the pairs (Φ, φ), with*
   $\tilde{\mathcal{F}}_1 \xrightarrow{\Phi} \tilde{\mathcal{F}}_2$ a morphism of $\tilde{\mathcal{O}}$-modules,
   $\mathcal{M}_1 \xrightarrow{\varphi} \mathcal{M}_2$ a morphism of $\mathcal{O}/\mathcal{J}$-modules, such that the following diagram is commutative

$$\begin{array}{ccccc} \tilde{\mathcal{F}}_1 & \longrightarrow & \tilde{\mathcal{F}}_1 \otimes_{\tilde{\mathcal{O}}} \tilde{\mathcal{O}}/\mathcal{J} & \xleftarrow{i_1} & \mathcal{M}_1 \\ \downarrow \Phi & & \downarrow \bar{\Phi} & & \downarrow \varphi \\ \tilde{\mathcal{F}}_2 & \longrightarrow & \tilde{\mathcal{F}}_2 \otimes_{\tilde{\mathcal{O}}} \tilde{\mathcal{O}}/\mathcal{J} & \xleftarrow[i_2]{} & \mathcal{M}_2 . \end{array}$$

Now we formulate the main theorem of this section

THEOREM 2.2. *The functor*

$$\mathrm{VB}_X \xrightarrow{\mathbf{F}} \mathrm{T}_X$$
$$\mathcal{F} \longrightarrow (\tilde{\mathcal{F}}, \mathcal{M}, i),$$

where $\tilde{\mathcal{F}} := \mathcal{F}_{\otimes_{\mathcal{O}}} \tilde{\mathcal{O}}$, $\mathcal{M} := \mathcal{F}/\mathcal{J}\mathcal{F}$, and $i: \mathcal{F}/\mathcal{J}\mathcal{F} \longrightarrow \mathcal{F} \otimes_{\mathcal{O}} \tilde{\mathcal{O}}/\mathcal{J}(\mathcal{F} \otimes_{\mathcal{O}} \tilde{\mathcal{O}})$ is an equivalence of categories. $\tilde{\mathcal{F}}$ is called the normalization *of $\mathcal{F}$.*

PROOF. We construct the quasi-inverse functor

$$\mathrm{T}_X \xrightarrow{\mathbf{G}} \mathrm{VB}_X$$

as follows. Let $(\tilde{\mathcal{F}}, \mathcal{M}, i)$ be some triple. Consider the pull-back diagram

$$\begin{array}{ccccccccc} 0 & \longrightarrow & \mathcal{J}\tilde{\mathcal{F}} & \longrightarrow & \mathcal{F} & \longrightarrow & \mathcal{M} & \longrightarrow & 0 \\ & & \downarrow id & & \downarrow & & \downarrow i & & \\ 0 & \longrightarrow & \mathcal{J}\tilde{\mathcal{F}} & \longrightarrow & \tilde{\mathcal{F}} & \xrightarrow{\pi} & \tilde{\mathcal{F}}/\mathcal{J}\tilde{\mathcal{F}} & \longrightarrow & 0 \end{array}$$

in the category of $\mathcal{O}$-modules. Since the pull-back is functorial, we get a functor $\mathrm{T}_X \xrightarrow{\mathbf{G}} \mathrm{Coh}_X$. One has to show that

1. the pull-back of π and i is a vector bundle
2. the functors Φ and Ψ are quasi-inverse.

We refer to (Drozd and Greuel) for details of the proof. □

3. Vector bundles on a rational curve with one node

Since the idea to reduce the classification of vector bundles on singular curves to a matrix problem seems to be new and since the technique of matrix problems appear to be unfamiliar to algebraic geometers and, moreover, since the general procedure is not so easy to understand, we consider in this section a rather simple case. We treat it in some detail in order to clarify the ideas.

Let X be a plane curve, given by the equation $zy^2 - x^3 - zx^2 = 0$. Then its normalization is $\mathbb{P}^1 = \tilde{X} \xrightarrow{\pi} X$. Without loss of generality we may suppose that the pre-images of the singular point are $0 = (0:1)$ and $\infty = (1:0)$.

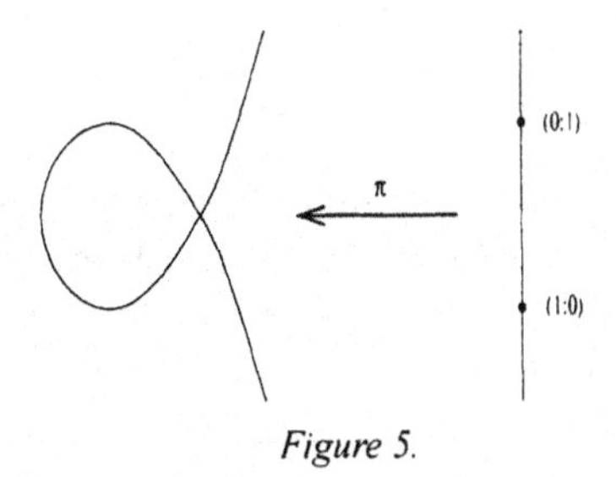

Figure 5.

What does the result of the previous section mean? A vector bundle $\mathcal{F}$ on the curve X is uniquely determined by some triple $(\tilde{\mathcal{F}}, \mathcal{M}, i)$. $\tilde{\mathcal{F}}$ is a locally free $\tilde{\mathcal{O}}$-module, or equivalently, a locally free $\mathcal{O}_{\mathbb{P}^1}$-module. By the theorem of Grothendieck $\tilde{\mathcal{F}} \cong \bigoplus_{n \in Z} \tilde{\mathcal{O}}(n)^{m_n}$.

Since $\mathcal{O}/\mathcal{J} = k$, $\mathcal{M}_0$ is nothing but a k-vector space and $i : \mathcal{M}_0 \longrightarrow (\tilde{\mathcal{F}}/\mathcal{J})_0$ can be viewed as a k-linear map. But $(\tilde{\mathcal{F}}/\mathcal{J})_0$ is a $k \times k$-module, hence the map i is given by two matrices $i(0:1)$ and $i(1:0)$. The canonical map $(\mathcal{O}/\mathcal{J})_0 \longrightarrow (\tilde{\mathcal{O}}/\mathcal{J})_0$ is the diagonal map $k \longrightarrow k \times k$, so the condition that $\tilde{i} : \mathcal{M} \otimes_{\mathcal{O}/\mathcal{J}} \tilde{\mathcal{O}}/\mathcal{J} \longrightarrow \tilde{\mathcal{F}}/\mathcal{J}$ is an isomorphism means that both matrices $i(0:1)$ and $i(1:0)$ are invertible square matrices.

Fix a direct sum decomposition $\tilde{\mathcal{F}} \cong \bigoplus_{n \in Z} \tilde{\mathcal{O}}(n)^{m_n}$ and choose trivializations of $\tilde{\mathcal{F}}$ at the points 0 and ∞. They induce a basis of the $k \times k$-module $\tilde{\mathcal{F}}/\mathcal{J}\tilde{\mathcal{F}}$. Choose also a basis of $\mathcal{M}_0$. With respect to these choices i is given by some matrix, divided into horizontal blocks, as in Figure 6.

Now we have to answer the main question: when do two triples $(\tilde{\mathcal{F}}_1, \mathcal{M}_1, i_1)$ and $(\tilde{\mathcal{F}}_2, \mathcal{M}_2, i_2)$ define isomorphic vector bundles? Surely, we have to require $\tilde{\mathcal{F}}_1 \cong \tilde{\mathcal{F}}_2$, $\mathcal{M}_1 \cong \mathcal{M}_2$. But what condition should be satisfied by the matrices defining i_1 and i_2 in order to give isomorphic vector bundles? The answer follows from the definition of the morphism in the category of triples. Namely, there should be isomorphisms $\Phi : \tilde{\mathcal{F}}_1 \longrightarrow \tilde{\mathcal{F}}_2$, $\varphi : \mathcal{M}_1 \longrightarrow \mathcal{M}_2$ such that $\bar{\Phi} i_1 = i_2 \varphi$.

Let $\tilde{\mathcal{F}} = \bigoplus_{n \in Z} \tilde{\mathcal{O}}(n)^{m_n}$. An endomorphism of $\tilde{\mathcal{F}}$ can be written in a matrix form: $\Phi = (\Phi)_{ij}$, where Φ_{ij} is a $n_j \times n_i$-matrix with coefficients in the vector space $\mathrm{Hom}(\tilde{\mathcal{O}}(j), \tilde{\mathcal{O}}(i))$. But since $\mathrm{Hom}(\tilde{\mathcal{O}}(i), \tilde{\mathcal{O}}(j)) = k[x_0, x_1]_{i-j}$, our matrix is lower

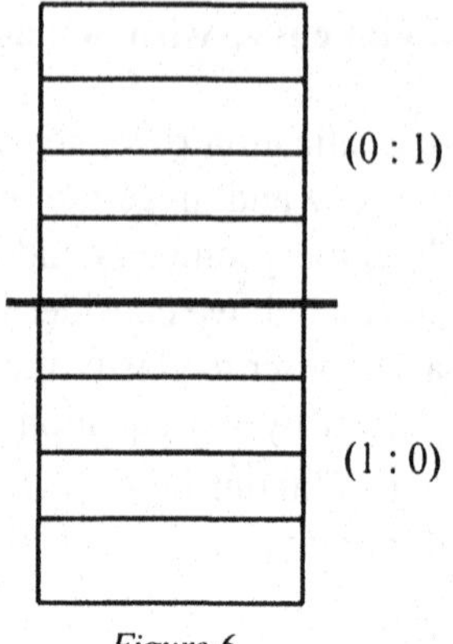

Figure 6.

triangular. Moreover, the diagonal $n_i \times n_i$-matrices Φ_{ii} are just matrices over k. Φ is an isomorphism if and only if all Φ_{ii} are invertible. What is a map $\bar{\Phi} : \tilde{\mathcal{F}}/\mathcal{J} \longrightarrow \tilde{\mathcal{F}}/\mathcal{J}$? Let $N = \mathrm{rank}(\mathcal{F})$. Then $\bar{\Phi} : k^{2N} \longrightarrow k^{2N}$ is given by the diagonal block matrix $\mathrm{diag}(\Phi(0:1), \Phi(1:0))$. Note, that the matrices $\Phi_{ij}(1:0)$ and $\Phi_{ij}(0:1)$, $(i > j)$ can be *arbitrary*. As a result we get a matrix problem:

We have two matrices $i(1:0)$ and $i(0:1)$. We require them to be square and nondegenerate. Each of them is divided into horizontal blocks labeled by integer numbers (they are called sometimes weights). Blocks of $i(0:1)$ and $i(1:0)$, labeled by the same integer, have to have the same size. We can perform the following transformations:

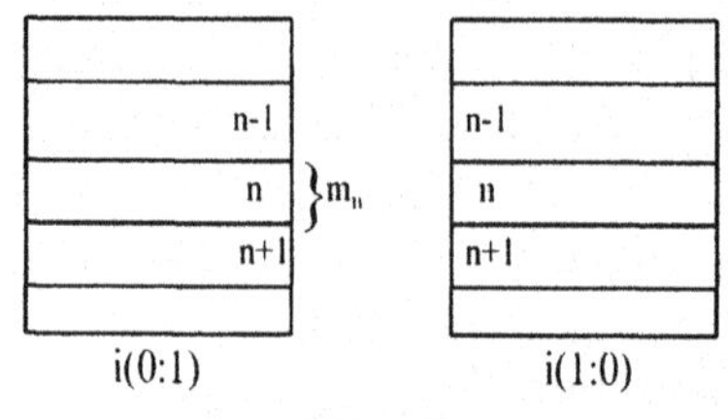

Figure 7.

1. We can simultaneously do any elementary transformations of columns of $i(0:1)$ and $i(1:0)$.
2. We can simultaneously do any invertible elementary transformations of rows inside of the conjugated horisontal blocks.
3. We can independentely add in each of the matrices $i(0:1)$ and $i(1:0)$ a scalar multiple of any row with lower weight to any row with higher weight.

These types of matrix problems are well-known in representation theory. First they appeared in the work of Nazarova and Roiter (1969) about the classification of $k[[x,y]]/(xy)$-modules. They are sometimes called Gelfand problems in honour of I. M. Gelfand, who formulated a conjecture at the International Congress of Mathematics in Nice (1970) about the structure of Harish-Chandra modules at the singular point of $SL_2(\mathbb{R})$. This problem was reduced to some matrix problem of this type in (Nazarova and Roiter, 1973). The idea to apply this technique is that we can write the matrix i in some canonical form which is quite analogous to the Jordan normal form.

EXAMPLE 3.1. *The following data define an indecomposable vector bundle of rank 2 on X: the normalization $\tilde{\mathcal{O}} \oplus \mathcal{O}(n), n \neq 0$, together with the matrices shown in figure 8.*

$$\begin{array}{|c|c|l} 1 & & \mathbf{n} \\ \hline & 1 & 0 \end{array} \qquad \begin{array}{|c|c|l} & \lambda & \mathbf{n} \\ \hline 1 & & 0 \end{array}$$

Figure 8.

A Gelfand matrix problem can be coded by some partially ordered set, together with some equivalence relation on it. For example, the problem of classifying vector bundles on a rational curve with one node, corresponds to the following partially ordered set. There are two infinite sets $E_0 = \{E_0(i) | i \in \mathbb{Z}\}$ and $E_\infty = \{E_\infty(i) | i \in \mathbb{Z}\}$ with total order induced by the order on $\mathbb{Z}$, and two one-point sets F_0 and F_∞. On the set

$$\mathbf{E} \bigcup \mathbf{F} = (E_0 \cup E_\infty) \bigcup (F_0 \cup F_\infty)$$

we introduce an equivalence relation: $E_0(i) \sim E_\infty(i), i \in \mathbb{Z}, F_0 \sim F_\infty$.

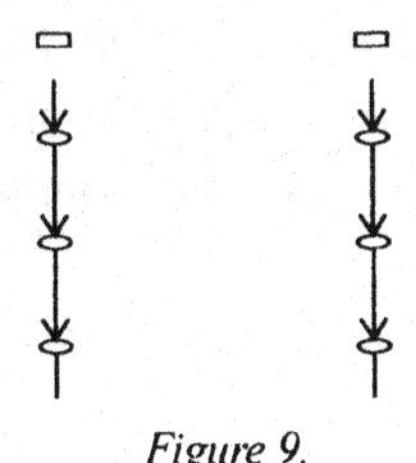

Figure 9.

From this picture we can easily recover the corresponding matrix problem: $E_0 \cup F_0$ and $E_\infty \cup F_\infty$ correspond to $i(0:1)$ and $i(1:0)$ respectively. $F_0 \sim F_\infty$ means that we have to do elementary transformations with columns of $i(0:1)$ and $i(1:0)$ simultaneously, the partial order on E_0 and E_∞ implies the division of matrices into horizontal blocks, where each block has some weight. $E_0(i) \sim E_\infty(i), i \in \mathbb{Z}$, means the conjugation of blocks: they have the same number of rows and elementary transformations inside of them should be done simultaneously.

Let now X be a configuration of projective lines of type A_n or $\tilde{A}_n$. We can proceed with constructing the category of triples in the similar way as we have done for a rational curve with one node. The matrix problem we get is coded by the Bondarenko's partially ordered set as follows: consider the set of pairs $\{(L,a)\}$, where L is an irreducible component of X, $a \in L$ a singular point. To each such pair corresponds a totally ordered set $E_{(L,a)} = \{E_{(L,a)}(i) | i \in \mathbb{Z}\}$ and a one-point set $F_{(L,a)}$. On the set

$$\mathbf{E} \bigcup \mathbf{F} = \bigcup_{(L,a)} (E_{(L,a)} \cup F_{(L,a)})$$

we introduce an equivalence relation:

1. $F_{(L,a)} \sim F_{(L',a)}$,
2. $E_{(L,a)}(i) \sim F_{(L,a')}(i), i \in \mathbb{Z}$.

This means that we have a set of matrices $M(L,a)$, where (L,a) runs through all possible pairs $(L,a), a \in L$, and each of the matrices is divided into horizontal blocks with respect to the partial order on $E_{(L,a)}$. The principle of conjugation of blocks is the same as for a rational curve with one node.

What is the combinatorics of indecomposable objects in this case? A Gelfand problem has two types of indecomposable objects: *bands* and *strings* (see (Bondarenko, 1992) and Appendix A in (Drozd and Greuel)). If X is a configuration of projective lines of type A_n, then each indecomposable vector bundle has to be a line bundle. Let X be either a rational curve with one node or a configuration of projective lines of type $\tilde{A}_n$. The condition on matrices $M(L,a)$ to be square and nondegenerate implies that vector bundles correspond to band representations.

DEFINITION 3.2. *Let X be either a rational projective curve with one node or a cycle of s projective lines. Let $\{a_1, a_2, \ldots, a_s\}$ be the set of singular points of X, $\tilde{X} \xrightarrow{\pi} X$ the normalization of X, i.e. $\tilde{X}$ is the disjoint union of s copies $L_1, \ldots, L_s$ of the projective line, and $\{a'_i, a''_i\} = \pi^{-1}(a_i)$. Suppose that $a'_i, a''_{i+1} \in L_i$, where $a''_{s+1} = a''_1$.*

A band $\mathcal{B}(\mathbf{d}, m, \lambda) = (\tilde{\mathcal{B}}, \mathcal{M}, i)$ is defined by the following parameters:

1. *$\mathbf{d} = d_1 d_2 \ldots d_s d_{s+1} d_{s+2} \ldots d_{2s} \ldots d_{rs-s+1} d_{rs-s+2} \ldots d_{rs}$ is a sequence of degrees on the normalized curve $\tilde{X}$. This sequence should not be of the form $\mathbf{e}^l$, where $\mathbf{e}$ is another sequence (i. e. $\mathbf{d}$ is not a self-concatenation of some other sequence).*
2. *m is the size of the elementary block of the matrix defining the glueing. The first two properties mean that the resriction of the normalized vector bundle on the l-th component of $\tilde{X}$ is*

$$\bigoplus_{i=1}^{r} \mathcal{O}_{L_l}(d_{l+is})^m.$$

3. *$\lambda \in k^*$ is a continuous parameter.*

We have 2s matrices $M(L_i,a'_i)$ and $M(L_i,a''_{i+1}), i=1,\ldots,s$, occurring in the triple, corresponding to $\mathcal{B}(\mathbf{d},m,\lambda)$. Each of them has size $mr \times mr$. Divide these matrices into $m \times m$ square blocks.

Consider then sequences $\mathbf{d}(i) = d_i d_{i+s} \ldots d_{i+(r-1)s}$ and label the horizontal strips of $M(L_i,a'_i)$ and $M(L_i,a''_{i+1})$ with respect to occurence of integers in the sequence $\mathbf{d}(i)$. If some integer d occured k times in $\mathbf{d}(i)$ then the horizontal strip corresponding to d consists of k substrips with m rows each. Recall now an algorithm of writing the components of the matrix i in normal form:

1. *Write a sequence*

$$(L_1,a''_2) \xrightarrow{1} (L_2,a'_2) \xrightarrow{1} (L_2,a''_3) \xrightarrow{1} \ldots$$
$$\xrightarrow{1} (L_1,a'_1) \xrightarrow{1} (L_1,a''_1) \xrightarrow{2} \cdots \xrightarrow{r} (L_1,a'_1).$$

2. *We unroll the sequence $\mathbf{d}$. This means that we write over each (L_i,a) the corresponding term of the subsequence $\mathbf{d}(i)$ together with the number as often as this term already occured in $\mathbf{d}(i)$:*

$$(L_1,a''_2)^{(d_1,1)} \xrightarrow{1} (L_2,a'_2)^{(d_2,1)} \xrightarrow{1} (L_2,a''_3)^{(d_2,1)} \xrightarrow{1} \ldots$$
$$\xrightarrow{r} (L_s,a''_1)^{(d_{rs},*)} \xrightarrow{r} (L_1,a'_1)^{(d_1,1)}.$$

3. *We are ready to fill the entries of the matrices $M(L,a)$:*
   a) *Consider each arrow above: $(L,a)^{(d,i)} \xrightarrow{k}$. Then we put the matrix I_m in the block $((d,i),k)$ of the matrix $M(L,a)$ (the block which is an intersection of the i-th substrip of the horizontal strip with label d' and the k-th vertical strip).*
   b) *We put on the $((d_1,1),r)$-th place of $M(L_1,a'_1)$ the Jordan block $J_m(\lambda)$ (our continous parameter λ appears at this moment).*

EXAMPLE 3.3. *Let X be a union of two projective lines intersecting transversally it two different points ($\tilde{A}_1$-configuration), $d = 01131\text{–}2$. Then $\mathcal{B}(\mathbf{d},m,\lambda)$ is a vector bundle of rank $3m$ given by its normalization*

$$(\mathcal{O}^m_{L_1} \oplus \mathcal{O}_{L_1}(1)^{2m}) \oplus (\mathcal{O}_{L_2}(-2)^m \oplus \mathcal{O}_{L_2}(1)^m \oplus \mathcal{O}_{L_2}(3)^m)$$

and the matrices in figure 10.

EXAMPLE 3.4. *A vector bundle from example 3.1 is just $\mathcal{B}(0n,1,\lambda)$.*

REMARK 3.5. *One can ask the following question: if bands corresponds to vector bundles, what do correspond to strings that are degenerated series in the Gelfand problem? As one can guess, some of the strings correspond to torsion-free sheaves which are not locally free. For this purpose one should modify a little bit the*

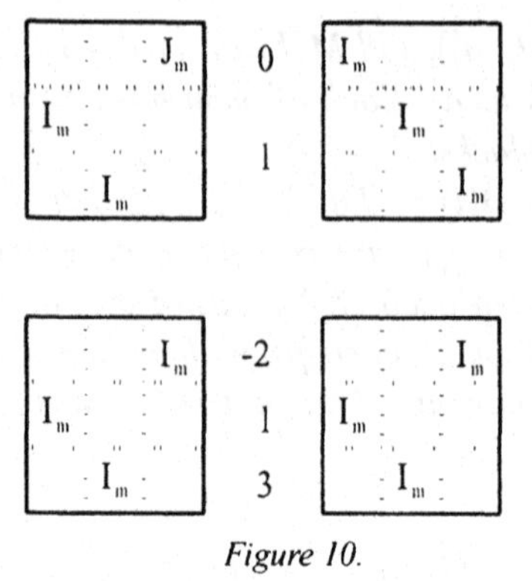

Figure 10.

definition of the category of triples. For example, if X is a rational curve with one node, then the triple $(\tilde{\mathcal{O}}_{\mathbb{P}^1}, k^2, i)$, where i is given by two matrices

$$0 \begin{array}{|c:c|} \hline 1 & 0 \\ \hline \end{array} \qquad\qquad 0 \begin{array}{|c:c|} \hline 0 & 1 \\ \hline \end{array}$$

defines $\tilde{\mathcal{O}} = \pi_*(\mathcal{O}_{\tilde{X}})$.

In what follows we shall see that this way of representing of vector bundles on rational curves is indeed a convenient one. It allows in particular to determine the dual bundle, the decomposition of tensor products and the computation of cohomology groups.

4. Dual vector bundle

THEOREM 4.1. *Let X be either a projective curve with one node or a configuration of projective lines of type $\tilde{A}_n$, let $\mathcal{B} = \mathcal{B}(\mathbf{d}, m, \lambda)$ be a vector bundle on X. Then $\mathcal{B}^\vee \cong \mathcal{B}(-\mathbf{d}, m, \lambda^{-1})$.*

PROOF. Let $\mathcal{B}$ be a vector bundle on X. By the adjoint property we have:

$$\mathcal{H}om_{\tilde{\mathcal{O}}}(\mathcal{B} \otimes_{\mathcal{O}} \tilde{\mathcal{O}}, \tilde{\mathcal{O}}) \cong \mathcal{H}om_{\mathcal{O}}(\mathcal{B}, \mathcal{H}om_{\tilde{\mathcal{O}}}(\tilde{\mathcal{O}}, \tilde{\mathcal{O}})) \cong \mathcal{H}om_{\mathcal{O}}(\mathcal{B}, \tilde{\mathcal{O}}).$$

But we have a canonical map $\mathcal{H}om_{\mathcal{O}}(\mathcal{B}, \mathcal{O}) \otimes_{\mathcal{O}} \tilde{\mathcal{O}} \longrightarrow \mathcal{H}om_{\mathcal{O}}(\mathcal{B}, \tilde{\mathcal{O}})$, which is an isomorphism in case $\mathcal{B}$ is locally-free. Hence, if the normalization of $\mathcal{B}$ is $\tilde{\mathcal{B}}$ then the normalization of $\mathcal{B}^\vee$ is $\tilde{\mathcal{B}}^\vee$. Now, let $\mathcal{B}$ be given by a triple $(\tilde{\mathcal{B}}, \mathcal{M}, i)$. Then we have a commutative diagram

$$\begin{array}{ccccccccc} 0 & \longrightarrow & \mathcal{J}\tilde{\mathcal{B}} & \longrightarrow & \mathcal{B} & \longrightarrow & \mathcal{M} & \longrightarrow & 0 \\ & & \Big\downarrow{\scriptstyle id} & & \Big\downarrow & & \Big\downarrow{\scriptstyle i} & & \\ 0 & \longrightarrow & \mathcal{J}\tilde{\mathcal{B}} & \longrightarrow & \tilde{\mathcal{B}} & \xrightarrow{\pi} & \tilde{\mathcal{B}}/\mathcal{J}\tilde{\mathcal{B}} & \longrightarrow & 0\,. \end{array}$$

Apply the functor $\mathcal{H}om_{\mathcal{O}}(\ \ ,\mathcal{O})$ to obtain the glueing matrices of $\mathcal{H}om_{\mathcal{O}}(\mathcal{B},\mathcal{O})$. Then we get

$$\begin{array}{ccccccc} 0 \longrightarrow \mathcal{H}om_{\mathcal{O}}(\tilde{\mathcal{B}},\mathcal{O}) & \longrightarrow & \mathcal{H}om_{\mathcal{O}}(\mathcal{J}\tilde{\mathcal{B}},\mathcal{O}) & \longrightarrow & \mathcal{E}xt^1_{\mathcal{O}}(\tilde{\mathcal{B}}/\mathcal{J}\tilde{\mathcal{B}},\mathcal{O}) & \longrightarrow & \mathcal{E}xt^1_{\mathcal{O}}(\tilde{\mathcal{B}},\mathcal{O}) \\ \uparrow & & \uparrow & & \uparrow & & \uparrow \\ 0 \longrightarrow \mathcal{H}om_{\mathcal{O}}(\tilde{\mathcal{B}},\mathcal{O}) & \longrightarrow & \mathcal{H}om_{\mathcal{O}}(\mathcal{B},\mathcal{O}) & \longrightarrow & \mathcal{E}xt^1_{\mathcal{O}}(\tilde{\mathcal{B}}/\mathcal{B},\mathcal{O}) & \longrightarrow & \mathcal{E}xt^1_{\mathcal{O}}(\tilde{\mathcal{B}},\mathcal{O}) \end{array}$$

But since X has nodal singularities, $\mathcal{E}xt^1_{\mathcal{O}}(\tilde{\mathcal{O}},\mathcal{O}) = 0$, and we get, as a corollary, $\mathcal{E}xt^1_{\mathcal{O}}(\tilde{\mathcal{B}},\mathcal{O}) = 0$. Moreover, since our curve X is Gorenstein, $\mathcal{K}/\mathcal{O}$ is an injective $\mathcal{O}$-module ($\mathcal{K}$ denotes a sheaf of rational functions). Hence,

$$0 \longrightarrow \mathcal{O} \longrightarrow \mathcal{K} \longrightarrow \mathcal{K}/\mathcal{O} \longrightarrow 0$$

is an injective resolution of $\mathcal{O}$. If $\mathcal{N}$ is a skyscraper sheaf, then $\mathcal{H}om_{\mathcal{O}}(\mathcal{N},\mathcal{K}) = 0$ and $\mathcal{E}xt^1_{\mathcal{O}}(\mathcal{N},\mathcal{O})$ is naturally isomorphic to $\mathcal{H}om_{\mathcal{O}}(\mathcal{N},\mathcal{K}/\mathcal{O})$. So, we have a commutative diagram

$$\begin{array}{ccc} \mathcal{E}xt^1_{\mathcal{O}}(\tilde{\mathcal{B}}/\mathcal{J}\tilde{\mathcal{B}},\mathcal{O}) & \xrightarrow{\cong} & \mathcal{H}om_{\mathcal{O}}(\tilde{\mathcal{B}}/\mathcal{J}\tilde{\mathcal{B}},\tilde{\mathcal{O}}/\mathcal{O}) \\ \uparrow & & \uparrow \\ \mathcal{E}xt^1_{\mathcal{O}}(\tilde{\mathcal{B}}/\mathcal{B},\mathcal{O}) & \xrightarrow{\cong} & \mathcal{H}om_{\mathcal{O}}(\tilde{\mathcal{B}}/\mathcal{B},\tilde{\mathcal{O}}/\mathcal{O}) \end{array}$$

where $\mathcal{H}om_{\mathcal{O}}(\tilde{\mathcal{B}}/\mathcal{B},\tilde{\mathcal{O}}/\mathcal{O}) \longrightarrow \mathcal{H}om_{\mathcal{O}}(\tilde{\mathcal{B}}/\mathcal{J}\tilde{\mathcal{B}},\tilde{\mathcal{O}}/\mathcal{O})$ is the map induced from the exact sequence

$$0 \longrightarrow \mathcal{M} \xrightarrow{i} \tilde{\mathcal{B}}/\mathcal{J} \longrightarrow \tilde{\mathcal{B}}/\mathcal{B} \longrightarrow 0.$$

Let us see what are $\mathcal{H}om_{\mathcal{O}}(\tilde{\mathcal{B}},\mathcal{O})$ and $\mathcal{H}om_{\mathcal{O}}(\mathcal{J}\tilde{\mathcal{B}},\mathcal{O})$.

1. We have canonical $\mathcal{O}$-module homomorphisms

$$\mathcal{J}\mathcal{H}om_{\tilde{\mathcal{O}}}(\tilde{\mathcal{B}},\tilde{\mathcal{O}}) \longrightarrow \mathcal{H}om_{\tilde{\mathcal{O}}}(\tilde{\mathcal{B}},\mathcal{J}) \longrightarrow \mathcal{H}om_{\mathcal{O}}(\tilde{\mathcal{B}},\mathcal{O}).$$

   Since all modules occurring in the sequence are coherent $\mathcal{O}$-modules, we can check that these homomorphisms are isomorphisms just looking at stalks. For the stalks however, this is true since all singular points of X are nodes.
2. We have canonical $\mathcal{O}$-module homomorphisms

$$\mathcal{H}om_{\mathcal{O}}(\mathcal{J}\tilde{\mathcal{B}},\mathcal{O}) \longrightarrow \mathcal{H}om_{\tilde{\mathcal{O}}}(\mathcal{J}\tilde{\mathcal{B}},\mathcal{J}) \longleftarrow \mathcal{H}om_{\tilde{\mathcal{O}}}(\tilde{\mathcal{B}},\tilde{\mathcal{O}})$$

   which are isomorphisms on stalks.

Finally we get a commutative diagram:

$$\begin{array}{ccccccc} 0 \longrightarrow \mathcal{J}\mathcal{H}om_{\tilde{\mathcal{O}}}(\tilde{\mathcal{B}},\tilde{\mathcal{O}}) & \longrightarrow & \mathcal{H}om_{\tilde{\mathcal{O}}}(\tilde{\mathcal{B}},\tilde{\mathcal{O}}) & \longrightarrow & \mathcal{H}om_{\mathcal{O}}(\tilde{\mathcal{B}}/\mathcal{J}\tilde{\mathcal{B}},\tilde{\mathcal{O}}/\mathcal{O}) & \longrightarrow & 0 \\ \uparrow{\scriptstyle id} & & \uparrow & & \uparrow & & \\ 0 \longrightarrow \mathcal{J}\mathcal{H}om_{\tilde{\mathcal{O}}}(\tilde{\mathcal{B}},\tilde{\mathcal{O}}) & \longrightarrow & \mathcal{H}om_{\mathcal{O}}(\mathcal{B},\mathcal{O}) & \longrightarrow & \mathcal{H}om_{\mathcal{O}}(\tilde{\mathcal{B}}/\mathcal{B},\tilde{\mathcal{O}}/\mathcal{O}) & \longrightarrow & 0. \end{array}$$

But this is a diagram we are looking for. It remains only to describe a map giving an inclusion of the kernel of the map of $\mathcal{O}/\mathcal{J}$-modules

$$\mathcal{H}om_{\mathcal{O}/\mathcal{J}}(\tilde{\mathcal{B}}/\mathcal{B},\tilde{\mathcal{O}}/\mathcal{O}) \xrightarrow{i^*} \mathcal{H}om_{\mathcal{O}/\mathcal{J}}(\mathcal{M},\tilde{\mathcal{O}}/\mathcal{O}).$$

Let x be a singular point, $\{u_1,u_2,\dots,u_n\}$ a basis of $\mathcal{M}_x$, $\{v_1,v_2,\dots,v_n;w_1,w_2,\dots,w_n\}$ a basis of $(\tilde{\mathcal{B}}/\mathcal{J}\tilde{\mathcal{B}})_x$ and i_x is given by

$$\begin{cases} i_x(v_1) &= \sum a_{i1}v_i + \sum b_{i1}w_i \\ i_x(v_2) &= \sum a_{i2}v_i + \sum b_{i2}w_i \\ &\vdots \\ i_x(v_n) &= \sum a_{in}v_i + \sum b_{in}w_i. \end{cases}$$

Let $(\tilde{\mathcal{O}}/\mathcal{J})_x = \langle\alpha,\beta\rangle$, $(\mathcal{O}/\mathcal{J})_x = \langle\gamma\rangle$. Since $\tilde{\mathcal{O}}/\mathcal{J}$ is supplied with an $\mathcal{O}/\mathcal{J}$-module structure by the diagonal mapping, $(\tilde{\mathcal{O}}/\mathcal{O})_x = \langle\alpha,\beta\rangle/\langle\gamma,\gamma\rangle$. So we may suppose that the isomorphism $(\tilde{\mathcal{O}}/\mathcal{O})_x \longrightarrow k$ is given by $[\alpha]\mapsto 1$, $[\beta]\mapsto -1$.

The space $\mathcal{H}om_{\mathcal{O}/\mathcal{J}}(\tilde{\mathcal{B}}/\mathcal{B},\tilde{\mathcal{O}}/\mathcal{O})$ has a basis v_1^*, v_2^*, …, v_n^*; w_1^*, w_2^*, …, w_n^*, where $v_i^*(v_j) = \delta_{ij}[\alpha]$, $v_i^*(w_j) = 0$, $w_i^*(v_j) = 0$, $w_i^*(w_j) = \delta_{ij}[\beta]$. Therefore we get

$$i_x^*(v_i^*)(u_j) = a_{ij}[\alpha] = a_{ij}$$

and on the other hand

$$i_x^*(w_i^*)(u_j) = b_{ij}[\beta] = -b_{ij}.$$

So, if i_x was $(\frac{A}{B})$, then i_x^* is given by $(A^T,-B^T)$. Now we may suppose that the matrix i has a canonical form. Then we can easily compute the matrix giving an embedding of the kernel of i^*. As a corollary we get the claim of the theorem. □

5. Cohomology groups, tensor products and homomorphism spaces

5.1. COMPUTATION OF COHOMOLOGY GROUPS OF VECTOR BUNDLES

Let $\mathcal{B} = \mathcal{B}(\mathbf{d},m,\lambda)$. The developed technique allows us to compute the cohomology groups of $\mathcal{B}$ in terms of the combinatorics of the sequence $\mathbf{d}$. Let $(\tilde{\mathcal{B}},\mathcal{M},i)$ be a triple corresponding to $\mathcal{B}$. We have an exact sequence

$$0 \longrightarrow \mathcal{B} \longrightarrow \tilde{\mathcal{B}} \longrightarrow \tilde{\mathcal{B}}/\mathcal{B} \longrightarrow 0.$$

After taking cohomology we get the long exact sequence:

$$0 \longrightarrow H^0(\mathcal{B}) \longrightarrow H^0(\tilde{\mathcal{B}}) \xrightarrow{0} H^0(\tilde{\mathcal{B}}/\mathcal{B}) \longrightarrow H^1(\mathcal{B}) \longrightarrow H^1(\tilde{\mathcal{B}}).$$

The map $f: H^0(\tilde{\mathcal{B}}) \longrightarrow H^0(\tilde{\mathcal{B}}/\mathcal{B})$ can be computed explicitly; it is just the composition

$$H^0(\tilde{\mathcal{B}}) \longrightarrow H^0(\tilde{\mathcal{B}}/\mathcal{J}\tilde{\mathcal{B}}) \longrightarrow H^0(\tilde{\mathcal{B}}/\mathcal{J}\tilde{\mathcal{B}})/H^0(\mathcal{B}/\mathcal{J}\mathcal{B}).$$

But $\mathcal{B}/\mathcal{J}\mathcal{B} \cong \mathcal{M}$, and the embedding $H^0(\mathcal{B}/\mathcal{J}\mathcal{B}) \longrightarrow H^0(\tilde{\mathcal{B}}/\mathcal{J}\tilde{\mathcal{B}})$ is given by the matrix i. So we can compute cohomology groups as the kernel and cokernel of the map f. We refer to (Drozd, Greuel and Kashuba) for more details and just give the result:

$$\dim_k H^0(\mathcal{E}) = m\left(\sum_{i=1}^{rs}(d_i+1)^+ - \theta(\mathbf{d})\right) + \delta(\mathbf{d},\lambda),$$

$$\dim_k H^1(\mathcal{E}) = m\left(\sum_{i=1}^{rs}(d_i+1)^- + rs - \theta(\mathbf{d})\right) + \delta(\mathbf{d},\lambda),$$

where $\delta(\mathbf{d},\lambda) = 1$ if $\mathbf{d} = (0,\ldots,0)$, $\lambda = 1$ and 0 otherwise; $k^+ = k$ if $k > 0$ and zero otherwise, $k^- = k^+ - k$. Call a subsequence $p = (d_{k+1},\ldots,d_{k+l})$, where $0 \leq k < rs$ and $1 \leq l \leq rs$, a *positive part* of d if all $d_{k+j} \geq 0$ and either $l = rs$ or both $d_k < 0$ and $d_{k+l+1} < 0$. For such a positive part put $\theta(p) = l$ if either $l = rs$ or $p = (0,\ldots,0)$ and $\theta(p) = l+1$ otherwise. Then $\theta(\mathbf{d}) = \sum\theta(p)$, where we take a sum over all positive subparts of d.

5.2. TENSOR PRODUCT OF VECTOR BUNDLES

All the results of this subsection are taken from (Yudin). Let $\mathcal{B}_1$ and $\mathcal{B}_2$ be two indecomposable vector bundles either on a rational curve with one node or on a configuration of projective lines of type $\tilde{A}_n$. What is $\mathcal{B}_1 \otimes_{\mathcal{O}} \mathcal{B}_2$?

Let $(\tilde{\mathcal{B}}_1, \mathcal{M}_1, i_1)$ and $(\tilde{\mathcal{B}}_2, \mathcal{M}_2, i_2)$ be triples corresponding to $\mathcal{B}_1$ and $\mathcal{B}_2$ respectively. Then it is not difficult to see that $\mathcal{B}_1 \otimes_{\mathcal{O}} \mathcal{B}_2$ corresponds to $(\tilde{\mathcal{B}}_1 \otimes_{\tilde{\mathcal{O}}} \tilde{\mathcal{B}}_2, \mathcal{M}_1 \otimes_{\mathcal{O}/\mathcal{J}} \mathcal{M}_2, i_1 \otimes i_2)$. Moreover $i_1 \otimes i_2$ is given by the Kronecker product of matrices, but the problem is that we have to care of the horizontal block division of these matrices.

Let us suppose for simplicity that $\mathrm{char}(k) = 0$ (for fields of prime characteristic an answer is analogous but more sophisticated). The first two steps in describing the decoposition of the tensor product are the following lemmas

LEMMA 5.1. *Let s be the number of components of X, $\mathbf{0} = 00\ldots0$ a sequence of 0's of length s. Then*

$$\mathcal{B}(\mathbf{d},m,\lambda) \cong \mathcal{B}(\mathbf{d},1,\lambda) \otimes_{\mathcal{O}} \mathcal{B}(\mathbf{0},m,1).$$

LEMMA 5.2. *Moreover*

$$\mathcal{B}(\mathbf{0},m,1) \otimes_{\mathcal{O}} \mathcal{B}(\mathbf{0},n,1) \cong \bigoplus_{j=1}^{m} \mathcal{B}(\mathbf{0},n-m+1+2j,1).$$

It remains to describe the tensor product of vector bundles of type $\mathcal{B}(\mathbf{d},1,\lambda)$. For simplicity let us use the following notation. Let $\mathbf{d}^l = \mathbf{dd}\ldots\mathbf{d}$ (l times), then

$$\mathcal{B}(\mathbf{d}^l,m,\lambda) = \bigoplus_{i=1}^{l} \mathcal{B}(\mathbf{d},m,\xi^i\sqrt[l]{\lambda}),$$

where ξ is a l-th primitive root of 1.

Let $\mathcal{B}(\mathbf{d}, 1, \lambda)$ and $\mathcal{B}(\mathbf{e}, 1, \mu)$ be two vector bundles of rank k and l respectively. The tensor product of this two bundles has to be of rank kl. Let $\mathbf{d} = \vec{d}_1\vec{d}_2\dots\vec{d}_k$ (each $\vec{d}_i = d_{(i-1)s+1}d_{(i-1)s+2}\dots d_{is}$ defines a sequence of degrees on $L_1, L_2, \dots, L_s$; one should not mix $\vec{d}_i$ with $\mathbf{d}(i)$ from section 2), and $\mathbf{e} = \vec{e}_1\vec{e}_2\dots\vec{e}_l$. Write

$$\tilde{\mathbf{d}} = \underbrace{\vec{d}_1\vec{d}_2\dots\vec{d}_k}_{1}\underbrace{\vec{d}_1\vec{d}_2\dots\vec{d}_k}_{2}\dots\underbrace{\vec{d}_1\vec{d}_2\dots\vec{d}_k}_{l}, \quad \tilde{\mathbf{e}} = \underbrace{\vec{e}_1\vec{e}_2\dots\vec{e}_l}_{1}\underbrace{\vec{e}_1\vec{e}_2\dots\vec{e}_l}_{2}\dots\underbrace{\vec{e}_1\vec{e}_2\dots\vec{e}_l}_{k}.$$

We have (k,l) sequences $\mathbf{f}_1, \mathbf{f}_2, \dots, \mathbf{f}_{(k,l)}$ (where (k,l) is the greatest common divisor of k and l):

$$\mathbf{f}_1 = \vec{d}_1 + \vec{e}_1, \vec{d}_2 + \vec{e}_2, \dots, \vec{d}_k + \vec{e}_l$$

(the lenght of $\mathbf{f}_1$ is $[k,l]$, the smallest common multiple of k and l),

$$\mathbf{f}_2 = \vec{d}_1 + \vec{e}_2, \vec{d}_2 + \vec{e}_3, \dots, \vec{d}_k + \vec{e}_1,$$

$$\mathbf{f}_{(k,l)} = \vec{d}_1 + \vec{e}_{(k,l)}, \vec{d}_2 + \vec{e}_{(k,l)+1}, \dots, \vec{d}_k + \vec{e}_{(k,l)-1}.$$

Then

$$\mathcal{B}(\mathbf{d}, 1, \lambda) \otimes_{\mathcal{O}} \mathcal{B}(\mathbf{e}, 1, \mu) \cong \bigoplus_{i=1}^{(k,l)} \mathcal{B}(\mathbf{f}_i, 1, \lambda^{\frac{l}{k,l}}\mu^{\frac{k}{k,l}}).$$

5.3. COMPUTATION OF EXTENSION AND HOMOMORPHISM SPACES

Now we are able to compute the dimension of the homomorphism space and of the first Ext-group between two vector bundles,

$$\mathrm{Hom}_{\mathcal{O}}(\mathcal{E}, \mathcal{F}) \cong H^0(\mathcal{E}^\vee \otimes_{\mathcal{O}} \mathcal{F}),$$

$$\mathrm{Ext}^1_{\mathcal{O}}(\mathcal{E}, \mathcal{F}) \cong H^1(\mathcal{E}^\vee \otimes_{\mathcal{O}} \mathcal{F}).$$

We have formulas for computing the dual of a vector bundle and for decomposing a tensor product into a direct sum of indecomposables.

Let us describe the simple vector bundles (i.e. those bundles $\mathcal{E}$, for which $\mathrm{End}_{\mathcal{O}}(\mathcal{E}) = k$. In other words, each automorphism of $\mathcal{E}$ is a scalar multiple of the identity automorphism). This question is motivated by the recent work of (A. Polishchuk), which relates spherical objects (Seidel and Thomas) with solutions of the classical Yang-Baxter equation. Suppose $\mathcal{E} = \mathcal{B}(\mathbf{d}, m, \lambda)$ a simple vector bundle. Since an automorphism of a Jordan block defines also an automorphism of $\mathcal{B}(\mathbf{d}, m, \lambda)$, we can conclude that $m = 1$. Further, it is easy to see that each $\mathbf{d}(i)$ should not contain degrees with difference greater than one (otherwise there will be an automorphism of $\tilde{\mathcal{E}}$ which induces the identity map modulo conductor). Since it is enough to describe simple bundles modulo the action of the Picard group, we may suppose that each $\mathbf{d}_i (i = 1, 2, \dots, s)$ consists of 0's and 1's. Now we apply the machinery developed above and get

THEOREM 5.3. *$\mathcal{B}(\mathbf{d}, m, \lambda)$ is a simple vector bundle if and only if*

1. *$m = 1$;*
2. *the difference between degrees of any two vector bundles on the same component of the normalized curve is at most 1;*
3. *consider all possible differences $\mathbf{d} - \mathbf{d}[t]$, where t is a shift of the sequence $\mathbf{d}$ ($\mathbf{d}[1] = d_{s+1}d_{s+2}\dots d_{rs}d_1d_2\dots d_s$); then each of these differences does not contain a subsequence of type $10\dots01$.*

References

Atiyah, M. (1957) Vector bundles over an elliptic curve, *Proc. London Math. Soc.* **7**, 414–452.

Bass, H. (1963) On the ubiquity of Gorenstein rings, *Math. Zeitsch.* **82**, 8–27.

Bondarenko, V. V. (1988) Bundles of semi-chains and their representations, preprint of the Kiev Institute of mathematics.

Bondarenko, V. V. (1992) Representations of bundles of semi-chains and their applications, *St. Petersburg Math. J.* **3**, 973–996.

Bondarenko, V. V., Nazarova, L. A., Roiter, A. V., and Sergijchuck, V. V. (1972), Applications of the modules over a diad to the classification of finite p-groups, having an abelian subgroup of index p, *Zapiski Nauchn. Seminara LOMI* **28**, 69–92.

Drozd, Yu. A. (1972) Matrix problems and categories of matrices, *Zapiski Nauchn. Seminara LOMI* **28**, 144–153.

Drozd, Yu. A., and Greuel, G.-M. (1999) On the classification of vector bundles on projective curves, Max-Plank-Institut für Mathematik Preprint Series 130.

Drozd, Yu. A., Greuel, G.-M., and Kashuba, I. M. (2000) On Cohen-Macaulay Modules on Surface Singularities, preprint, Max-Planck-Institut für Mathematik Bonn.

Gelfand, I. M. (1970) Cohomology of the infinite dimensional Lie algebras; some questions of the integral geometry, *International congress of mathematics*, Nice.

Gelfand, I. M., and Ponomarev, V. A. (1968) Indecomposable representations of the Lorenz group, *Uspehi Mat. Nauk* **140**, 3–60.

Grothendieck, A. (1956) Sur la classification des fibres holomorphes sur la sphère de Riemann, *Amer. J. Math.* **79**, 121–138.

Hartshorne, R. (1977) *Algebraic Geometry*, Springer.

Nazarova, L. A., and Roiter, A. V. (1969) Finitely generated modules over diad of two discrete valuation rings, *Izv. Akad. Nauk USSR* **33**, 65–89.

Nazarova, L. A., and Roiter, A. V. (1973) About one problem of I. M. Gelfand, *Functional analysis and its applications* **4**, 54–69.

Polishchuk, A. (2000) Classical Yang-Baxter equation and the A_∞-constraint, preprint, arXiv: math.AG/0008156.

Scharlau, W. (2001) On the classification of vector bundles and symmetric bilinear forms over projective varieties, preprint, Universität Münster.

Seidel, P., and Thomas, R. P. (2000) Braid group actions on derived categories of coherent sheaves, preprint, arXiv: math.AG/0001043.

Yudin, I. (2001) Diploma thesis, Kaiserslautern.

ON DOUBLE PLANES WITH KODAIRA DIMENSION ZERO

A. CALABRI*
University of Rome "Tor Vergata", Italy

Abstract. In this paper we report on a work in progress about the classification of birational equivalence classes of double planes which are surfaces of Kodaira dimension zero, namely $K3$, Enriques and bielliptic surfaces.

Key words: double planes, birational equivalence, $K3$ surfaces, Enriques surfaces, bielliptic surfaces.

Mathematics Subject Classification (2000): 14J28, 14E99.

1. Introduction

In the last few decades, birational geometry has seen a new spring, after some decades of a kind of lethargy, during which it seemed that many classical problems had fallen into oblivion.

In particular Mori theory started a new development of birational geometry, especially in higher dimensional geometry, but giving a new perspective and insight also to several problems in surface theory.

For example very recently Bayle and Beauville gave an elegant and short proof of Bertini's birational classification of plane involutions, that is equivalent to the classification of birational equivalence classes of rational double planes (i.e. double coverings of $\mathbb{P}^2$ which are rational surfaces), that was known already to Noether more than one century ago and had been considered "proven" by Castelnuovo and Enriques.

In a previous paper, at the same time and independently from Bayle and Beauville, the author of the present contribution showed how to emend and to improve Castelnuovo and Enriques' techniques in order to complete their "proof" of birational classification of rational double planes. We stress the fact that the argument we used is algorithmic and constructive, while the proof of Bayle and Beauville relies on the Mori cone theorem, which is theoretical in nature.

* Partially supported by E.C. project EAGER, contract n. HPRN-CT-2000-00099.

C. Ciliberto et al. (eds.),
Applications of Algebraic Geometry to Coding Theory, Physics and Computation, 17–22.

Therefore it seems interesting to attack with similar techniques also the problem of birational classification of double planes which are not rational, partially answered in the past only by Enriques, though with arguments that sometimes appear rather obscure and difficult to us.

Recall that irrational ruled double planes (i.e. non-rational double planes with Kodaira dimension $-\infty$) are easily classified by applying a theorem of De Franchis and the birational classification of pencils of rational plane curves.

In this contribution we report on a work in progress towards the birational classification of double planes with Kodaira dimension zero, namely double planes which are $K3$, Enriques, or bielliptic surfaces (abelian surfaces cannot be birationally equivalent to double planes).

Before stating the results in section 3, we introduce some notation and preliminaries in section 2. However details of the proofs will be presented elsewhere.

We finish this introduction with a few general remarks.

Note that the birational classification of double planes which are $K3$ surfaces (we mean with all plurigenera equal to 1) is equivalent to the classification of hyperelliptic linear systems of dimension larger than 1 on a (minimal) $K3$ surface. It turns out that such a linear system is always a subsystem of a *complete* linear system of hyperelliptic curves, already classified by Saint-Donat, cf. section 3.

This leads to think about other interesting questions: is it possible to say anything about pencils of hyperelliptic curves on $K3$ surfaces? And about rigid hyperelliptic curves?

Moreover relevant results have been established regarding the invariance of the gonality and the Clifford index of linearly equivalent smooth curves on a minimal $K3$ surface, see e.g. (Ciliberto and Pareschi) and the references therein. Is it possible to generalize these results to singular curves?

2. Notation

A double plane $\pi : X \to \mathbb{P}^2$ is a double covering over the projective plane $\mathbb{P}^2$, i.e. π is a finite morphism $\pi : X \to \mathbb{P}^2$ of degree 2. We work over the complex numbers, so by $\mathbb{P}^2$ we mean $\mathbb{P}^2(\mathbb{C})$. We remark that we do not require X to be smooth.

Two double planes π and $\rho : Y \to \mathbb{P}^2$ are said *birationally equivalent* if there exists two birational maps $\gamma : \mathbb{P}^2 \dashrightarrow \mathbb{P}^2$ and $\varphi : Y \dashrightarrow X$ such that $\pi \circ \varphi = \gamma \circ \rho$.

Recall that a double covering over a smooth rational surface S (in particular a double plane) is uniquely determined by its branch curve in S, because the Picard group of S has no torsion. We will say that a double covering $\rho : Y \to S$ is smooth (resp. singular, rational, ruled, $K3$, Enriques, etc.) if Y is.

Since we deal with birationality problems, let us say that a surface Y is:

ruled if Y has Kodaira dimension $\mathrm{kod}(Y) = -\infty$;

K3 if the minimal model of Y is a $K3$ surface, or equivalently all plurigenera of Y are equal to 1;

Enriques if the minimal model of Y is an Enriques surface, or equivalently if its irregularity is 0 and all odd (resp. even) plurigenera equal 0 (resp. 1);

bielliptic if the minimal model of Y is a bielliptic surface.

Recall that bielliptic surfaces have irregularity 1 and split in several families: the m-genus is $P_m = 1$ for all m multiple of 2 (resp. 3, 4, or 6) and zero otherwise. For standard definitions in surface theory see (Barth, Peters and Van de Ven).

Let us go back to a double plane $\pi : X \to \mathbb{P}^2$. We may assume X to be normal with no loss in generality. Indeed if X is not normal (i.e. if its branch curve $C \subset \mathbb{P}^2$ is not reduced) then the normalization $\nu : X^\nu \to X$ induces a double covering $X^\nu \to \mathbb{P}^2$ (branched along the reduced curve obtained from C by removing even multiplicity components and reducing odd multiplicity components) that is clearly birationally equivalent to π.

If X is not smooth (i.e. if C is not smooth), we consider *the canonical resolution*: there exists a birational morphism $\sigma : S \to \mathbb{P}^2$ such that the induced normal double covering $\tilde{\pi} : \tilde{X} \to S$, where X^σ is the normalization of $X \times_S T$, is smooth. Indeed the canonical resolution consists simply in blowing-up a singular point of the branch curve and normalizing, then going on until the double covering is smooth. For explicit descriptions of the canonical resolution (of an isolated double point singularity, but generalizing trivially to a whole double covering) we refer to (Laufer; Calabri and Ferraro).

Let us say that the branch curve B of $\tilde{\pi}$ is the *virtual* branch curve of the double plane π, while C is its *effective* one. Note that B is the strict transform of C in S plus possibly some exceptional curves.

The birational morphism $\sigma : S \to \mathbb{P}^2$ is the composition of finitely many blow-ups, say $\sigma = \sigma_0 \circ \sigma_1 \circ \cdots \circ \sigma_n$, where $\mathbb{P}^2 = S_{-1}$, $S = S_n$ and $\sigma_i : S_i \to S_{i-1}$ is the blow-up at a point $x_i \in S_{i-1}$. We will use the classical notions of infinitely near points, see (Enriques and Chisini; Casas-Alvero) for general theory or (Calabri) for a brief account in this setting.

We denote by E_i (resp. E_i^*) the strict (resp. total) transform in S of the exceptional curve $\sigma_i^{-1}(x_i) \subset S_i$ of σ_i. Recall that $E_i = E_i^* - \sum_j q_{ij} E_j^*$ in $\mathrm{Pic}(S)$, where $q_{ij} = 1$ if x_j is proximate to x_i and $q_{ij} = 0$ otherwise.

In $\mathrm{Pic}(S)$ we may write $C = 2dL - \sum_i c_i E_i^*$ and $B = 2dL - \sum_i b_i E_i^*$, where $2d = \deg(C)$, $c_i = \mathrm{mult}_{x_i}(C_i)$ and L is the total transform of a general line in $\mathbb{P}^2$. It easy to prove that $A = B/2$ is well defined in $\mathrm{Pic}(S)$, namely all the b_i's are even.

Although B is smooth, let us say that b_i is the *multiplicity* of B at x_i (we hope not to confuse readers with this terminology). Curiously (at a first sight), it may happen that x_j is infinitely near of the first order to x_i and $b_j > b_i$. If this occurs, then $b_j = b_i + 2$, $c_i = c_j$ and we say that x_i (resp. x_j) is a *defective* (resp. *excessive*) point. Equivalently x_i is defective if and only if $\tilde{\pi}^{-1}(E_i)$ is a (-1)-curve in $\tilde{X}$. In particular if x_i is proper (i.e. not infinitely near to any point), then c_i must be odd, see (Calabri; Calabri and Ferraro) for more details about defective points.

Then the plurigenera of X are easily computed:

$$P_m(X) = h^0(S, mA + mK_S) + h^0(S, (m-1)A + mK_S), \qquad \text{for all } m,$$

while its irregularity is $q(X) = p_g(X) - p_a(A)$.

3. On the birational classification of double planes

Let us first recall the classification of ruled double planes.

THEOREM 3.1. *A double plane is ruled of genus $g > 0$ if and only if it is birationally equivalent to a double plane branched along $2g+2$ distinct lines through a point.*

PROOF. It follows from De Franchis' theorem and the birational classification of pencils of rational plane curves, see (Calabri). □

THEOREM 3.2. *A double plane is rational if and only if it is birationally equivalent to a double plane branched along one of the following curves:*

1. *a smooth conic;*
2. *a smooth quartic;*
3. *an irreducible sextic with two infinitely near triple points;*
4. *an irreducible curve of degree $2d$ with a point of multiplicity $2d-2$ and either no other singularities or at most one additional node.*

Moreover the above types are birationally distinct.

PROOF. (Bayle and Beauville; Calabri). □

Using the terminology of section 2, type (3) and (4) of the above theorem are double planes whose virtual branch curve has degree $2d$ and a point of multiplicity $2d-2$ (and in (4) possibly a node), because the triple point of the sextic is defective.

The idea of the proof is the following. Let $\pi : X \to \mathbb{P}^2$ be a double plane branched along a reduced curve C, as in the above notation. The vanishing of plurigenera of X implies the vanishing of the linear systems $|B + mK_S|$, for $m > 1$, and this condition, together with $p_a(A) = 0$, is sufficient to find a Cremona tranformation $\gamma : \mathbb{P}^2 \dashrightarrow \mathbb{P}^2$ such that the induced normal double plane $\pi_\gamma : X_\gamma \to \mathbb{P}^2$ (where X_γ is the normalization of the double plane branched along the total transform $\gamma^*(C)$ of C), birationally equivalent to π, is branched along a curve in the statement of Theorem 3.2.

Improving the techniques of (Calabri), we are able to prove the following:

THEOREM 3.3 (ENRIQUES, 1896). *A double plane is a K3 surface if and only if it is birationally equivalent to a normal double plane branched along one of the following curves:*

1. *a sextic;*
2. *a curve of degree 8 with two proper quadruple points;*
3. *a curve of degree 8 with two infinitely near quadruple points;*
4. *a curve of degree 10 with a point x_0 of multiplicity 7 and two triple points each infinitely near of the first order to x_0;*
5. *a curve of degree 12 with a point x_0 of multiplicity 9 and three triple points each infinitely near of the first order to x_0.*

The virtual branch curve of double planes of type (3), (4), (5) has degree $2d$, a point x_0 of multiplicity $2d-4$ and $d-3$ quadruple points in the first infinitesimal neighbourhood of x_0. This may happen only if $c_0 = 2d-3$ is odd and $2d < 14$.

This time we show that the condition $P_m = 1$ for all m implies that $|mA + mK_S|$ is a rigid curve (namely the rigid curve $A + K_S$ counted m times), $|B + mK| = \emptyset$ for $m > 2$ and $p_a(A) = 1$. Again these conditions are sufficient to find a Cremona trasformation which shows the birational equivalence of a given $K3$ double plane with one in the list of statement 3.3.

We remark that the five types of $K3$ double planes in Theorem 3.3 corresponds respectively to types (a), (b), (c.3), (c.2), (c.1) in (Saint-Donat, p. 622–623), namely the double plane (2) corresponds to (b) a double covering of $\mathbb{P}^1 \times \mathbb{P}^1$ branched along a curve of bidegree (4,4); the double plane (3) corresponds to (c.3) the quadric cone $\mathbb{F}_2^* \subset \mathbb{P}^3$; type (4) is (c.2) the cone $\mathbb{F}_3^*$ over a rational normal twisted cubic curve in $\mathbb{P}^3$ and finally type (5) is (c.1) the cone $\mathbb{F}_4^*$ over a rational normal twisted quartic curve in $\mathbb{P}^4$.

Similarly, one can reach the following result about Enriques double planes:

A double plane is an Enriques surface if and only if it is birationally equivalent to a normal double plane branched along a curve of degree 8 that splits in two lines L_1, L_2 and a sextic C_1 with a node in $x_0 = L_1 \cap L_2$ and two tacnodes in two points lying respectively on L_1, L_2 where L_1, L_2 are the tacnodal tangents.

We remark that the sextic can be reducible. The virtual branch curve of the double plane in the statement has three points of multiplicity 4. Two of them are excessive, each infinitely near of the first order to a defective double point (tacnode of the sextic) and lying on the tacnodal tangent.

This type of double plane has been first described by Enriques in 1906. Later Averbuh proved that an Enriques surface is birationally equivalent to a surface that is a double plane with such a branch curve, which is a weaker result than the above statement.

In order to achieve this classification result, we show that an Enriques double plane is characterized by the following conditions: $|A + K_S| = \emptyset$, $p_a(A) = 0$, $|B + 2K_S|$ is a single rigid curve and $|B + mK_S| = \emptyset$ for $m > 2$.

Regarding bielliptic surfaces, it is easy to show that the bi-genus must be 1 (therefore only two of seven families of bielliptic surfaces can be double planes) and one gets the following result:

A double plane is a bielliptic surface if and only if it is birationally equivalent to a normal double plane branched along a curve of degree 10 that splits in four lines $L_1,\dots,L_4$ through a point x_1 and either

1. *an irreducible sextic with a node in x_1 and four tacnodes in points lying on $L_1,\dots,L_4$ where $L_1,\dots,L_4$ are the tacnodal tangents, or*
2. *two smooth cubics each with a flex in x_1 and tangent to $L_1,\dots,L_4$ respectively in x_1,x_2,x_3,x_4, such that L_1 is the flex tangent and x_2,x_3,x_4 are aligned.*

The virtual branch curves of the double plane in the statement have multiplicity 6 at x_1 and four excessive points of multiplicity 4, each infinitely near of the first order to a defective double point and lying respectively on $L_1,\dots,L_4$.

The bielliptic double planes have been described also by Bagnera and De Franchis, and the above statement can be considered implicit in their work.

The conditions that we use for the above result are the following: $|A+K_S| = \emptyset$, $p_a(A) = -1$, $|B+2K_S|$ is a single rigid curve and $|B+mK_S| = \emptyset$ for $m > 2$.

References

Averbuh, B.G. (1965) Chapter X and Appendix, in I.R. Šafarevič et al., Algebraic surfaces, *Proc. Steklov Inst. Math.* **75**.

Bagnera, G., and De Franchis, M. (1908) Le superficie algebriche le quali ammettono una rappresentazione parametrica mediante funzioni iperellittiche di due argomenti, *Mem. Soc. Italiana delle Scienze detta dei XL* **15**, 251–343.

Barth, W., Peters, C., and Van de Ven, A. (1984) *Compact complex surfaces*, Ergebnisse der Math., 3. Folge, Band 4, Springer, Berlin.

Bayle, L., and Beauville, A. (2000) Birational involutions of $\mathbb{P}^2$, *Asian J. Math.* **4**, no. 1, 11–17.

Calabri, A. (2000) On rational and ruled double planes, preprint, submitted.

Calabri, A., and Ferraro, R. (2000) Explicit resolution of double point singularities of surfaces, preprint, submitted.

Casas-Alvero, E. (2000) *Singularities of plane curves*, London Math. Soc. Lect. Note Series 276, Cambridge University Press, Cambridge.

Castelnuovo, G., and Enriques, F. (1900) Sulle condizioni di razionalità dei piani doppi, *Rend. Circ. Mat. Palermo* **14**, 290–302.

Ciliberto, C., and Pareschi, G. (1995) Pencils of minimal degree on curves on a K3 surface, *J. reine angew. Math.* **460**, 15–36.

De Franchis, M. (1904) I piani doppi dotati di due o più differenziali totali di prima specie, *Rend. Accad. Lincei* **13**, 688–695.

Enriques, F. (1896) Sui piani doppi di genere uno, *Mem. Soc. Italiana delle Scienze detta dei XL* **10**, 201–224.

Enriques, F. (1906) Sopra le superficie algebriche di bigenere uno, *Mem. Soc. Italiana delle Scienze detta dei XL* **14**.

Enriques, F., and Chisini, O. (1920) *Lezioni sulla teoria geometrica delle equazioni e delle funzioni algebriche*, 5 voll., Zanichelli.

Laufer, H. (1978) On normal two-dimensional double point singularities, *Israel J. Math.* **31**, 315–334.

Saint-Donat, B. (1974) Projective models of K-3 surfaces, *Amer. J. Math.* **96**, 602–639.

COMPUTING MINIMAL GENERATORS OF IDEALS OF ELLIPTIC CURVES

L. CHIANTINI
University of Siena, Italy

F. CIOFFI and F. ORECCHIA
University of Naples "Federico II", Italy

Abstract. In this paper results about the Hilbert function and about the number of minimal generators stated in (Orecchia) for disjoint unions of rational smooth curves are generalized to disjoint unions of distinct smooth non special curves. Hence, the maximal rank and the minimal generation of such curves are studied. In particular, we consider elliptic curves and we describe a method to compute their Hilbert functions in any dimension and for every choice of the degrees. Applications to the study of elliptic curves on threefolds are shown.

Key words: Elliptic curves, Hilbert function.

Mathematics Subject Classification (2000): 14H52, 14Q05.

1. Introduction

In this paper we study the maximal rank and the minimal generation of disjoint unions of distinct non special curves, with applications to elliptic curves. We construct elliptic curves in projective spaces $\mathbb{P}^n$ of any dimension $n \geq 3$ and for every choice of the degree via their rational points.

Remind that a curve $C \subset \mathbb{P}^n$ is said to have maximal rank if for every integer $t > 0$ the restriction map $\rho_C(t) : H^0(\mathbb{P}^n, \mathcal{O}_{\mathbb{P}^n}(t)) \to H^0(C, \mathcal{O}_C(t))$ has maximal rank as a map of vector spaces, i.e., if it is injective or surjective. Standard semicontinuity arguments show that the maximal rank property is open in the Hilbert scheme (Hartshorne and Hirschowitz, p. 2); hence if it holds for one curve, then it holds also for the general curve in the same irreducible component. The maximal rank for general curves has been deeply studied by E. Ballico and P. Ellia who proved that a general non degenerate and non special curve of genus $g \geq 0$ and degree $d \geq g+n$ in $\mathbb{P}^n$ has maximal rank, where $n \geq 3$ (see (Ballico and Ellia) and the references therein). In (Orecchia) disjoint unions of rational curves with maximal

C. Ciliberto et al. (eds.),
***Applications of Algebraic Geometry to Coding Theory, Physics and Computation,* 23–35.**

rank are characterized by means of the Hilbert function; moreover results about the number of minimal generators for such unions of rational curves are described. In section 2 we generalize these results to disjoint unions of non special curves, i.e. of curves C whose index of speciality $e_C := \max\{t | H^1(\mathcal{O}_C(t)) \neq 0\}$ is null.

In section 3 we describe how to construct an elliptic curve of any degree $d \geq 3$ via its rational points (Remark 3.1). So, recall that a point $[y_0 : \ldots : y_n]$ of $\mathbb{P}^n_{\bar{K}}$ is a "rational point" over K if there exist $y'_0, \ldots, y'_r$ of K such that $[y_0 : \ldots : y_r] = [y'_0 : \ldots : y'_r]$. The Mordell-Weil group of rational points of an elliptic curve has been studied in many different contests (for example, see (Cremona; Darmon)). We have implemented our construction of elliptic curves of given degree in the object-oriented language C^{++} by using the NTL library of (V. Shoup) in a software called "Points" that is available at http://cds.unina.it/~orecchia/ gruppo/EPoints.html.

In section 4, by our method we check the maximal rank and the minimal generation for disjoint unions of degree $d \leq 30$ of $s \leq 6$ distinct elliptic curves in $\mathbb{P}^n$, where $3 \leq n \leq 30$, and we show the existence of smooth elliptic quintic curves on a general threefold in $\mathbb{P}^4$. The question about which curves lie in a smooth general threefold recently grew in interest because of its connections with hyperbolic geometry (see e.g. (Johnsen and Kleiman; Clemens; Kley)); the problem is also related with the study of Hilbert schemes of curves (see (Kleppe and Miró-Roig)). We show in section 4 how one can use the equations of an elliptic curve, generated by our algorithm, to obtain the existence of elliptic normal curves on a general quartic threefold in $\mathbb{P}^4$. The method was used in (Madonna), where a complete classification of rank 2 bundles without intermediate cohomology on general quartic threefolds is achieved. We also hope to apply it for the study of rank 2 bundles over the general (Calabi-Yau) quintic threefold.

In the following $S = K[x_0, \ldots, x_n]$ is a ring of polynomials over a field K, $\bar{K}$ is the algebraic closure of K and the characteristic of K is not 2. By projective algebraic variety we mean the set of points of $\mathbb{P}^n_{\bar{K}}$ that are the zeros of the polynomials of a homogeneous ideal of S. In particular, by a (projective) curve in $\mathbb{P}^n_K$ we mean a projective variety of pure dimension 1.

2. Maximal rank and minimal generation of disjoint unions of non special curves

We will use freely the common notations of sheaf cohomology in $\mathbb{P}^n$. We say that a variety is generated in degree t if its ideal I can be generated by forms of degree less than or equal to t.

REMARK 2.1. We recall briefly some well-known fact about the generators of a homogeneous ideal and its Castelnuovo-Mumford regularity.

Let m be an integer. A coherent sheaf $\mathcal{F}$ is m-regular if $H^q(\mathcal{F}(m-q)) = 0$ for all $q > 0$ (Mumford).

If I is a saturated homogeneous ideal, then we say that I is m-regular when its sheafification I is. It turns out that I is m-regular if its jth module of syzygies is generated in degree $\leq m-j$, for each $j>0$ (see for example (Green, Prop. 2.6)). This last definition can be extended to any homogeneous ideal I.

Let $Y \subset \mathbb{P}^n$ be a projective variety. We say that Y is m-regular when its associated homogeneous ideal I is.

It can be proved that if I is m-regular, then it is also $(m+t)$-regular, for every $t \geq 0$ (Mumford). The regularity reg(I) of I is defined as the smallest integer m for which I is m-regular. Regularity provides a stop for an algorithm that finds generators of a saturated ideal. In fact if I is m-regular, then it is generated in degree m (Mumford, Lecture 14).

Let $I = I(Y) = \bigoplus_{t\geq 0} I(Y)_t$ be the ideal of a closed subvariety Y and $A = S/I$ be the coordinate ring of Y. Set

$$H_Y(t) = \dim_K A_t = \dim_K S_t - \dim_K I_t = \binom{t+n}{n} - \dim_K I_t$$

for the Hilbert function of Y and call $P_Y(t)$ the Hilbert polynomial of Y. Let

$$\Delta H_Y(t) := H_Y(t) - H_Y(t-1), \Delta^i H_Y(t) := \Delta^{i-1} H_Y(t) - \Delta^{i-1} H_Y(t-1)$$

be the difference functions.

The Poincaré series of Y

$$HP_Y(z) = \sum_{t\geq 0} H_Y(t) z^t$$

can be expressed as a rational function $h(z)/(1-z)^{d+1}$; let δ be the degree of $h(z)$.

DEFINITION 2.1. *Let Y be a closed (possibly reducible) subvariety of $\mathbb{P}^n$; define $\rho_Y(t)$ as the natural restriction map*

$$\rho_Y(t) : H^0(\mathbb{P}^n, \mathcal{O}_{\mathbb{P}^n}(t)) \to H^0(Y, \mathcal{O}_Y(t)).$$

We say that Y has maximal rank if, for every integer $t>0$, $\rho_Y(t)$ has maximal rank as a map of vector spaces, i.e., it is injective or surjective.

DEFINITION 2.2. *The index of speciality of a curve C is*

$$e_C := \max\{t | H^1(\mathcal{O}_C(t)) \neq 0\}$$

and the sheaf $\mathcal{O}_C(t)$ is called non special if $H^1(\mathcal{O}_C(t)) = 0$.

Using the previous notation, we can restate the notion of regularity in the case of homogeneous ideals associated to curves:

LEMMA 2.1. *Let C be a curve. For all $t \geq 0$, C is t-regular if and only if $\mathcal{O}_C(t-2)$ is non special and $\rho_C(t-1) : H^0(\mathbb{P}^n, \mathcal{O}_{\mathbb{P}^n}(t-1)) \to H^0(C, \mathcal{O}_C(t-1))$ is surjective. Furthermore, if $m \geq \operatorname{reg}(C)$, then*

(1) $H_C(t) = P_C(t)$, *for* $t \geq m-1$;
(2) $P_C(t) = \Delta H_C(m)t + H_C(m-1) - \Delta H_C(m)(m-1)$;
(3) $HP_C(z) = \sum_{i=1}^{m} h_i z^i/(1-z)^2$, *with* $h_i = \Delta^2 H_C(i)$.

PROOF. The first assertion is obvious. For (1) see, for example, Lemma 4(i) in (Nagel). (2) Since the Hilbert polynomial $P_C(t)$ has degree 1, it is sufficient to determine the unique polynomial of degree 1 in a variable t which assumes the values $H_C(m)$ and $H_C(m-1)$ respectively for $t = m$ and $t = m-1$. (3) See, for example, section 1.4 in (Migliore). □

The following results extend to non-special curves properties that have been already proved in (Orecchia) for rational smooth curves.

THEOREM 2.1. *Let $C \subset \mathbb{P}^n$ ($n \geq 3$) be a non degenerate disjoint union of s distinct irreducible smooth curves C_i of genus g and of degrees d_i. Assume that $\mathcal{O}_{C_i}(t)$ is non special for every $t > 0$ and for every $i = 1, \ldots, s$, and let I be the ideal of C.*

Then $H_C(t) \leq P_C(t) = dt + s(1-g)$, for every $t > 0$. Furthermore:

(1) for every $t > 0$ one has $H_C(t) \leq \min\left\{ \binom{t+n}{n}, dt + s(1-g) \right\}$;
(2) in (1) the equality holds for all $t > 0$ if and only if $\rho_C(t)$ has maximal rank;
(3) $\operatorname{reg}(I) = \min\{t \geq 3 \mid H_C(t-1) = P_C(t-1)\}$.

PROOF. Since $\mathcal{O}_C(t) = \oplus \mathcal{O}_{C_i}(t)$ and the cohomology commutes with direct sums, then $H^1(\mathcal{O}_C(t)) = 0$ for all $t > 0$. By applying sheafification to the short exact sequence $0 \to I \to S \to S/I \to 0$, we determine the exact sequence

$$0 \to \mathcal{I}_C(t) \to \mathcal{O}_{\mathbb{P}^r}(t) \to \mathcal{O}_C(t) \to 0 \tag{2.1}$$

to which we apply cohomology, getting

$$0 \to H^0(\mathcal{I}_C(t)) \to H^0(\mathcal{O}_{\mathbb{P}^r}(t)) \xrightarrow{\rho_C(t)} H^0(\mathcal{O}_C(t)) \to H^1(\mathcal{I}_C(t)) \to 0.$$

Since $H^1(\mathcal{O}_C(t)) = 0$, for every $t > 0$,

$$H_C(t) = \dim_K(\operatorname{Im}(\rho_C(t))) \leq h^0(\mathcal{O}_C(t)) = h^0(\mathcal{O}_C(t)) - h^1(\mathcal{O}_C(t)) = P_C(t).$$

The first assertion follows since clearly it is $H_C(t) \leq \dim H^0(\mathcal{O}_{\mathbb{P}^r}(t)) = \binom{t+n}{n}$, for any t. Also equality means that $\rho_C(t)$ has maximal rank.

To see (3), observe that sequence (2.1) and our assumptions imply $H^2(\mathcal{I}_C(t-2)) = H^1(\mathcal{O}_C(t-2)) = 0$ for all $t \geq 3$. If $H_C(\bar{t}-1) = P_C(\bar{t}-1)$ for some $\bar{t} \geq 3$, then the map $\rho_C(\bar{t}-1)$ surjects, so $H^1(\mathcal{I}_C(\bar{t}-1)) = 0$. Moreover, by Grothendieck, it is $H^i(\mathcal{O}_C(t)) = 0$, for every $i > 1$ and for every t. Since $H^q(\mathcal{O}_{\mathbb{P}^n}(t)) = 0$ for every $q > 0$ ($q \neq n$) and every t, then from the same exact long sequence in cohomology one sees that $0 = H^{q-1}(\mathcal{O}_C(t)) = H^q(\mathcal{I}_C(t))$, for every $q > 2$ ($q \neq n$) and t. If $q = n$, then $H^q(\mathcal{O}_{\mathbb{P}^n}) = 0$ for every $t > -(n+1)$. Since $\bar{t} \geq 3$, thus $H^q(\mathcal{I}_C(t-q)) = 0$ for every $q > 2$ and $t \geq \bar{t}$. It follows that I is $\bar{t}$-regular. Then, by Lemma 2.1(1) we are done. □

Next proposition provides our main tool for determining when we get an embedding of disjoint union of smooth non-special curves.

PROPOSITION 2.1. *Let $\bar{C}$ be a disjoint union of s smooth curves $\bar{C}_1, \ldots, \bar{C}_s$ of geometric genus g. Let $\phi : \bar{C} \to \mathbb{P}^n$ be a map of $\bar{C}$ to some projective space and call $C = \bigcup_{i=1}^s C_i$ the image. Assume that the pull-back H_i of $\mathcal{O}_{\mathbb{P}^n}(1)$ to any $\bar{C}_i$ is a very ample and non special divisor of degree d_i. Put $d = \sum_{i=1}^s d_i$.*

If $H_C(\bar{t}) = d\bar{t} + s(1-g)$ for some $\bar{t} > 0$, then ϕ is an embedding, i.e. C is a disjoint union of smooth curves of degrees $d_1, \ldots, d_s$.

PROOF. By Riemann-Roch, on each $\bar{C}_i$ one has $h^0(\mathcal{O}_{\bar{C}_i}(tH_i)) = d_i t - g + 1$ for all $t > 0$. Now we have $H^0(\mathcal{O}_{\bar{C}}) = \bigoplus_{i=1}^s H^0(\mathcal{O}_{\bar{C}_i})$ so that $\dim H^0(\mathcal{O}_{\bar{C}}(tH)) = dt - sg + s$ for all $t > 0$, where H is the pull-back to $\bar{C}$ of $\mathcal{O}_{\mathbb{P}^n}(1)$.

Assume now $H_C(\bar{t}) = d\bar{t} + s(1-g)$. The map ϕ induces a map $\mathcal{O}_{C_i} \to \mathcal{O}_{\bar{C}_i}$ which is injective, for the kernel cannot have torsion and $\phi_{|\bar{C}_i}$ is not trivial. It follows that the composition $H^0(\mathcal{O}_{\mathbb{P}^r}(\bar{t})) \to H^0(\mathcal{O}_{\bar{C}}(\bar{t}H))$ is surjective; so it induces on each $\bar{C}_i$ the complete linear series associated to H_i; by assumptions it follows that ϕ restricts to an embedding of any $\bar{C}_i$. Furthermore, by assumptions since the restriction map surjects onto $H^0(\mathcal{O}_C(t))$, then there are surfaces of degree $\bar{t}$ which separate the points of two different pieces $\bar{C}_i$, $\bar{C}_j$ of $\bar{C}$. The claim follows. □

PROPOSITION 2.2. *Let $C = \bigcup_{i=1}^s C_i$ be a non degenerate curve such that $C_1, \ldots, C_s$ are irreducible curves of geometric genus g and of degrees $d_1, \ldots, d_s$, where $d_i \geq 2g+1$ for every $i = 1, \ldots, s$. Let $d = \sum_{i=1}^s d_i$ and*

$$\alpha = \min\left\{t \in N \mid \binom{t+n}{n} > dt + s(1-g)\right\}.$$

Then the curves C_i are disjoint, smooth, and C has maximal rank if, and only if,

$$H_C(\alpha-1) = \binom{\alpha-1+n}{n} \quad \text{and} \quad H_C(\alpha) = d\alpha + s(1-g).$$

Moreover, in this situation $I(C)$ is generated by forms of degrees α and $\alpha+1$ because $\operatorname{reg}(I(C)) = \alpha + 1$.

PROOF. It is sufficient to apply Theorem 2.1 and Proposition 2.1. Indeed, in the hypotheses that $H_C(\alpha-1) = \binom{\alpha-1+n}{n}$ and $H_C(\alpha) = d\alpha + s(1-g)$, let $\Phi_i : \tilde{C}_i \to C_i$

be the normalization of C_i. Hence $\check{C}_i$ is a smooth curve of degree d_i and genus g and $H_{C_i}(t) \leq H_{\check{C}_i}(t)$. Since $d_i \geq 2g+1$, $\mathcal{O}_{\check{C}_i}(1)$ is a very ample and non special divisor of degree d_i. By Theorem 2.1 it is $H_{\check{C}_i}(t) \leq d_i t + 1 - g$, for every $t \geq 0$, and hence

$$d\alpha + s(1-g) = H_C(\alpha) \leq \sum H_{C_i}(\alpha) \leq \sum H_{\check{C}_i}(\alpha) \leq d\alpha + s(1-g).$$

If $\bigcup \check{C}_i$ were not a disjoint union, it would be $\sum H_{\check{C}_i}(\alpha) < d\alpha + s(1-g)$, a contradiction. Hence we can apply Proposition 2.1 and so C is a disjoint union of smooth curves. The maximal rank follows by Theorem 2.1(2). Vice versa, apply Theorem 2.1 (1) and (2). The result about the regularity follows from Theorem 2.1(3) and Lemma 2.1(1). □

Notice that since the index of speciality of an elliptic curve E is null and, so, $H^1(\mathcal{O}_E(t)) = 0$ for every $t > 0$, then Theorem 2.1 and Propositions 2.1 and 2.2 can be applied for every elliptic curve and for non degenerate unions of distinct elliptic curves.

Let C be a non degenerate disjoint union of s distinct irreducible smooth curves C_i of degrees d_i and of the same genus g, with $\mathcal{O}_{C_i}(t)$ non special, for every $t > 0$. Moreover, assume that C has maximal rank. By Proposition 2.2 we know that these properties are characterized by the behaviour of the Hilbert function of C.

In this situation, the ideal $I(C)$ is generated by forms of degrees α and $\alpha+1$, where $\alpha = \min\{t \mid \binom{t+n}{n} > dt + s(1-g)\}$, since $\operatorname{reg}(I) = \alpha + 1$. So it is interesting to study the expected number for the minimal generators of $I(C)$ (see (Orecchia) for rational curves).

DEFINITION 2.3. *The curve C is minimally generated if a minimal set of homogeneous generators of $I(C)$ has cardinality*

$$\dim_K(I_\alpha) + \dim_K(I_{\alpha+1}) - \min\{(n+1)\dim_K(I_\alpha), \dim_K(I_{\alpha+1})\}$$

that is

$$\sigma(\alpha) : H^0(I_C(\alpha)) \otimes H^0(\mathcal{O}_{\mathbb{P}^n}(1)) \rightarrow H^0(I_C(\alpha+1))$$

is of maximal rank.

Since the kernel of $\sigma(\alpha)$ is $H^0(I_C(\alpha) \otimes \Omega(1))$, see (Idá), then for curves of maximal rank, by semicontinuity, minimal generation is an open property hence if it holds for one curve, then it holds also for the general curve in the same irreducible component of the Hilbert scheme.

It turns out that the maximal rank and the minimal generation can be checked by computing the generators of a suitable number of points on C.

LEMMA 2.2. *Let C be a curve, I its ideal and $C_1, \ldots, C_h$ its irreducible components of degrees $d_1, \ldots, d_h$ respectively, so that $d = \sum_{i=1}^{h} d_i$ is the degree of C. Let $m \geq \mathrm{reg}(I)$. Suppose that every curve C_i, $i = 1, \ldots, h$, contains $d_i \cdot m + 1$ distinct rational points and let J be the ideal of these $d \cdot m + h$ points of C. Then, the polynomials of degree $\leq m$ of a minimal set of generators for J form a minimal set of generators for I. Hence $H(I,t) = H(J,t)$, for $t \leq m$.*

PROOF. (Albano, Cioffi, Orecchia and Ramella, Lemma 2.1) Clearly $I \subset J$ and it is well known that a hypersurface of degree $t \leq m$ containing $d_i m + 1$ points of C_i must contain C_i. Hence, $I_t = J_t$ for each degree $t \leq m$. It remains to observe that, since I is generated in degree m, then I is generated by $\bigoplus_{t \leq m} I_t = \bigoplus_{t \leq m} J_t$. □

Hence we may use suitable sets of points on C to control that C has maximal rank and is minimally generated.

THEOREM 2.2. *Let C be a non degenerate disjoint union of s distinct irreducible smooth curves C_i of the same genus g, resp. of degree d_i, and with $\mathcal{O}_{C_i}(t)$ non special, for every $t > 0$. Let $d = \sum_{i=1}^{s} d_i$ and $\alpha = \min\{t \in N | \binom{t+n}{n} > dt + s(1-g)\}$. Let V_t^i be a set of $d_i t + 1$ distinct points of C_i and $V_t = \cup V_t^i$. Let G_t be the matrix whose columns are the vectors of the evaluations of the terms of degree t over the points of V_t. Then the curves $C_1, \ldots, C_s$ are smooth and disjoint and C has maximal rank if, and only if, $H_V(\alpha - 1) = \mathrm{rank}(G_{\alpha-1}) = \binom{\alpha-1+n}{n}$ and $H_V(\alpha) = \mathrm{rank}(G_\alpha) = d\alpha + s(1-g)$. Moreover, if $g_1, \ldots, g_e$ is a basis for $I(C)_\alpha$ and $H_{\alpha+1}$ is the matrix whose columns are the vectors of the coefficients of $x_i g_j$ (for every $i = 0, \ldots, n$ and $j = 1, \ldots, e$), then C is minimally generated if, and only if,*

$$\mathrm{rank}(H_{\alpha+1}) = \min\{(n+1)\dim_K I(C)_\alpha, \dim_K I(C)_{\alpha+1}\} =$$

$$= \min\left\{(n+1)\left(\binom{\alpha+n}{n} - d\alpha - s\right), \binom{\alpha+1+n}{n} - d(\alpha+1) - s\right\}.$$

Note that Theorem 2.2 can be applied to elliptic curves.

3. How to construct an elliptic curve in $\mathbb{P}^n$

In this section we show an effective method to compute elliptic curves by exploiting the availability of algorithms that construct minimal generators of ideals of points in polynomial time (for example, see (Cioffi, 1999) and the references therein). This construction allows to check the minimal rank and the minimal generation of elliptic curves. The described algorithm has been implemented in the object oriented language C^{++} by using the NTL library of (Shoup) in a software called "Points", available at http://cds.unina.it/˜orecchia/gruppo/EPoints.html.

Recall that a point $[y_0 : \ldots : y_n]$ of $\mathbb{P}^n_{\bar{K}}$ is a "rational point" over K if there exist $y'_0, \ldots, y'_r$ of K such that $[y_0 : \ldots : y_r] = [y'_0 : \ldots : y'_r]$. Over a field K of characteristic different from 2 every elliptic curve E can be trasformed by an isomorphism over $\bar{K}$ into a plane curve with an equation of the following form

$$y^2 = x(x-1)(x-\lambda) \tag{3.1}$$

where $\lambda \neq 0, 1$, see (Silverman, III, Prop. 1.7). The equation (3.1) is a Weierstrass representation of an elliptic plane curve in the Legendre form. The condition $\lambda \neq 0, 1$ is equivalent to the fact that E is smooth, see for example (Silverman and Tate, IV, section 3). We suppose that $\lambda \in K$.

Every field K contains a copy of $\mathbb{Q}$ or of $\mathbb{Z}_p$ for some prime p and so we can assume that $K = \mathbb{Q}$ or $\mathbb{Z}_p$. To study an algebraic object over $\mathbb{Q}$ it is useful to study the reduction of the object modulo primes. Hence, let K be the field $\mathbb{F}_q$ with q elements, where q is a power of a prime $p \geq 3$. Moreover, let $E(\mathbb{F}_q)$ be the Mordell-Weil group of the rational points over $\mathbb{F}_q$ of an elliptic curve E.

LEMMA 3.1. *Let E be an elliptic curve defined over the field $\mathbb{F}_q$ with q elements, where q is a power of p. Then $|\#E(\mathbb{F}_q) - q - 1| \leq 2\sqrt{q}$.*

PROOF. This result was conjectured by Artin and was proved by Hasse in 1930. See for example (Silverman, V, Theorem 1.1). □

REMARK 3.1. Now we can describe how from any equation of type (3.1) it is possible to construct a non degenerate elliptic curve E of degree $d \geq 3$ in $\mathbb{P}^n$ over a field K of characteristic $p \geq 3$ such that $p + 1 - 2\sqrt{p} > d(d+2-n) + 1$. The regularity of an irreducible non degenerate curve of degree d in $\mathbb{P}^n$ is lower than or equal to $d + 2 - n$ (Gruson, Lazarsfeld and Peskine). Hence, by Lemma 2.2 and by the availability of algorithms that compute generators of ideals of points, it is possible to determine the ideal of the curve from a set of $d(d+2-n)+1$ distinct points of the curve. So, it is enough to construct $d(d+2-n)+1$ distinct points of E. Note that by Lemma 3.1, $\#E(K) \geq p + 1 - 2\sqrt{p}$. Hence over a field K of characteristic p such that $p + 1 - 2\sqrt{p} > d(d+2-n)+1$, an elliptic curve represented by (3.1) has at least $d(d+2-n)+1$ distinct rational points. So, if we evaluate the polynomial $x(x-1)(x-\lambda)$ on every $\bar{x}$ of F_p and then we compare the result with the evaluation of y^2 on every $\bar{y}$ of F_p, we must find at least $d(d+2-n)+1$ points $(\bar{x}, \bar{y})$ of $\mathbb{A}^2$ that belong to E. Let $f_0(y_0, y_1, y_2), \ldots, f_n(y_0, y_1, y_2)$ be $n+1$ polynomials of degree $\bar{d}$ with no common zeros in $\mathbb{P}^2$; they define a birational map $\phi : \mathbb{P}^2 \to X \subset \mathbb{P}^n$ into a surface X of $\mathbb{P}^n$, when they are sufficiently generic (one can check that the image is an elliptic curve by the Hilbert polynomial that must be of type dz, where z is a variable). The image of the projective closure of a curve (3.1) by ϕ is a curve of geometric genus 1. We put $\bar{d} = d/3$ if $d \equiv 0 \pmod 3$, otherwise $\bar{d} = d/3 + 1$. Moreover, if $d \equiv 2 \pmod 3$, we make all the polynomials f_i to have a null coefficient of $y_2^{\bar{d}}$ (the unique point at infinity of the curve (3.1) becomes a base point of the map), and if $d \equiv 1 \pmod 3$ also the coefficient of $y_0^{\bar{d}}$ is null (the origin becomes a base point of the map). In this way, when it is necessary we put enough base points in the rational map so that the curve through the images of the computed points of the curve (3.1) is elliptic of degree d.

Hence, if we want to determine elliptic curves of any degree d in $\mathbb{P}^n$ we need to compute enough rational points on a curve of type (3.1) and to take a suitable rational map. About the computation of the points, it is enough to choose the characteristic p of the field as it is suggested in Lemma 3.1. About the rational map, we take random coefficients for the polynomials f_i and then we impose the conditions on base points described in Remark 3.1. So now we can formulate the following algorithm, where E_i is an elliptic curve, n_i is the minimum degree of a linear variety that contains the curve E_i and m_i is an upper bound for the regularity of E_i, as it is proved in (Gruson, Lazarsfeld and Peskine).

REMARK 3.2. Let M be the matrix whose entries are the coordinates of $d+1$ distinct points of the curve E_i. By Bèzout Theorem, the minimum degree n_i of a linear variety that contains the curve E_i is equal to rank$(M)-1$.

ALGORITHM

INPUT: number s of the irreducible components $E_1,\dots,E_s$ of a curve E and their degrees $d_1,\dots,d_s$; characteristic $p\geq 3$ of the field K.

OUTPUT: if the variable q is false, then the characteristic of the field is too low for a successful computation. Otherwise, the output consists of a minimal set of homogeneous generators of the ideal of the curve $I(E)$, the Hilbert function $H_E(t)$, the Hilbert polynomial and the Poincaré series of $I(E)$. If the Hilbert polynomial is the expected one, then the properties of maximal rank and of minimally generation of E are checked.

BEGIN

1. set $d=\sum_{j=1}^{s} d_i$, $q=true$ and $i=0$;
2. WHILE $i<s$ and $q=true$ DO
   - 2.1. set $i=i+1$ and $\bar{d}=d_i/3$
   - 2.2. random choice of an equation of type (3.1)
   - 2.3. IF $d_i\not\equiv 0 \pmod 3$ THEN $\bar{d}=\bar{d}+1$ ENDIF
   - 2.4. random choice of a rational map $x_0=f_0(y_0,y_1,y_2)$, $x_1=f_1(y_0,y_1,y_2),\dots,$ $x_n=f_n(y_0,y_1,y_2)$ where every f_j is a homogeneous polynomial of degree $\bar{d}$
   - 2.5. IF $d_i\equiv 2 \pmod 3$ THEN

     set the coefficient of $y_2^{\bar{d}}$ equal to zero in all the polynomials f_i

     ENDIF
   - 2.6. IF $d_i\equiv 1 \pmod 3$ THEN

     set the coefficients of $y_2^{\bar{d}}$ and of $y_2^{\bar{d}}$ equal to 0 in all the polynomials f_i

ENDIF

2.7. computation of a set of $d_i + 1$ points that satisfy the equation (3.1) (as it is described in Remark 3.1) and of their images by the rational map; hence, determination of n_i (as it is described in Remark 3.2) and of $m_i = d_i + 2 - n_i$

2.8. IF $p+1-2\sqrt{p} > d_i m_i + 1$ THEN

computation of $d_i m_i + 1$ points that satisfy the equation of type (3.1) that was determined in step 2.2 (see Remark 3.1) and of the set X_i of their images by the rational map

ELSE

set $q = false$

ENDIF

ENDWHILE

3. IF $q = true$ THEN

3.1. set $X = \cup_{i=1}^{h} X_i^i$ and $m = \sum_{i=1}^{h} m_i$

3.2. construction of the minimal generators of the ideal $I(X)$ and computation of the Hilbert function, the Hilbert polynomial and the Poincaré series of $I(E)$ by applying the algorithm of (Cioffi, 1999) up to degree $\alpha + 1$, if the curve has maximal rank, or with the same termination criterion as algorithm of (Albano, Cioffi, Orecchia and Ramella, 2000)

3.3. if the Hilbert polynomial is $dt + s$, then check of the properties of maximal rank and of minimally generation by Theorem 2.2

ELSE

3.4. print: "the chosen characteristic is too low for a successful computation"

ENDIF

END

4. Some applications

EXAMPLE 4.1. *Disjoint union of elliptic curves.*
By running the implementation of the above algorithm over $\mathbb{F}_p$ with $p = 32003$ we have observed the behaviour of disjoint union of space elliptic curves, see (Ballico, specially the introduction). It turns out that if at least one of the irreducible components is degenerate, then there are many examples of union of elliptic curves that are not minimally generated.

If we consider only non degenerate irreducible components, for every $d \leq 30$ and every n such that $3 \leq n \leq 10$ there exists always a union of $s \leq 6$ distinct non degenerate elliptic curves with maximal rank and minimally generated, except for $s = 2$ and $d_1 = d_2 = 4$ or $d_1 = d_2 = 5$ in $\mathbb{P}^3$.

It is easy to explain the first exception. In fact, if $d_1 = d_2 = 4$, each curve E_1, E_2 has maximal rank for every $t \geq 2$ (each curve is complete intersection, hence projectively normal). Then $\alpha_{E_i} = 2$ and $I(E_i)$, $i = 1,2$, has 2 minimal generators of degree 2. Hence $I(E_1 \cup E_2)$ has 4 minimal generators of degree 4. But if the maximal rank holds, then there are no more than 3 generators of degree 4.

The second exception can be proved directly by observing that any elliptic quintic curve E_i sits in 5 independent cubic surfaces, so the union E is contained in 25 sextics (1 more than the expected): if the 25 sextics are independent, the exception is proved; in fact the independence of the 25 sextics can be checked directly by computer.

One would like to get an easy geometrical description of the reason why a general union E of 2 disjoint elliptic curves of degree 5 in $\mathbb{P}^3$ has necessarily one minimal generator of degree 6. Observe that E has no 6-secant lines, for 3-secants to any of its components describe a surface.

EXAMPLE 4.2. *Existence of elliptic curves on quartic threefolds.*
As an application, we show how one can use the previous algorithm to prove the existence of smooth elliptic quintic curves on a general quartic threefold in $\mathbb{P}^4$. This algorithm was used in (Madonna), where curves over a general quartic threefold are studied extensively.

Start with a plane elliptic curve Γ as in the beginning of the algorithm and take 5 general polynomials of degree 2 with one base point on Γ; these polynomial define a map $\Gamma \to \mathbb{P}^4$; call C the image. One uses the algorithm to compute the Hilbert function of C and the generators of its ideal in $\mathbb{P}^4$. As in Proposition 2.1 one can deduce from the Hilbert function that the map is in fact an embedding of Γ as an elliptic quintic curve of maximal rank in $\mathbb{P}^4$.

The ideal of C contains some quartic element F, corresponding to a smooth quartic threefold X, see (Gruson, Lazarsfeld and Peskine). Next one considers the component $\mathcal{H}$ of the Hilbert scheme in $\mathbb{P}^4$, containing C, whose elements are quintic elliptic curves of maximal rank; then one looks at the subvariety $J \subset \mathcal{H} \times H^0(\mathcal{O}_{\mathbb{P}^4}(4))$ of all pairs (C', F') such that C' lies in the threefold defined by F'. The goal is to prove that J dominates $H^0(\mathcal{O}_{\mathbb{P}^4}(4))$. J dominates $\mathcal{H}$, with fibers of dimension 50, so one computes $\dim J = 75$. It is enough to prove that a general fiber of the map $J \to H^0(\mathcal{O}_{\mathbb{P}^4}(4))$ has dimension 5, for $H^0(\mathcal{O}_{\mathbb{P}^4}(4))$ has dimension 70. So one needs to extimate the dimension of H^0 of the normal bundle N of C on X. By Riemann-Roch one gets $\dim H^0(N) = 5$ if and only if $\dim H^1(N) = 0$.

In order to see the vanishing of $H^1(N)$, it is enough to prove the vanishing of $H^1(T_{X|C})$, where $T_{X|C}$ is the restriction to C of the tangent bundle of X. Consider the exact sequence of bundles on C

$$0 \to T_{X|C} \to T_{\mathbb{P}^4|C} \to \mathcal{O}_C(4) \to 0;$$

the image of the induced map $\alpha : H^0(T_{\mathbb{P}^4|C}) \to H^0(\mathcal{O}_C(4))$ is generated by the products of the 5 derivatives of the equation F with linear forms, modulo the ideal of C (i.e. it is a homogeneous piece of the Jacobian ideal of F mod C). Since $H^1(T_{\mathbb{P}^4|C}) = 0$, it is enough to show that α surjects. Since $\dim(H^0(\mathcal{O}_C(4)) = 20$, one just needs to prove that the ideal of C intersects the ideal generated by the derivatives of F, in degree 4, in a vector space of dimension 5.

Once the algorithm provides the equations of C, this last computation is a straightforward application of Gröbner bases calculus that can be made by many softwares dedicated to symbolic computation. All the computations were performed over the finite field $\mathbb{F}_p$ with $p = 32003$ and the output was as expected.

Observe that all the previous computations can be performed in positive characteristic and the output is valid over the complex field. Indeed the curve Γ exists in any characteristic, so does C. We only compute the generators for the ideal of C in some positive characteristic p, but they correspond, mod p, to the equations of C in characteristic 0.

REMARK 4.1. Problems arise in attempting to apply directly the procedure that we described in section 3 to obtain results also over the rational field $\mathbb{Q}$.

Indeed we need to check that the elliptic curve E has "enough" rational points over $\mathbb{Q}$. Recall that the set $E(K)$ of the rational points of an elliptic curve over $K = \mathbb{Q}$ can be structured as an abelian algebraic group that is always finitely generated by a theorem of Mordell, see for example (Silverman). In fact, it is $E(\mathbb{Q}) = E(\mathbb{Q})_{\text{tors}} \oplus \mathbb{Z}^r$, where r is called rank of the curve. Since the number of rational points with finite order is at most 16 (Mazur; Cohen, Th. 7.1.11), the curve has infinite rational points if, and only if, its rank is at least 1. The question of deciding whether the curve has positive rank is still open. However, there are many results about the rank of an elliptic curve and there are many examples of curves with infinitely many rational points. So, to study elliptic curves over $\mathbb{Q}$ it is enough to apply the described algorithm to a plane elliptic curve that is known to have positive rank (for some example of such a curve see (Silverman) and for a very recent result see (Yamagishi)).

References

Albano, G., Cioffi, F., Orecchia, F., and Ramella, I. (2000) Minimally generating ideals of rational parametric curves in polynomial time, *J. Symbolic Computation* **30**, n. 2, 137–149.

Ballico, E. (1986) On the postulation of disjoint rational curves in a projective space, *Rend. Sem. Mat. Univers. Politecn. Torino* **44**, 207–249.

Ballico, E., and Ellia, Ph. (1987) The Maximal Rank Conjecture for Non-Special Curves in $\mathbb{P}^n$, *Math. Z.* **196**, 355–367.

Cioffi, F. (1999) Minimally generating ideals of points in polynomial time using linear algebra, *Ricerche di Matematica* **XLVIII**, Fasc. 1, 55–63.

Cohen, H. (1995) *A Course in Computational Algebraic Number Theory*, GTM 138, Springer Verlag.

Cremona, J. (1992) *Algorithms for Modular Elliptics Curves*, Cambridge University Press.

Clemens, H. (1986) Curves in generic hypersurfaces, *Ann. Sup. Sc. Ecole Norm. Sup.* **19**, 629–636.

Darmon, H. (1999) A Proof of the Full Shimura-Taniyama-Weil Conjecture is Announced, Research News in *Notices of AMS* **46**, n. 11, 1397–1401.

Green, M. (1996) Generic Initial Ideals, in *Six Lectures on Commutative Algebra*, Progress in Mathematics 166, Birkhäuser Verlag.

Gruson, L., Lazarsfeld, R., and Peskine, C. (1983) On a Theorem of Castelnuovo and the Equations Defining Space Curves, *Invent. math.* **72**, 491–506.

Hartshorne, R., and Hirschowitz, A. (1982) Droites en position générale dans l'espace projectif, *Lect. Notes Math.* **961**, 169–189.

Idá, M. (1990) On the homogeneous ideal of the generic union of s-lines in $\mathbb{P}^3$, *J. Reine Angew. Math.* **403**, 67–153.

Johnsen, T., and Kleiman, S. (1996) Rational curves of degree at most 9 on a general quintic threefold, *Comm. Alg.* **24**, 2721–2753.

Kley, H.P. (2000) Rigid curves in complete intersection Calabi-Yau threefolds, *Compositio Mathematica* **123**, no. 2, 185–208.

Kleppe, J.O., and Miró-Roig, M.R. (1998) The dimension of the Hilbert scheme of Gorenstein codimension 3 subschemes, *J. Pure Appl. Alg.* **127**, 73–82.

Madonna, C. (2000) Rank 2 vector bundles on general quartic hypersurfaces in $\mathbb{P}^4$, preprint.

Mazur, B. (1978) Rational isogenies of prime degree, *Invent. Math.* **44**, 129–162.

Migliore, J. (1998) *Introduction to Liaison Theory and Deficiency Modules*, Progress in Mathematics 165, Birkhäuser.

Mumford, D. (1966) *Lectures on curves on an algebraic surface*, Ann. of Math. Studies 59.

Nagel, U. (1990) On Castelnuovo's regularity and Hilbert functions, *Compositio Mathematica* **76**, 265–275.

Orecchia, F. (2001) The ideal generation conjecture for s general rational curves in $\mathbb{P}^r$, *Journal of Pure and Applied Algebra* **155**, 77–89.

Shoup, V. (1999) NTL: a Library for doing Number Theory, ver. 3.9b, at http://www.shoup.net/ntl.

Silverman, J. (1986) *The Arithmetic of Elliptic Curves*, GTM 106, Springer Verlag.

Silverman, J., and Tate, J. (1992) *Rational Points on Elliptic Curves*, UTM 56, Springer Verlag.

Yamagishi, H. (1999) A unified method of construction of elliptic curves with high Mordell-Weil rank, *Pacific J. Math.* **191**, n. 1, 189–200.

THE SEGRE AND HARBOURNE-HIRSCHOWITZ CONJECTURES

C. CILIBERTO*
University of Rome "Tor Vergata", Italy

R. MIRANDA
Colorado State University, Fort Collins, U.S.A.

Abstract. In this paper we prove the equivalence of two conjectures, one by B. Segre, the other by Harbourne and Hirschowitz, concerning linear systems of plane curves with general assigned base points of given multiplicities. Then we deduce, under the assumption that either one of the two conjectures is true, some general results about the structure of the aforementioned linear systems.

Key words: linear systems, plane curves.

Mathematics Subject Classification (2000): 14J26, 14C20.

1. Introduction

Consider the linear system of plane curves of a given degree d having at n specified, but general, points of the plane, singularities of multiplicities say $m_1, \ldots, m_n$. The classical multivariate polynomial interpolation problem consists in evaluting the dimension of the linear system in question. Since the conditions one imposes, by imposing the n multiple points, are $\sum_{i=1}^{n} \frac{m_i(m_i+1)}{2}$ linear and since the plane curves of degree d depend on $\frac{d(d+3)}{2}$ parameters, i.e., up to a scalar, the coefficients of their equation, we see that the expected dimension of the linear system we are dealing with is the maximum between $\frac{d(d+3)}{2} - \sum_{i=1}^{n} \frac{m_i(m_i+1)}{2}$ and -1, the latter being the dimension of the empty system. A system for which the dimension is bigger than the expected one is called *special*, and it is a long-standing problem to classify all special linear systems of the above type.

One way speciality of linear systems can arise, is if they have as a multiple base curve a so called (-1)-*curve*, i.e. a rational curve which, on the blow up of the plane at the multiple base points, ends up having self-intersection -1. An important conjecture due to Harbourne (1986) and Hirschowitz (1989) says that this is

* Partially supported by E.C. project EAGER, contract n. HPRN-CT-2000-00099.

C. Ciliberto et al. (eds.),
Applications of Algebraic Geometry to Coding Theory, Physics and Computation, 37–51.

the only way speciality can arise (see §3). Some authors, among them Gimigliano (1987), also pointed out an earlier conjecture, formulated by B. Segre (1961), to the effect that if a system is special then it has some multiple curve in its base locus (see §2). Clearly Harbourne-Hirschowitz conjecture implies Segre's one. In this paper we prove that also the converse holds (see §4). This is also implied by some results of Nagata's (1959), but the proof presented here, which relies only on basics of surface theory, is really very elementary and faster than Nagata's. We also classify, up to Cremona transformations and under the assumption that one of the conjectures holds, all systems as above whose general curve is reducible (see §5) as well as all special system for which the multiplicities of the assigned base points are the same, i.e. the so called *homogeneous* linear system see §6). Finally, in §7, we observe that Segre's conjecture implies a famous conjecture of Nagata also concerning plane curves.

The multivariate polynomial interpolation problem has, of course, relations and connections with numerical analysis. For this reason some of these issues have been addressed by numerical analysts too. For general references we refer the reader to the expository article (Ciliberto, 2000).

The authors wish to thank H. Clemens for useful discussions on the subject of this paper.

2. The Segre Conjecture

Let $p_1, \ldots, p_n$ be general points in the complex projective plane $\mathbb{P}^2$ and let $m_1, \ldots, m_n$ be positive integers. We let $\mathcal{L}_d$ be the linear system of plane curves of degree d and $\mathcal{L}_d(p_1^{m_1}, \ldots, p_n^{m_n})$ the sub-linear system of $\mathcal{L}_d$ of curves having multiplicity at least m_i at the point p_i, $i = 1, \ldots, n$. If $m_i = 1$ we suppress the superscript m_i for p_i in $\mathcal{L}_d(p_1^{m_1}, \ldots, p_n^{m_n})$.

Let $\pi : S \to \mathbb{P}^2$ be the blow-up of $\mathbb{P}^2$ at the points $p_1, \ldots, p_n$. Let $\mathcal{L}$ be a line bundle on S, or, by abusing notation, the corresponding complete linear system. One defines the *virtual dimension* of $\mathcal{L}$ to be:

$$\nu(\mathcal{L}) := \chi(\mathcal{L}) - 1 = \frac{\mathcal{L} \cdot (\mathcal{L} - K_S)}{2}$$

where K_S is the canonical class on S.

If C is any divisor on S, we similarly define $\nu(C) := \chi(\mathcal{O}_S(C)) - 1$. Riemann-Roch Theorem says that if $\mathcal{L}$ is effective, then

$$\dim(\mathcal{L}) = \nu(\mathcal{L}) + h^1(S, \mathcal{L}) \tag{2.1}$$

since $h^2(\mathcal{L}) = 0$. One also defines the *expected dimension* of $\mathcal{L}$ to be

$$\varepsilon(\mathcal{L}) := \max\{\nu(\mathcal{L}), -1\}.$$

If C is any divisor on S we can accordingly define $\varepsilon(C) := \max\{\nu(C), -1\}$.

One says that a linear system $\mathcal{L}$ on S is *non-special* if its dimension equals the expected dimension. This is equivalent to saying that $\mathcal{L}$ is non-special if and only if $h^0(S,\mathcal{L})\cdot h^1(S,\mathcal{L})=0$. In other words $\mathcal{L}$ is non special if and only if either it is empty or it is *regular*, namely not empty and with $h^1(S,\mathcal{L})=0$.

Let now H be the pull-back via π of a general line of the plane and let $E_1,\dots,E_n$ be the exceptional divisors contracted by π to the points $p_1,\dots,p_n$. The *proper transform* of $\mathcal{L}_d(p_1^{m_1},\dots,p_n^{m_n})$ on S is the complete linear system $\mathcal{L}:=|dH-m_1E_1-\dots-m_nE_n|$. By abusing notation, we will denote by $\mathcal{L}$ also the line bundle associated to this linear system.

We apply the language of virtual and expected dimension to the linear system $\mathcal{L}_d(p_1^{m_1},\dots,p_n^{m_n})$ on the plane also, by using the corresponding notions of the proper transform. In particular, the *virtual dimension* of $\mathcal{L}_d(p_1^{m_1},\dots,p_n^{m_n})$ is

$$\nu(\mathcal{L}_d(p_1^{m_1},\dots,p_n^{m_n})):=\nu(\mathcal{L})=\frac{d(d+3)}{2}-\sum_{i=1}^{n}\frac{m_i(m_i+1)}{2}$$

and the *expected dimension* of $\mathcal{L}_d(p_1^{m_1},\dots,p_n^{m_n})$ is

$$\varepsilon(\mathcal{L}_d(p_1^{m_1},\dots,p_n^{m_n})):=\varepsilon(\mathcal{L}).$$

One says that a system $\mathcal{L}_d(p_1^{m_1},\dots,p_n^{m_n})$ of plane curves is *non-special* if the proper transform $\mathcal{L}$ on S is such.

A linear system $\mathcal{L}$ on S, which is not empty, is called *reducible* [resp. *reduced*] if its general curve C is reducible [resp. reduced]. Bertini's theorem tells us that, if $\mathcal{L}$ is reducible, then either it has some fixed components or it is composed with a rational pencil $\mathcal{P}$, i.e. the movable part of $\mathcal{L}$ consists of the sum of $h\geq 2$ curves of $\mathcal{P}$. The following conjecture is due to B. Segre (1961):

CONJECTURE 2.2 (SEGRE'S CONJECTURE). *Suppose that $\mathcal{L}$ as above is non-empty and reduced. Then $\mathcal{L}$ is non-special.*

Since a plane curve is reduced if and only if it has isolated singularities, another way of phrasing Segre's Conjecture is: if the general member of $\mathcal{L}$ has isolated singularities, then $H^1(S,\mathcal{L})=0$. In this form it may generalize to higher dimensions (Ciliberto, 2000).

3. The Harbourne-Hirschowitz Conjecture

Next we need some more definitions. A smooth irreducible rational curve E on S such that $E^2=-1$ is called a (-1)-*curve*.

Let C be the general curve of the linear system $\mathcal{L}$. The linear system $\mathcal{L}$ is said to be (-1)-*reducible*, if $\mathcal{L}$ is not empty and there is some (-1)-curve E such that $E\cdot C<0$. In this case E is a fixed component of $\mathcal{L}$.

A linear system $\mathcal{L}$ on S, which is not empty, is called (-1)-*special*, if there is some (-1)-curve E such that $E\cdot C<-1$, where, again, C is the general curve of $\mathcal{L}$. If $E\cdot C=-h, h>0$, then hE appears in the fixed part of $\mathcal{L}$. Moreover $E\cdot(C-hE)=0$ and $h^1(S,\mathcal{O}_S(C))\geq\binom{h-1}{2}$ (Ciliberto and Miranda, 1998). This motivates the above definition, inasmuch as a (-1)-special system is indeed special.

Similar definitions can be given for a linear system $\mathcal{L}_d(p_1^{m_1},\ldots,p_n^{m_n})$ of plane curves, inheriting them from the proper transform system $\mathcal{L}$ on S. In particular we will speak of the *self-intersection* of the curves of the system $\mathcal{L}_d(p_1^{m_1},\ldots,p_n^{m_n})$ referring to the self-intersection of the curves in the proper transform system $\mathcal{L}$ on S, etc.

The following conjecture is essentially due to Harbourne (1986) and Hirschowitz (1989) and is related to an earlier conjecture by Gimigliano (1987). We will formulate it as follows.

CONJECTURE 3.1. *In the above situation one has:*

(i) *a linear system $\mathcal{L}$ on S is special if and only if it is (-1)-special;*
(ii) *if a linear system $\mathcal{L}_d(p_1^{m_1},\ldots,p_n^{m_n})$ is non-empty then the general curve in it has singular points of multiplicities exactly $m_1,\ldots,m_n$ at $p_1,\ldots,p_n$;*
(iii) *if a linear system $\mathcal{L}$ on S is non-empty and non-special, then it is irreducible unless either it is (-1)-reducible or it consists of a unique multiple irreducible curve of genus 1 and self-intersection 0 or it is composed of a pencil of rational curves with self-intersection zero;*
(iv) *if a linear system $\mathcal{L}_d(p_1^{m_1},\ldots,p_n^{m_n})$ is non-empty and non-special, and if the general curve in the system is irreducible, then it has ordinary singular points of multiplicities $m_1,\ldots,m_n$ at $p_1,\ldots,p_n$ and it has no other singularities.*

Notice that part (i) of the conjecture in particular applies to systems $\mathcal{L}$ corresponding to systems $\mathcal{L}_d(p_1^{m_1},\ldots,p_n^{m_n})$ of plane curves. Hence, if the conjecture is true, then a system $\mathcal{L}_d(p_1^{m_1},\ldots,p_n^{m_n})$ is special if and only if it is (-1)-special.

4. Segre Implies Harbourne-Hirschowitz

Note that Harbourne-Hirschowitz Conjecture immediately implies Segre's Conjecture, since (-1)-special systems are not reduced. In this section we will show that in fact Segre's Conjecture is not weaker that the Harbourne-Hirschowitz.

We keep the notation from the previous section. We will mainly work with linear systems on the blown-up surface S rather than with systems $\mathcal{L}_d(p_1^{m_1},\ldots,p_n^{m_n})$. First we notice that Segre's Conjecture implies the following:

THEOREM 4.1. *Suppose that Segre's Conjecture is true. Let $p_1,\ldots,p_n$ be general points in the plane and let S be the blow-up of the plane at $p_1,\ldots,p_n$. Let $\mathcal{L}$ be a complete linear system on S which is not empty and such that its general curve C is irreducible and reduced of arithmetic genus g. Then:*

(i) $H^1(S, \mathcal{O}_S(C)) = 0$, $\dim(\mathcal{L}) = \nu(\mathcal{L})$, *and* $\mathcal{L}$ *is non-special;*
(ii) $C^2 \geq g-1$ *and therefore* $C^2 < 0$ *if and only if* C *is a* (-1)*-curve.*

PROOF. The assertion about $H^1(S, \mathcal{O}_S(C))$ is exactly Segre's Conjecture; the rest of (i) follows from (2.1). Part (ii) follows by Riemann-Roch and $\nu(C) \geq 0$: we have $C^2 + C \cdot K_S = 2g-2$, and $2\nu(C) = C^2 - C \cdot K_S \geq 0$; summing these gives the result. □

From this theorem we will deduce the following result, which relates to parts (i), (ii) and (iii) of the Harbourne-Hirschowitz Conjecture.

THEOREM 4.2. *Suppose that Segre's Conjecture is true. Let* $p_1, \ldots, p_n$ *be general points in the plane and let* S *be the blow-up of the plane at* $p_1, \ldots, p_n$. *Let* $\mathcal{L}$ *be a complete, nonempty linear system on* S. *Then:*

(i) If $\mathcal{L}$ *is special then* $\mathcal{L}$ *is* (-1)*-special.*
(ii) If $\mathcal{L}$ *is non-special and the general element of* $\mathcal{L}$ *is not reduced, then* $\dim(\mathcal{L}) = 0$ *and* $\mathcal{L}$ *consists of a unique curve of the form* $mC + D_1 + \ldots + D_h$, *where* C *is an elliptic curve with self intersection* 0 *and* $D_1, \ldots, D_h$ *are* (-1)*-curves which are mutually disjoint and disjoint from* C.
(iii) If $\mathcal{L}$ *is non-special, the general curve in* $\mathcal{L}$ *is reduced, and there are fixed components of* $\mathcal{L}$, *then the fixed components of* $\mathcal{L}$ *are disjoint curves which are* (-1)*-curves, plus, at most one more curve of positive genus. If there is such a fixed component of positive genus, then* $\mathcal{L}$ *has dimension* 0.
(iv) If $\mathcal{L}$ *is non-special of positive dimension, then its general curve is reduced and its movable part is disjoint from the fixed part, consisting of a union of disjoint* (-1)*-curves. The movable part, if reducible, is composed of a base point free pencil of rational curves.*

We will prove Theorem 4.2 in several steps. We start with the following:

LEMMA 4.3. *Suppose that Segre's Conjecture is true. Let* $p_1, \ldots, p_n$ *be general points in the plane and let* S *be the blow-up of the plane at* $p_1, \ldots, p_n$. *Suppose that* C *and* D *are general members of their linear systems* $|C|$ *and* $|D|$ *on* S *respectively, and that they are both irreducible and disjoint. If* C *has positive arithmetic genus, then* D *has arithmetic genus zero and* $\dim|D| = 0$.

PROOF. Consider the exact sequence

$$0 \to \mathcal{O}_S(D-C) \to \mathcal{O}_S(D) \to \mathcal{O}_C(D) \cong \mathcal{O}_C \to 0$$

the final isomorphism holding since C and D are disjoint. By Theorem 4.1,(i), the higher cohomology of $\mathcal{O}_S(D)$ is zero, so that $h^2(\mathcal{O}_S(D-C)) = h^1(\mathcal{O}_C) \geq 1$. Hence by Serre duality $|K_S + C - D|$ is not empty. Reversing the roles of C and D we see by symmetry that $h^2(\mathcal{O}_S(C-D)) = h^1(\mathcal{O}_D)$. If D has positive genus, then this is also nonzero; hence $|K_S + D - C|$ is not empty, and we conclude that $|2K_S|$ is not empty, a contradiction. Therefore D has genus zero.

If $\dim|D| \geq 1$, then $h^0(\mathcal{O}_S(D)) \geq 2$, so the above exact sequence implies that $h^0(\mathcal{O}_S(D-C)) \geq 1$, i.e., $|D-C|$ is not empty. Since we have seen that $|K_S+C-D|$ is also not empty, we conclude that $|K_S|$ is not empty, which is a contradiction. Therefore $\dim|D| = 0$ as claimed. □

The following lemma analyzes the fixed components of a linear system $\mathcal{L}$ on a general blowup S of the plane.

LEMMA 4.4. *Suppose that Segre's Conjecture is true. Let S and $\mathcal{L}$ be as above, with $\mathcal{L}$ not empty, and let C be the general curve in $\mathcal{L}$. Let us write:*

$$C = h_1C_1 + \ldots + h_kC_k + D_1 + \ldots + D_h + M, \quad k \geq 0, \quad h \geq 0$$

where $C_1, \ldots, C_k, D_1, \ldots, D_h$ are the distinct, irreducible, fixed components of $\mathcal{L}$, and M is the general curve in the movable part of the system (which may be empty). Also we assume $h_1 \geq \ldots \geq h_k \geq 2$. Then one has:

(i) $C_i^2 = C_i \cdot K_S$, $C_i^2 \leq 0$, $i = 1, \ldots, k$; $D_j^2 = D_j \cdot K_S$, $j = 1, \ldots, h$;
(ii) $C_i \cdot C_j = 0$, $1 \leq i < j \leq k$; $D_i \cdot D_j = 0$, $1 \leq i < j \leq h$; $C_i \cdot D_j = 0$, $i = 1, \ldots, k$, $j = 1, \ldots, h$;
(iii) if F is any irreducible component of M, then $F \cdot C_i = F \cdot D_j = 0$, $i = 1, \ldots, k$, $j = 1, \ldots, h$.

PROOF. One has

$$\dim|C_i| = \dim|D_j| = 0, \text{ for } i = 1, \ldots, k, \ j = 1, \ldots, h$$

since they are fixed components. By Theorem 4.1, this implies that:

$$\nu(C_i) = \nu(D_j) = 0, \text{ for } i = 1, \ldots, k, \ j = 1, \ldots, h$$

i.e.

$$C_i^2 = C_i \cdot K_S, \text{ for } i = 1, \ldots, k; \ \ D_j^2 = D_j \cdot K_S, \text{ for } j = 1, \ldots, h.$$

Also

$$\dim|C_i + C_j| = 0, \text{ for } 1 \leq i < j \leq k,$$

and hence (using (2.1)) we have

$$\nu(C_i + C_j) \leq 0, \text{ for } 1 \leq i < j \leq k.$$

This yields:

$$C_i \cdot C_j = 0, 1 \leq i < j \leq k$$

and, in a completely similar manner, we prove the rest of (ii).

Let F be an irreducible component of M. Since F must move in its linear system, in this case we have

$$\dim|F| = \nu(F) > 0.$$

Since $\dim|F+A| = \dim|F|$ for $A = C_i$ or $A = D_j$, we must have

$$\nu(F+C_i) \leq \nu(F), \text{ for } i = 1,\dots,k \text{ and } \nu(F+D_j) \leq \nu(F), \text{ for } j = 1,\dots,h$$

which yields (iii), since for any curve A we have

$$2(\nu(F+A) - \nu(F)) = 2A\cdot F + A^2 - A\cdot K_S$$

and applying this to either $A = C_i$ or $A = D_j$ and using the proved part of (i) gives $F\cdot C_i \leq 0$ and $F\cdot D_j \leq 0$. Since F, C_i and D_j are actual curves on S, we must have $F\cdot C_i = F\cdot D_j = 0$.

Finally we have:

$$\dim|2C_i| = 0, \text{ for } i = 1,\dots,k,$$

hence

$$\nu(2C_i) \leq 0, \text{ for } i = 1,\dots,k.$$

This reads:

$$2C_i^2 \leq C_i\cdot K_S, \text{ for } i = 1,\dots,k,$$

which, together with $C_i^2 = C_i\cdot K_S$ forces $C_i^2 \leq 0$. This finishes the proof of (i). □

Next we analyze the nonreduced systems, and prove the:

PROPOSITION 4.5. *Suppose that Segre's Conjecture is true. With the same assumptions and notation as in Lemma 4.4, assume that the general member of the system is nonreduced, i.e., that* $k \geq 1$. *Then either one of the curves* C_i *is a* (-1)*-curve, in which case the system is* (-1)*-special, or* $k = 1$, $C = C_1$ *is elliptic,* $D_1,\dots,D_h$ *are* (-1)*-curves,* $M = 0$ *and the system is non-special.*

PROOF. By the adjunction formula, we have $2C_i^2 = C_i^2 + C_i\cdot K_S \geq -2$ for $i = 1,\dots,k$. Hence, if $C_i^2 < 0$, then one has $C_i^2 = C_i\cdot K_S = -1$ and C_i is a (-1)-curve. Since $h_i \geq 2$ the system is special in this case.

Suppose that $C_i^2 = 0$ for all $i = 1,\dots,k$, i.e. each C_i is an elliptic curve. If $k \geq 2$ we have a contradiction, applying Lemma 4.3, since the C_i's are disjoint. Therefore we must have $k = 1$ and we set $C_1 = C$, $h_1 = h$.

Let F be an irreducible component of M; since $F\cdot C = 0$, we apply Lemma 4.3 again and conclude that F has genus zero and $\dim|F| = 0$. This gives a contradiction, since F moves in its linear system. We conclude that $M = 0$.

Finally for every $j = 1,\dots,h$, we apply Lemma 4.3 to C and D_j and conclude that each D_j is rational. Then Lemma 4.4,(i) implies that it is a (-1)-curve (since $D_j^2 = D_j\cdot K_S$ and the sum is -2 by rationality).

The assertion about the non-speciality of the system is an easy computation: $\nu(h_1C_1 + D_1 + \ldots + D_h) = 0$ and the system also has dimension zero. □

We next analyze systems with no fixed components:

PROPOSITION 4.6. *Suppose that Segre's Conjecture is true. With the same assumptions and notation as in Lemma 4.4, assume that $k = h = 0$, i.e. $\mathcal{L}$ has no fixed components. Then either M is irreducible or $\mathcal{L}$ is composed of a base point free pencil of rational curves; in either case the system is non-special.*

PROOF. If M is irreducible, then the system is non-special by Theorem 4.1,(i). Assume M is reducible. Then $\mathcal{L}$ is composed of a pencil, namely $M \equiv \mu F$, where $|F|$ is a pencil and $\mu \geq 2$. By Theorem 4.1 we have

$$\dim|F| = \nu(F) = 1,$$

i.e. $F^2 - F \cdot K_S = 2$. On the other hand one has

$$\nu(\mu F) \leq \dim|\mu F| = \dim|M| = \mu$$

which gives $\mu^2 F^2 - \mu F \cdot K \leq 2\mu$, or $\mu F^2 - F \cdot K \leq 2$. Subtracting $F^2 - F \cdot K_S = 2$ from this inequality gives $(\mu - 1)F^2 \leq 0$. Since F moves in a pencil, we must have $F^2 \geq 0$, which forces $F^2 = 0$; then $F \cdot K_S = -2$, which implies that $|F|$ is a base-point-free rational pencil as claimed. Moreover in this case then $\nu(\mu F) = \mu$, so that $M \equiv \mu F$ is non-special also. □

Now we can analyse the reduced linear systems with fixed components and prove the following:

PROPOSITION 4.7. *Suppose that Segre's Conjecture is true. With the same assumptions and notation as in Lemma 4.4, assume that $k = 0$ and $h > 0$. Then either all the curves D_j are (-1)-curves or all of them, but one, say D_1, are (-1)-curves. In the latter case $M = 0$ and in both cases the system is non-special.*

PROOF. The proof imitates the one of Proposition 4.5. Since the D_j's are disjoint, we may apply Lemma 4.3 and conclude that at most one of them is non-rational. Any rational D_j must be a (-1)-curve (see Lemma 4.4).

Suppose that one of them, say D_1, is not rational and consider an irreducible component F of M. Since $F \cdot D_1 = 0$, we apply Lemma 4.3 and conclude that F is rational and does not move in its linear system, a contradiction. Therefore $M = 0$.

If $M = 0$, the assertion about the non-speciality of the system is a straightforward computation: $\nu(D_1 + \cdots + D_h) = 0$, and the dimension is also zero. If $M \neq 0$, then each D_j is a (-1)-curve. Again one checks that

$$\nu(D_1 + \cdots + D_h + M) = \nu(M)$$

so that since these systems have the same dimension, $D_1 + \cdots + D_h + M$ is non-special if and only if M is nonspecial. The result now follows from the previous Proposition. □

The above Propositions prove Theorem 4.2. Next we see how Conjecture 3.1, (i, ii, iii) follows from the above results.

THEOREM 4.8. *Suppose that Segre's Conjecture is true. Then parts (i), (ii) and (iii) of Conjecture 3.1 hold.*

PROOF. Part (i) of the Conjecture is nothing but Theorem 4.2, (i). Part (iii) also follows from Theorem 4.2, (ii, iii, iv).

As for part (ii), let $\mathcal{L}$ be the proper transform of $\mathcal{L}_d(p_1^{m_1},...,p_n^{m_n})$ on S. Let C be the general curve in $\mathcal{L}$. Suppose that the exceptional divisor E_i contracted to p_i is an irreducible component of multiplicity $h \geq 1$ of C. Then, by Lemma 4.4, (ii, iii) one has $E \cdot (C - hE) = 0$ hence $m_i = C \cdot E = hE^2 = -h < 0$, a contradiction. Hence E_i is not a component of C. Since $C \cdot E_i = m_i$, then 3.1, (ii) also follows. □

5. Reducible, Non-special Linear Systems

In this section we will describe non-special, reducible linear systems which we met in Conjecture 3.1, (iii) and in Theorem 4.2. First a remark is in order. We express it in the following Lemma which is a corollary of Theorem 4.1.

LEMMA 5.1. *Suppose that Segre's Conjecture is true. Let $p_1,...,p_n$ be general points in the plane and let S be the blow-up of the plane at $p_1,\ldots,p_n$. Let X be a minimal rational surface obtained from S by a sequence of contractions of (-1)-curves. Then either $X = \mathbb{P}^2$ or $X = \mathbb{P}^1 \times \mathbb{P}^1$. In the former case the map $\sigma : S \to X = \mathbb{P}^2$ is a sequence of contractions of n disjoint (-1)-curves of S to general points $q_1,\ldots,q_n$ of the plane; in the latter case the map $\sigma : S \to X = \mathbb{P}^1 \times \mathbb{P}^1$ is a sequence of contractions of $n-1$ disjoint (-1)-curves of S to general points $r_1,\ldots,r_{n-1}$ of $\mathbb{P}^1 \times \mathbb{P}^1$. In the latter case S also contracts to $\mathbb{P}^2$, as in the former case.*

PROOF. The surface X is either $\mathbb{P}^2$ or a Hirzebruch surface $\mathbb{F}_n$, with either $n \geq 2$ or $n = 0$. But $\mathbb{F}_n$ contains a smooth rational curve with self-intersection $-n$; in this case Theorem 4.1,(ii) implies that $n = 0$. Hence either $X = \mathbb{P}^2$ or $X = \mathbb{F}_0 = \mathbb{P}^1 \times \mathbb{P}^1$.

Suppose $X = \mathbb{P}^2$. The map $\sigma : S \to X = \mathbb{P}^2$ is a sequence of n contractions of (-1)-curves $\sigma_i : S_i \to S_{i+1}$, $i = 0,\ldots,n$, where $S_0 = S$ and $S_n = \mathbb{P}^2$. Clearly, each surface S_i shares with S the property of not containing any irreducible curve of negative self-intersection, except for (-1)-curves. One moment of reflection shows that this implies that σ is obtained by contracting n disjoint (-1)-curves of S to points $q_1,...,q_n$ of the plane. The fact that $q_1,...,q_n$ must be general follows by an obvious parameter count.

A similar argument works in the case $X = \mathbb{P}^1 \times \mathbb{P}^1$.

Finally the last assertion follows from noting that one blow-up of $\mathbb{P}^1 \times \mathbb{P}^1$ contracts to $\mathbb{P}^2$. □

With the above notation, suppose $X = \mathbb{P}^2$ (which we may assume by the last assertion). Let $\sigma : S \to X$ be a birational map obtained by a sequence of contractions of n disjoint (-1)-curves of S as in the statement of Lemma 5.1 above. Then $\sigma \circ \pi^{-1} : \mathbb{P}^2 \to \mathbb{P}^2$ is a *Cremona transformation.*

We will use such Cremona transformations to understand the linear systems $\mathcal{L}_d(p_1^{m_1}, \ldots, p_n^{m_n})$ which are non-special but reducible. Given a non-special system $\mathcal{L}_d(p_1^{m_1}, \ldots, p_n^{m_n})$, we consider its proper transform $\mathcal{L}$ on S. Then $\sigma_*(\mathcal{L})$ is a linear system of the type $\mathcal{L}_\delta(q_1^{n_1}, \ldots, q_r^{n_r})$, where $q_1, \ldots, q_r$ are general points in $\mathbb{P}^2$, and actually $r \leq n$. We will define this system to be the *Cremona transform* of $\mathcal{L}_d(p_1^{m_1}, \ldots, p_n^{m_n})$ via the given Cremona transformation $\sigma \circ \pi^{-1}$. Of course the dimension of the two systems are the same.

If $\mathcal{L}'$ is the system associated to $\mathcal{L}_\delta(q_1^{n_1}, \ldots, q_r^{n_r})$ on S, it is clear that $\mathcal{L} \otimes \mathcal{L}'^{-1} \cong \mathcal{O}_S(E)$ where $E = E_1 + \ldots + E_h$ and $E_1, \ldots, E_h$ are the distinct (-1)-curves contracted by σ which are in the base locus of $\mathcal{L}$. Hence also $\nu(\mathcal{L}) = \nu(\mathcal{L}')$.

Having all this in mind, we can state our next theorem:

THEOREM 5.2. *Suppose that Segre's Conjecture is true. Let $p_1, \ldots, p_n$ be general points in the plane and let S be the blow-up of the plane at $p_1, \ldots, p_n$. Let $\mathcal{L}_d(p_1^{m_1}, \ldots, p_n^{m_n})$ be a non-special, reducible linear system. Then it can be Cremona transformed either to:*

(i) the linear system $\mathcal{L}_d(p^d)$ whose general member consists of d lines through a point p, or;

(ii) the linear system $\mathcal{L}_{3m}(q_1^m, \ldots, q_9^m)$ that consists of the unique cubic curve containing the nine general points $q_1, \ldots, q_9$, counted with multiplicity m.

The main tool for proving this result is the following classical result, apparently already known to M. Noether:

THEOREM 5.3. *Let $p_1, \ldots, p_n$ be general points in the plane and consider a linear system $\mathcal{L}_d(p_1^{m_1}, \ldots, p_n^{m_n})$ with $m_1 \geq m_2 \geq \ldots \geq m_n$ and $m_3 \geq 1$. Let S be the blow-up of the plane at the points $p_1, \ldots, p_n$ and let $\mathcal{L}$ be the proper transform of $\mathcal{L}_d(p_1^{m_1}, \ldots, p_n^{m_n})$ on S. Assume that the general member C of $\mathcal{L}$ is irreducible and let g be its arithmetic genus. If $(m_3 - 1)C^2 \geq 2m_3(g-1)$ and either the strict inequality holds or $m_3 < m_1$, then $\mathcal{L}_d(p_1^{m_1}, \ldots, p_n^{m_n})$ can be Cremona transformed to a system $\mathcal{L}_\delta(q_1^{n_1}, \ldots, q_r^{n_r})$ with $\delta < d$.*

If $(m_3 - 1)C^2 = 2m_3(g-1)$ and $m_1 = m_2 = m_3 = m$, then $\mathcal{L}_d(p_1^{m_1}, \ldots, p_n^{m_n})$ can be Cremona transformed to a system $\mathcal{L}_\delta(q_1^{n_1}, \ldots, q_r^{n_r})$ with $\delta < d$ unless $d = 3m$.

PROOF. We want to prove that $d < m_1 + m_2 + m_3$, since then the quadratic transformation based at p_1, p_2, p_3 does the job of Cremona transforming the given linear system into a new one with lower degree.

One has:

$$d(d-3) - \sum_{i=1}^{n} m_i(m_i - 1) = 2g - 2 \quad \text{and} \quad d^2 - \sum_{i=1}^{n} m_i^2 = C^2.$$

Multiplying the first by m_3 and the second by $1-m_3$, summing, and re-arranging, yields

$$d(3m_3-d)+m_1^2+m_2^2-m_3(m_1+m_2)-[(m_3-1)C^2-2m_3(g-1)]$$
$$=m_3\sum_{i=4}^{n}m_i-\sum_{i=4}^{n}m_i^2\geq 0.$$

The term $d(3m_3-d)$ is maximized for $d=3m_3/2$, and monotonically decreases for larger d. Hence if $d\geq m_1+m_2+m_3$ we would have ***

$$(m_1+m_2+m_3)(2m_3-m_1-m_2)+m_1^2+m_2^2-m_3(m_1+m_2)$$
$$-[(m_3-1)C^2-2m_3(g-1)]\geq 0$$

which simplifies to the inequality

$$2(m_3^2-m_1m_2)-[(m_3-1)C^2-m_3(2g-2)]\geq 0.$$

Since $m_1\geq m_2\geq m_3$, the first term here is non-positive, and is strictly negative if $m_3<m_1$. Therefore by the first assumption we have a contradiction.

The last assertion is also clear. □

COROLLARY 5.4. *Suppose that Segre's Conjecture is true. Let $p_1,\ldots,p_n$ be general points in the plane. An irreducible linear system $\mathcal{L}_d(p_1^{m_1},\ldots,p_n^{m_n})$ whose general member has genus $g\leq 1$ can be Cremona transformed to one of the following types:*

- *a linear system of lines, dimension $l\leq 2$ and $g=0$*
- *the linear system of all conics, dimension 5 and $g=0$*
- *a linear system $\mathcal{L}_d(p^{d-1})$, $d\geq 2$, dimension $2d$ and $g=0$*
- *a linear system $\mathcal{L}_d(p^{d-1},q)$, $d\geq 2$, dimension $2d-1$ and $g=0$*
- *a linear system of cubics $\mathcal{L}_3(p_1,\ldots,p_h)$, $h\leq 9$, dimension $9-h$ and $g=1$*
- *a linear system $\mathcal{L}_4(p_1^2,p_2^2)$, dimension 8 and $g=1$.*

PROOF. Assume $g=0$. Then Theorem 5.3 tells us that $\mathcal{L}_d(p_1^{m_1},\ldots,p_n^{m_n})$ can be Cremona transformed to a system of lower degree, unless $n\leq 2$. In other words $\mathcal{L}_d(p_1^{m_1},\ldots,p_n^{m_n})$ can be Cremona transformed to a system of one of the forms $\mathcal{L}_\delta$, $\mathcal{L}_\delta(q^n)$, or $\mathcal{L}_\delta(q_1^{n_1},q_2^{n_2})$. It is then easy to see that the only possible cases are those listed in our statement.

Let $g=1$. Theorem 5.3 tells us that $\mathcal{L}_d(p_1^{m_1},\ldots,p_n^{m_n})$ can be Cremona transformed to a system of lower degree, unless either $n\leq 2$, or $m_1=m_2=m_3=m$, $C^2(m_3-1)=0$ and $d=3m$. Therefore the possibilities are:

(a) $n\leq 2$;
(b) $m_1=m_2=m_3=1$;
(c) $m_1=m_2=m_3=m\geq 2$, $C^2=0$, $d=3m$.

Cases (a) and (b) easily lead to the possibilities we listed in our statement. Case (c) cannot occur. In fact $d = 3m$ implies $n = 9$ and $m_1 = ... = m_9 = m$. Since there is a unique cubic E through $p_1, ..., p_9$ and since, by hypothesis, the original system contains an irreducible curve C, we see that C should vary in a pencil, the one generated by C and mE. Then we would have $\mathcal{O}_C(C) \simeq \mathcal{O}_C$, hence $h^1(C, \mathcal{O}_C(C)) = 1$ contradicting Theorem 4.1, (i). □

REMARK. The numerical case (c) which appears in the analysis of the case $g = 1$ in the above proof is impossible if $p_1, \ldots, p_9$ are general points, but can occur for special positions of the points, leading to the well known *Halphen pencils* (see (Cossec and Dolgachev, 1989)).

We are now in a position to give the:

PROOF OF THEOREM 5.2. Let $\mathcal{L}$ be the proper transform of $\mathcal{L}_d(p_1^{m_1}, ..., p_n^{m_n})$ on S, as usual. By blowing-down all the (-1)-curves that appear in the fixed part of $\mathcal{L}$, we may reduce ourselves, up to Cremona transformations, to assume that $\mathcal{L}$ is not (-1)-reducible. Then either $\mathcal{L}$ consists of a unique multiple elliptic curve with self-intersection 0, or it is composed of a pencil of rational curves with self intersection 0. The conclusion then follows by Corollary 5.4. □

6. Homogeneous Linear Systems

We want to address now the analysis of the so-called *homogeneous linear systems*, i.e. linear systems $\mathcal{L}_d(p_1^{m_1}, \ldots, p_n^{m_n})$ with $m_1 = \ldots = m_n = m$. We will denote such systems briefly by $\mathcal{L}_d(n^m)$.

Our result is the following:

THEOREM 6.1. *Suppose that Segre's Conjecture is true. One has:*

(i) a homogeneous linear system $\mathcal{L}_d(n^m)$ is special if and only if one of the following cases occurs:

1. $n = 2, m \leq d \leq 2m - 2$
2. $n = 3, \frac{3m}{2} \leq d \leq 2m - 2$
3. $n = 5, 2m \leq d \leq \frac{5m-2}{2}$
4. $n = 6, \frac{12m}{5} \leq d \leq \frac{5m-2}{2}$
5. $n = 7, \frac{21m}{8} \leq d \leq \frac{8m-2}{3}$
6. $n = 8, \frac{48m}{17} \leq d \leq \frac{17m-2}{6}$

(ii) a homogeneous, non-special linear system $\mathcal{L}_d(n^m)$ is (-1)-reducible if and only if one of the following cases occurs:

1. $n = 2, d = 2m - 1$
2. $n = 3, d = 2m - 1$

3. $n = 5, d = \frac{5m-1}{2}$
4. $n = 6, d = \frac{5m-1}{2}$
5. $n = 7, d = \frac{8m-1}{3}$
6. $n = 8, d = \frac{17m-1}{6}$

(iii) a homogeneous, non-special linear system $\mathcal{L}_d(n^m)$ is reducible but not (-1)-reducible if and only if one of the following cases occurs:

1. $n = 1, d = m$
2. $n = 4, d = 2m$
3. $n = 9, d = 3m$.

In particular, if $n \geq 9$ any homogeneous linear system $\mathcal{L}_d(n^m)$ is non-special and irreducible.

PROOF. Part (i) follows from Theorem (2.4) of (Ciliberto and Miranda, 2001) and from Theorem 4.8. Part (ii) follows by the same argument of Theorem (2.4) of (Ciliberto and Miranda, 2001).

In case (iii) we know that either:
(a) $\mathcal{L}_d(n^m)$ consists of a unique elliptic curve E of self-intersection 0, counted with multiplicity $\mu \geq 2$ or;
(b) it is composed of a pencil $\mathcal{P}$ of rational curves of self-intersection 0.

In case (a), by obvious monodromy arguments, E has to contain all the n base points of $\mathcal{L}_d(n^m)$. This forces $n = 9$ and $\mu = m$, thus $d = 3m$.

In case (b), one similarly sees that $\mathcal{P}$ can either be the pencil of lines through a point or the pencil of conics through four points. This finishes the proof of the theorem. □

7. Nagata's Conjecture

In this section we prove that Segre's Conjecture implies the following conjecture due to Nagata.

THEOREM 7.1. *Suppose that Segre's Conjecture is true. Let $p_1, \ldots, p_n$ be general points in the plane. Let C be a plane curve of degree d having multiplicity at least m_i at the point p_i. Then for all $n \geq 10$*

$$\sum_{i=1}^{n} m_i < d\sqrt{n}.$$

The theorem (without the Segre Conjecture hypothesis) was proved by M. Nagata (1959) for all perfect squares $n = p^2$, $p \geq 4$. The technique used is specializations of the points in the plane. As a consequence of this result Nagata produced a counterexample to Hilbert's 14th problem. More recently Z. Ran (1998) has given an affirmative answer for a broad selection of integers $n \geq 10$. The proof relies on degenerating the plane to certain reducible surfaces called mosaics. It is also worth mentioning that G. Xu (1994) proves a similar bound, namely that

$$d \geq \frac{\sqrt{n-1}}{n} \sum_{i=1}^{n} m_i.$$

Recently B. Harbourne (2001) has made progress on Nagata's Conjecture.

See also (Xu, 1994) and (McDuff and Polterovich, 1994) for applications to symplectic packings of $B^4(1)$ (the symplectic ball of radius 1), and (Xu, 1994) for applications to multiple-point Seshadri constants.

PROOF. If $\sum_{i=1}^{n} m_i \geq d\sqrt{n}$, then there exists an irreducible and reduced component C' of C of degree d' with multiplicity at least m'_i at the point p_i such that

$$\sum_{i=1}^{n} m'_i \geq d'\sqrt{n}.$$

We may therefore assume that the curve C itself is reduced and irreducible. By 4.1 the linear system $\mathcal{L}_d(p_1^{m_1}, \ldots, p_n^{m_n})$ is non empty and non-special. In particular

$$\dim(\mathcal{L}_d(p_1^{m_1}, \ldots, p_n^{m_n})) = \frac{d(d+3) - \sum_{i=1}^{n} m_i(m_i+1)}{2} \geq 0.$$

If $\sum_{i=1}^{n} m_i \geq d\sqrt{n}$, then

$$\begin{aligned} 0 &\leq d(d+3) - \sum_{i=1}^{n} m_i(m_i+1) \\ &\leq \frac{(\sum_{i=1}^{n} m_i)^2}{n} + \frac{3\sum_{i=1}^{n} m_i}{\sqrt{n}} - \sum_{i=1}^{n} m_i^2 - \sum_{i=1}^{n} m_i \end{aligned}$$

By Cauchy-Schwartz $(\sum_{i=1}^{n} m_i)^2 \leq n(\sum_{i=1}^{n} m_i^2)$. It follows that

$$0 \leq \sum_{i=1}^{n} m_i \left(\frac{3}{\sqrt{n}} - 1 \right).$$

This is clearly impossible for all $n \geq 10$. □

References

Cossec, F.R., and Dolgachev, I.V. (1989) *Enriques Surfaces I*, Progress in Mathematics 76, Birkhäuser, Boston.

Ciliberto, C. (2000) Geometric aspects of polynomial interpolation in more variables and of Waring's problem, in *Proceedings of the Third European Congress of Math.*, Barcelona 2000, Progress in Mathematics, Birkhäuser.

Ciliberto, C., and Miranda, R. (1998) Degenerations of Planar Linear Systems, *J. reine angew. Math.* **501**, 191–220.

Ciliberto, C., and Miranda, R. (2001) Linear Systems of Plane Curves with Base Points of Equal Multiplicity, to appear in the *Transactions of the AMS*.

Gimigliano, A. (1987) On linear systems of plane curves, Thesis, Queen's University, Kingston.

Harbourne, B. (1986) The Geometry of rational surfaces and Hilbert functions of points in the plane, *Can. Math. Soc. Conf. Proc.* **6**, 95–111.

Harbourne, B. (2001) On Nagata's Conjecture, *Journal of Algebra* **236**, 692–702.

Hirschowitz, A. (1989) Une conjecture pour la cohomologie des diviseurs sur les surfaces rationnelles generiques, *J. reine angew. Math.* **397 (1989)** 208-213.

McDuff, D., and Polterovich, L. (1994) Symplectic packings and algebraic geometry, *Inventiones Math.* **115**, 405–429.

Nagata, M. (1959) On the fourteenth problem of Hilbert, *American J. Math.* **81**, 766–772.

Segre, B. (1961) Alcune questioni su insiemi finiti di punti in geometria algebrica, *Atti Convegno Intern. di Geom. Alg. di Torino*, 15–33.

Xu, G. (1994) Curves in $\mathbb{P}^2$ and symplectic packings, *Math. Ann.* **299**, 609–613.

Xu, G. (1995) Ample line bundles on smooth surfaces, *J. Reine angew. Math.* **469**, 199–209.

Ran, Z. (1998) On the Nagata Problem, preprint, math.AG/9809101.

PILLOW DEGENERATIONS OF K3 SURFACES

C. CILIBERTO*
University of Rome "Tor Vergata", Italy

R. MIRANDA
Colorado State University, Fort Collins, U.S.A.

M. TEICHER
Bar-Ilan University, Ramat Gan, Israel

Abstract. In this paper we study a particular projective degeneration of general (not necessarily primitive) $K3$ surfaces into a union of planes.

Key words: $K3$ surfaces, degenerations.

Mathematics Subject Classification (2000): 14J28, 14D06.

1. Introduction

In this article we construct a specific projective degeneration of $K3$ surfaces of degree $2g-2$ in $\mathbb{P}^g$ to a union of $2g-2$ planes, which meet in such a way that the combinatorics of the configuration of planes is a triangulation of the 2-sphere. Abstractly, such degenerations are said to be Type III degenerations of $K3$ surfaces, see (Kulikov, 1977), (Persson and Pinkham, 1981), (Friedman and Morrison, 1983). Although the birational geometry of such degenerations is fairly well understood, the study of projective degenerations is not nearly as completely developed.

In (Ciliberto, Lopez and Miranda, 1993), projective degenerations of $K3$ surfaces to unions of planes were constructed, in which the general member was embedded by a primitive line bundle. The application featured there was a computation of the rank of the Wahl map for the general hyperplane section curve on the $K3$ surface.

In this article we construct degenerations for which the general member is embedded by a *multiple* of the primitive line bundle class. The construction depends

* Partially supported by E.C. project EAGER, contract n. HPRN-CT-2000-00099.

C. Ciliberto et al. (eds.),
Applications of Algebraic Geometry to Coding Theory, Physics and Computation, 53–63.

on two parameters, and we intend in follow-up work to use these degenerations to compute braid monodromy for Galois coverings, in the style of Moishezon and Teicher (1987 and 1994). We hope that the freedom afforded by the additional discrete parameters in the construction will yield interesting phenomena related to fundamental groups.

The specific degenerations which we construct can be viewed as two rectangular arrays of planes, joined along their boundary; for this reason we have given them the name "pillow" degenerations. They are described in Section 3. Following that, in Section 4, we study the degeneration of the general branch curve (for a general projection of the surfaces to a plane) to a union of lines (which is the "branch curve" for the union of planes). In particular when the general branch curve is a plane curve having only nodes and cusps as singularities, we describe the degeneration of the nodes and the cusps to the configuration of the union of lines. This is critical information in the application to the computation of the braid monodromy.

We are not aware of a modern reference for the statement that the general branch curve for a linear projection of a surface to a plane has only nodes and cusps as singularities. In this article we will operate under the assumption that this "folklore" statement is true and proceed. The reader may wish to consult (Kulikov and Kulikov, 2000) for further information. We have included a short section at the beginning of the article deriving the characters of a general branch curve (degree, number of nodes and cusps) for the convenience of the reader, under this assumption.

The authors are grateful to the NATO Scientific Affairs Division, the Ministry of Science of Israel, the Emmy Noether Research Institute of Mathematics at Bar-Ilan University, and EAGER (the European Union Research Network in Algebraic Geometry), for financial support for the Workshop in Eilat at which this article was completed.

2. Characters of a General Branch Curve

Here we briefly develop the formulas for the degree and number of nodes and cusps on a general branch curve B for a general projection of a smooth surface $S \subset \mathbb{P}^N$ to a general plane $\mathbb{P}^2$, assuming that these are the only singularities. These formulas are not new, see for example (Enriques, 1949), (Iversen, 1971), but these standard references do much more, in either outdated notation or with much more advanced techniques than are necessary for this more modest computation. Hence we thought it useful to include them here for completeness and for the convenience of the reader. The reader may also want to consult (Kulikov and Kulikov, 2000), (Moishezon and Teicher, 1987), (Moishezon, Teicher and Robb, 1996), and (Teicher and Robb, 1997) for additional insight.

Denote by $\pi : S \to \mathbb{P}^2$ such a general projection. Let K and H be the canonical and hyperplane classes of S respectively. Let d be the degree of S and $g(H)$ the genus of a smooth hyperplane divisor. The intersection numbers KH and H^2 are related to d and g by

$$d = H^2 \quad \text{and} \quad 2g(H) - 2 = H^2 + KH. \tag{2.1}$$

The degree of the finite map π is equal to the degree d of the surface S. The degree b of the branch curve may be easily computed by noting that the pull-back of a line in $\mathbb{P}^2$ is a hyperplane divisor; hence the Hurwitz formula gives

$$2g(H) - 2 = d(-2) + \deg(B)$$

from which it follows that

$$b = \deg(B) = 2d + 2g(H) - 2 = 3d + KH. \tag{2.2}$$

Let $R \subset S$ denote the ramification curve, and denote by R_0 the residual curve (equal set-theoretically to the closure of $\pi^{-1}(B) - R$). R is a smooth curve, and the mapping π, restricted to R, is a desingularization of B.

Suppose that B has n nodes and k cusps and no other singularities. Over a general smooth point of B, the map π has $d-1$ preimages, one on the ramification curve and $d-2$ on the residual curve. Over each node of B, the ramification curve R has two smooth branches, and over each cusp, R has one smooth branch. Over a node of B, the residual curve R_0 meets R once transversally at each branch of R, and otherwise has $d-4$ nodes of its own. Over a cusp of B, the residual curve R_0 meets R twice at the point of R lying over the cusp, and is smooth there; it otherwise has $d-3$ cusps of its own. In any case, over either a node or a cusp of B, there are only $d-2$ preimages, instead of the $d-1$ preimages over a general point of B. Therefore, computing Euler numbers, we see that

$$e(R \cup R_0) = e(\pi^{-1}(B)) = (d-1)e(B) - (n+k). \tag{2.3}$$

The genus of the ramification curve R, being a desingularization of the branch curve B, is

$$g(R) = (b-1)(b-2)/2 - (n+k)$$

using Plücker's formulas. Its Euler number is therefore

$$e(R) = 2 - 2g(R) = 2(n+k) - b^2 + 3b.$$

Since R and B differ, topologically, only over the nodes, we see that the Euler number of B is

$$e(B) = e(R) - n = n + 2k - b^2 + 3b.$$

Letting $e(S)$ be the Euler number of the surface S, we see that

$$e(S) = d[e(\mathbb{P}^2) - e(B)] + e(R \cup R_0)$$

$$= 3d - de(B) + (d-1)e(B) - (n+k) \quad \text{using} \quad (2.3)$$
$$= 3d - e(B) - n - k$$
$$= 3d - [n + 2k - b^2 + 3b] - n - k$$
$$= 3d + b^2 - 3b - 2n - 3k,$$

so that

$$2n + 3k = 3d + b^2 - 3b - e(S). \tag{2.4}$$

Pulling back 2-forms via π, we have the standard formula that

$$K_S = \pi^*(K_{\mathbb{P}^2}) + R = -3H + R$$

and since $bH = \pi^*(B)$, we see that

$$2R + R_0 = \pi^*(B) = bH,$$

so that, as classes on S,

$$R = K + 3H \quad \text{and} \quad R_0 = bH - 2R = -2K + (b-6)H.$$

Since R and R_0 meet transversally at each of the two points of R over a node, and meet to order two at the point of R lying over a cusp, we see that $R \cdot R_0 = 2(n+k)$. Therefore $2n + 2k = R \cdot R_0 = (K + 3H)(-2K + (b-6)H)$; multiplying this out gives

$$2n + 2k = -2K^2 + (b-12)KH + (3b-18)H^2. \tag{2.5}$$

Subtracting (2.5) from (2.4) gives

$$k = 3d + b^2 - 3b - e(S) + 2K^2 - (b-12)KH - (3b-18)d$$

and then one can solve either expression for the number of nodes. Simplifying the expressions somewhat leads to the following.

PROPOSITION 2.6. *Let S be a smooth surface of degree d in $\mathbb{P}^N$, and let $\pi : S \to \mathbb{P}^2$ be a general projection. Let K and H be the canonical and hyperplane classes of S, respectively. Let B be the branch curve of the projection π, which is assumed to be a plane curve of degree b with n nodes, k cusps, and no other singularities. Then:*

(a) $\deg(\pi) = \deg(S) = d = H^2$.
(b) The degree of the branch curve B is $b = 3d + KH$.
(c) The number of nodes of the branch curve B is

$$n = -3K^2 + e(S) + 24d + \frac{b^2}{2} - 15b.$$

(d) The number of cusps of the branch curve B is

$$k = 2K^2 - e(S) - 15d + 9b.$$

(e) Under a general projection of the branch curve B to a line, the number t of turning points (simple branch points) is

$$t = e(S) - 3d + 2b.$$

The last computation of turning points is obtained from the Hurwitz formula, applied to the ramification curve R, noting that there are simple branch points for such a projection at the points of R lying over the cusps of B also.

EXAMPLE 2.7 (VERONESE SURFACES). *Let S be the rth Veronese image of $\mathbb{P}^2$. In this case, if L denotes the line class of S, then $L^2 = 1$, $K = -3L$, and $H = rL$; hence $K^2 = 9$, $KH = -3r$, and $d = H^2 = r^2$. The Euler number $e(S) = 3$. Therefore*

$$b = 3r(r-1); \quad n = 3(r-1)(r-2)(3r^2+3r-8)/2;$$
$$k = 3(r-1)(4r-5); \quad t = 3(r-1)^2.$$

EXAMPLE 2.8 (RATIONAL NORMAL SCROLLS). *Let S be a rational normal scroll, e.g. $\mathbb{P}^1 \times \mathbb{P}^1$ embedded by the complete linear system H of type $(1,r)$. The canonical class is of type $(-2,-2)$, so that $K^2 = 8$, $KH = -2r-2$, and $d = H^2 = 2r$. The Euler number $e(S) = 4$. Therefore*

$$b = 4r - 2; \quad n = 4(r-1)(2r-3); \quad k = 6r - 6; \quad t = 2r.$$

EXAMPLE 2.9 (DEL PEZZO SURFACES). *Let S be a Del Pezzo Surface of degree d in $\mathbb{P}^d$, for $3 \leq d \leq 9$. Then S is isomorphic to the plane blown up at $9-d$ points; if L denotes the class of a line, and E the sum of the classes of the $9-d$ exceptional divisors, then $L^2 = 1$, $LE = 0$, and $E^2 = d-9$; also $K = -3L+E$, and $H = -K$, so that $K^2 = H^2 = d$, and $KH = -d$. The Euler number $e(S) = 12-d$. Therefore*

$$b = 2d; \quad n = 2(d-2)(d-3) = 2d^2 - 10d + 12; \quad k = 6(d-2); \quad t = 12.$$

EXAMPLE 2.10 ($K3$ SURFACES). *Let S be a $K3$ surface of degree $d = 2g-2$ in $\mathbb{P}^g$. The canonical class is trivial, so that $K^2 = KH = 0$. The Euler number $e(S) = 24$. Therefore*

$$b = 6g - 6; \quad n = 6(g-2)(3g-7) = 18g^2 - 78g + 84;$$
$$k = 24(g-2); \quad t = 6g + 18.$$

3. Construction of the Pillow Degeneration

A non-hyperelliptic $K3$ surface of genus $g \geq 3$ can be embedded by the sections of a very ample line bundle as a smooth surface of degree $2g-2$ in $\mathbb{P}^g$. When the line bundle generates the Picard group of the $K3$ surface, the embedded $K3$ surface can be degenerated to a union of $2g-2$ planes in a variety of ways (see for example (Ciliberto, Lopez and Miranda, 1993)). In this section we will describe a degeneration, which we call the *pillow* degeneration, which smooths to a $K3$ surface whose Picard group is generated by a sub-multiple of the hyperplane class.

Fix two integers a and b at least two; set $g = 2ab+1$. The number of planes in the pillow degeneration is then $2g-2 = 4ab$.

This projective space has $g+1 = 2ab+2$ coordinate points, and each of the $4ab$ planes is obtained as the span of three of these. The sets of three are indicated in Figure 1, which describes the bottom part of the "pillow" and the top part of the "pillow", which are identified along the boundaries of the two configurations. The reader will see that the boundary is a cycle of $2a+2b$ lines.

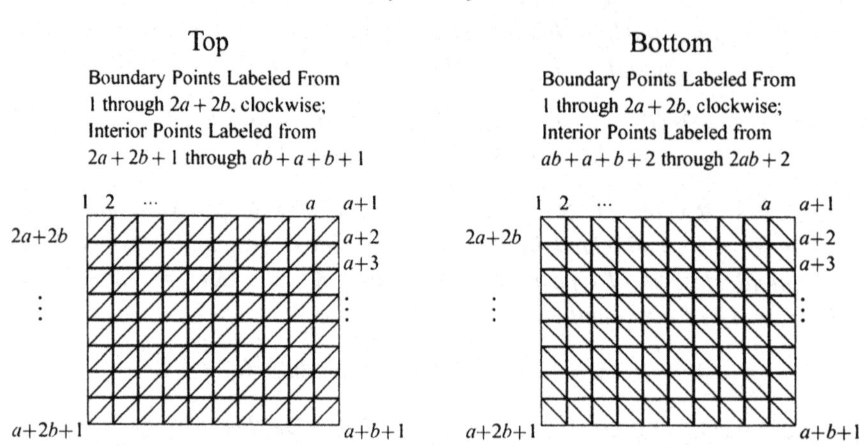

Figure 11. Configuration of Planes, Top and Bottom

Note that no three of the planes meet in a line. Also note that the set of bottom planes lies in a projective space of dimension $ab+a+b$, as do the set of top planes; these two projective spaces meet exactly along the span of the $2a+2b$ boundary points, which has dimension $2a+2b-1$. Finally note that the four corner points of the pillow degeneration (labeled 1, $a+1$, $a+b+1$, and $a+2b+1$) are each contained in three distinct planes, while all other points are each contained in six planes. This property, that the number of lines and planes incident on each of the points is bounded, is important for the later computations, and is a feature of the pillow degeneration that is not available in other previous degenerations.

We will call such a configuration of planes a *pillow of bidegree* (a,b).

THEOREM 3.1. *For any a and b at least 2, the pillow of bidgree (a,b) is a degeneration of a smooth $K3$ surface of degree $4ab$ in a projective space of dimension $g = 2ab+1$. If $c = g.c.d(a,b)$, then the general such $K3$ surface will have Picard group generated by a line bundle L such that cL is the hyperplane bundle.*

The proof of the Theorem will be made in various steps. First we will exhibit a degeneration of the $K3$ surface to a union of two rational surfaces, each isomorphic to $\mathbb{P}^1 \times \mathbb{P}^1$, embedded via the sections of the linear system of bidegree (a,b). These two rational surfaces will meet along an elliptic normal curve which is anticanonical in each. Secondly we will simultaneously degenerate each rational surface to a union of ab quadrics, resulting in a total of $2ab$ quadrics. Finally we will degenerate each quadric to a union of two planes.

PROOF. (Step One:) Note that the sections of the linear system of bidegree (a,b) embed $\mathbb{P}^1 \times \mathbb{P}^1$ as a surface in a projective space of dimension $ab+a+b$. Choose an anticanonical divisor (of bidegree $(2,2)$) which is a smooth elliptic curve; it is mapped by the above embedding to an elliptic normal curve in a subspace of dimension $2a+2b-1$.

In our original ambient space of dimension $2ab+1$, choose two subspaces of dimension $ab+a+b$ which meet along a subspace of dimension $2a+2b-1$. Make the above identical construction of the $\mathbb{P}^1 \times \mathbb{P}^1$ in each of the two subspaces, taking care to have the two elliptic normal curves identified in the intersection subspace.

This union R of the two rational surfaces is a degeneration of an embedded $K3$ surface, by an argument identical to that presented for Theorems 1 and 2 of (Ciliberto, Lopez and Miranda, 1993), which we will not repeat in detail here. Briefly, one first checks via standard calculations that $H^0(N_R)$ has dimension $g^2+2g+19$ and that $H^1(N_R) = H^2(N_R) = 0$. Secondly, the natural map from $H^0(N_R)$ to $H^0(T^1)$ is seen to be surjective. This is sufficient to prove that R represents a smooth point of its Hilbert scheme, whose general member is a smooth $K3$ surface of degree $2g-2$.

(Step Two:) The second step can be achieved as in (Moishezon and Teicher, 1987) by observing that each $\mathbb{P}^1 \times \mathbb{P}^1$ can be degenerated to a union of ab quadrics by degenerating the first coordinate $\mathbb{P}^1$ to a chain of a lines, and the second coordinate $\mathbb{P}^1$ to a chain of b lines. In this degeneration the elliptic curve degenerates to a cycle of $2(a+b)$ lines. This degeneration is made simultaneously for each of the two $\mathbb{P}^1 \times \mathbb{P}^1$'s, resulting in a degeneration to a union of $2ab$ quadrics. This configuration of $2ab$ quadrics meet as in Figure 1, without the diagonal lines: if one removes the diagonal lines from Figure 1 we obtain $2ab$ rectangles, each indicating a quadric. Each of these quadrics meets the others along a cycle of four lines (two vertical and two horizontal).

(Step Three:) Finally degenerate each quadric to a union of two planes, as in Figure 1. These degenerations can be executed completely independently of course, and it is elementary to see that this can be done keeping the four lines along which any one of the quadrics meet the others fixed.

(Step Four:) Finally note that if $c \neq 1$, the pillow degeneration of bidegree (a,b) is a degeneration of the c-uple embedding of the pillow degeneration of bidegree $(a/c, b/c)$. To see this, one uses the standard triangular degeneration of the Veronese embedding of the plane as described in (Moishezon and Teicher, 1994).

The final point to check is that the general $K3$ surface in this 19-dimensional family has Picard group generated by $(1/c)H$, where H is the hyperplane class. Since we have a 19-dimensional family of $K3$ surfaces, the only question to be decided is which sub-multiple of the hyperplane system is the generator of the general Picard group. The maximum possible is the g.c.d c. Since the pillow is a c-uple Veronese, the hyperplane class is at least a c-fold multiple, and since it cannot be any more, this shows that the Picard group is generated by $(1/c)H$. This completes the proof of the Theorem. □

Note that in this degeneration, the horizontal and vertical lines appear first, and the diagonal lines appear second.

4. The Degeneration of the Branch Curve

We assume that we are in a general enough situation that for a generic projection of a K3 surface of degree g in $\mathbb{P}^g$ to a plane, the branch curve is a curve of degree $6g-6$, having $6(g-2)(3g-7)$ nodes and $24(g-2)$ cusps and no other singularities; these numbers were presented in Section 2. If one projects this branch curve onto a general line, the projection will have $6g+18$ simple branch points.

It is our goal in this section to describe how these nodes, cusps, and branch points degenerate in a pillow degeneration.

Firstly, since the pillow degeneration consists entirely of planes, under a general projection each plane will map isomorphically onto the target plane. Therefore the degenerate branch curve is composed of the $3g-3$ planar lines which are the images of the $3g-3$ double lines of the pillow degeneration where two planes meet. Each of the $3g-3$ planar lines have multiplicity two in the limit branch curve.

We see therefore that the general branch curve (of degree $6g-6$) degenerates as a curve to the $3g-3$ planar lines, each doubled. Our next task is to describe the degeneration of the nodes, cusps and branch points of the general branch curve. In any case it is clear that these distinguished points of the general branch curve can only go to points of the $3g-3$ planar lines.

Secondly, it is elementary to compute that there are $(9/2)g^2-(51/2)g+39$ pairs of disjoint lines in the pillow degeneration. Each of these pairs of disjoint lines gives rise to an intersection of two planar line components of the limit branch curve. We refer to these points as *2-points* of the configuration of the $3g-3$ planar lines.

In addition to these 2-points, we have exactly four *3-points*, corresponding to the projection of the four points in the pillow degeneration where 3 planes (and 3 double lines) meet. Finally we have $g-3$ *6-points* corresponding to the projection of the $g-3$ points in the pillow degeneration where 6 planes (and 6 double lines) meet. At any one of these n-points ($n=2$, 3, or 6) exactly n of the $3g-3$ planar lines meet; moreover at no other point of the plane do any of these lines meet.

In the degeneration of the general branch curve to this configuration of $3g-3$ double lines, each of the nodes, cusps, and branch points can degenerate either to a 2-point, a 3-point, a 6-point, or a smooth point of one of the $3g-3$ lines. With the above terminology, we can now describe how many nodes, cusps, and branch points degenerate to each of these types of points.

THEOREM 4.1. *In the pillow degeneration of a K3 surface of degree g in* $\mathbb{P}^g$, *the nodes, cusps, and branch points of the general branch curve degenerate to the 2-points, 3-points, 6-points, and other smooth points of lines according to the following table:*

Object Type	*Number*	*Branch Points*	*Nodes*	*Cusps*
Lines	$3g-3$	0	0	0
3-points	4	9	0	6
6-points	$g-3$	6	24	24
2-points	$\frac{9}{2}g^2-\frac{51}{2}g+39$	0	4	0
Totals:		$6g+18$	$18g^2-78g+84$	$24(g-2)$

In particular no node, cusp, or branch point degenerates to a smooth point of any of the $3g-3$ *double lines of the limit branch curve.*

PROOF. We first look at the row of the table for the 2-points. Since each of the planar lines have multiplicity two in the branch curve, this crossing point actually is a limit of 4 nodes of the general branch curve (the 4 nodes appearing as the four intersection points of two pairs of lines). No cusp or branch point of the general branch curve has this crossing point as a limit in general, since these points are created by the projection of unrelated disjoint lines in the union of planes in $\mathbb{P}^g$.

We now turn our attention to the images of the multiple points of the pillow degeneration where n planes (and n double lines) meet at one point. We assume that $3 \leq n \leq 6$ in what follows. (In the pillow degeneration we have $n=3$ or $n=6$ only.) Under the generic planar projection, such points go to intersections of n of the corresponding planar lines. We will refer to these as *n-points* of the limit branch curve.

In order to analyze the number of nodes, cusps, and branch points of the general curve which go to these n-points, we make a local analysis near the multiple point of the union of planes. There are n planes incident to this multiple point, and they together span a $\mathbb{P}^n$. Locally this collection of n planes in $\mathbb{P}^n$ smooths to a Del

Pezzo surface of degree n. In a generic projection for such a Del Pezzo, the branch curve has degree $2n$, with $2(n-2)(n-3)$ nodes and $6n-12$ cusps; the number of simple branch points for this curve under generic projection to a line is 12.

The limit branch curve corresponding to the degeneration of the Del Pezzo to the union of n planes is a union of n lines concurrent at a point p, the images of the n lines through the multiple point.

A partial smoothing of the union of n planes may be obtained by taking two adjacent planes and smoothing them to a quadric surface. The corresponding smoothing of the limit branch curve smooths exactly one of the n lines to a conic, which is necessarily tangent to two adjacent lines. As the conic degenerates to the (double) line L, we see that no nodes of the general branch curve go to any point of L which is not p, and no cusps do either. The conic has two general branch points for a projection to a line, and one of these branch points goes to p and one does not.

This local analysis of this partial smoothing shows that in a complete smoothing to the Del Pezzo, no node can go to a point of any line except the concurrent point p, and neither can any cusp. Therefore all of the $2(n-2)(n-3)$ nodes degenerate to the concurrent point, and all of the $6n-12$ cusps do too. Moreover, of the 12 branch points for the general curve, all but n of them go to the concurrent point p. (The other n go to one on each line.)

In the cases $n=3$ and $n=6$ of interest in the pillow degeneration, the above analysis shows that arbitrarily close to a 3-point there are $9=12-3$ branch points, and no nodes and 6 cusps. Arbitrarily close to a 6-point there are $6=12-6$ branch points, and 24 nodes and 24 cusps. This gives the entries in the 3-point and 6-point rows of the table.

If we now total the number of branch points, nodes and cusps which degenerate to these multiple points, we obtain the values in the last row of the table. Since these are exactly the number of branch points, nodes, and cusps of the general curve, we must have accounted for all of the branch points, nodes, and cusps already. In particular there are none left to degenerate to smooth points of the double lines.

This completes the proof of the Theorem. □

References

Ciliberto, C., Lopez, A., and Miranda, R. (1993) Projective Degenerations of $K3$ Surfaces, Gaussian Maps, and Fano Threefolds, *Inv. math.* **114**, 641–667.

Enriques, F. (1949) *Le Superficie Algebriche*, Nicola Zanichelli Editore, Bologna.

Friedman, R., and Morrison, D. (1983) The Birational Geometry of Degenerations: An Overview, in R. Friedman and D. Morrison (eds.), *The Birational Geometry of Degenerations*, Progress in Mathematics 29, Birkhäuser Press.

Iversen, B. (1971) Critical points of an algebraic function, *Inv. math.* **12**, 210–224.

Kulikov, V. (1977) Degenerations of $K3$ Surfaces and Enriques Surfaces, *Math. USSR Izvestija* **11**, 957–989.

Kulikov, V., and Kulikov, Vik. (2000) Generic coverings of the plane with A-D-E singularities, *Izvestiya RAN* **64**, 65–106.

Moishezon, B., and Teicher, M. (1987) Simply Connected Algebraic Surfaces with Positive Index, *Inv. math.* **89**, 601–643.

Moishezon, B., and Teicher, M. (1994) Braid group techniques in complex geometry III: Projective degeneration of V_3, in *Classification of Algebraic Varieties*, Contemporary Mathematics 162, AMS, pp. 313–332.

Moishezon, B., Teicher, M., and Robb A. (1996) On Galois covers of Hirzebruch surfaces, *Math. Ann.* **305**, 493–539.

Persson, U., and Pinkham, H. (1981) Degeneration of surfaces with trivial canonical divisor, *Annals of Math.* **113**, 45–66.

Teicher, M., and Robb, A. (1997) Applications of Braid Group Techniques to the decomposition of moduli spaces, new examples, *Topology and its Applications* **78**, 143–151.

COMPUTATIONAL ALGEBRAIC GEOMETRY TODAY *

W. DECKER
Universität des Saarlandes, Germany

F.-O. SCHREYER
Universität Bayreuth, Germany

Abstract. In this article we give a survey on computational algebraic geometry today. It is not our goal to guide research in this area, but to help algebraic geometers to decide whether nowadays' computational techniques and computer algebra systems provide useful tools for their own research. To illustrate some of the main techniques we focus on rather small examples. But we also give hints on the actual size of the computations which can be carried through today, and we quote further survey articles for more information in this direction. With the exception of Section 10 all computations are done on a 400 MHz Pentium II processor.

Key words: Computer algebra systems, Gröbner bases, syzygies, ideal membership, Hilbert polynomials and Hilbert functions, elimination, Milnor numbers, D-modules, EXT and TOR, sheaf cohomology, Beilinson monads, primary decomposition, normalization, Puiseux expansion, rational parametrization, deformations, invariants rings, special varieties, intersection theory, syzygy conjectures, Zariski's conjecture, visualization, complexity, benchmarks.

Mathematics Subject Classification (2000): 13P, 14Q.

1. Computer algebra systems

Nowadays there is a large variety of computer algebra systems suiting different needs. User-friendly interfaces and comfortable help functions enable us to work with powerful computing tools by just looking up a few commands, without any knowledge in programming. But modern systems also come with a programming language which allows interested users to extend the system. Many user-written packages and libraries are publicly available, thus providing even more computing tools.

There are general purpose and special purpose computer algebra systems. Typically, one has to pay for a general purpose system whereas many of the special

* The authors would like to thank the MSRI in Berkeley for its support and for providing a stimulating working atmosphere.

C. Ciliberto et al. (eds.),
Applications of Algebraic Geometry to Coding Theory, Physics and Computation, 65–119.

purpose systems can be downloaded from the internet for free (use your favorite search engine to find the home-pages of the systems mentioned here). A general purpose system allows us to attack problems in many different areas. In addition to tools for *symbolic computation* such a system usually offers tools for *numeric computation* and for *visualization*. Well-established general purpose systems are REDUCE, MACSYMA, MAPLE, DERIVE, MATHEMATICA, MUPAD, AXIOM, and TI-92. We refer to Wester's practical guide (1999) for a comparison of some of their capabilities.

EXAMPLE 1.1. *In the following* MAPLE *session we factor the polynomial*

$$p = x^2y^4z - xy^9z^2 + xyz^3 + 2x - y^6z^4 - 2y^5z \in \mathbb{Z}[x,y,z].$$

```
> p:=x^2*y^4*z - x*y^9*z^2 + x*y*z^3 + 2*x - y^6*z^4 - 2*y^5*z:
> factor(p);
                         4         3              5
                       (y  z x + y z  + 2) (x - y  z)
```
□

For some of the more special and advanced applications general purpose systems are not powerful enough. Often the implementation of the required algorithms is not optimal with respect to speed and storage handling; in addition, some of the more advanced algorithms might not be implemented at all. Many of todays' special purpose systems have been created by people specializing in a field other than computer algebra, and having a desperate need for computational power in the context of some of their research problems. A pioneering and prominent example is Veltman's SCHOONSCHIP which helped to win a Nobel price in physics in 1999 (awarded to Veltman and t'Hooft "for having placed particle physics theory on a firmer mathematical foundation"). From a special purpose system we expect highly tuned implementations of the algorithms needed for the area in which the system is specializing. Examples are PARI, KANT, LIDIA, MAGMA (which is not free) and SIMATH for number theory, or GAP and MAGMA for group theory.

EXAMPLE 1.2. LIDIA *is "a* C++ *Library for Computational Number Theory". In the following* LIDIA *session we factor the 8th Fermat number* $F_8 = 2^{2^8} + 1$.

```
lc> factor(2^(2^8)+1);
$0 =  [(1238926361552897,1)
(93461639715357977769163558199606896584051237541638188580280321,1)]
```
□

EXAMPLE 1.3. GAP *(Groups, Algorithms and Programming) is a system "for computational discrete algebra with particular emphasis on computational group theory."* *We use* GAP to compute the character table of the Heisenberg group H_3.

In its Schrödinger representation, H_3 is the subgroup of $GL(3,\mathbb{Q}(\xi))$, $\xi = e^{\frac{2\pi i}{3}}$ a primitive 3rd root of unity, which is generated by

$$\sigma = \begin{pmatrix} 0 & 0 & 1 \\ 1 & 0 & 0 \\ 0 & 1 & 0 \end{pmatrix} \quad \text{and} \quad \tau = \begin{pmatrix} 1 & 0 & 0 \\ 0 & \xi & 0 \\ 0 & 0 & \xi^2 \end{pmatrix}.$$

```
gap> m1:=[[0,0,1],[1,0,0],[0,1,0]];;
gap> m2:=[[1,0,0],[0,E(3),0],[0,0,E(3)^2]];;
gap> G:=Group(m1,m2);;
gap> Size(G);
27
gap> Display(CharacterTable(G));
CT1

      3  3  2  2  2  2  2  2  2  2  3  3

        1a 3a 3b 3c 3d 3e 3f 3g 3h 3i 3j

X.1      1  1  1  1  1  1  1  1  1  1  1
X.2      1  A  A  A /A /A /A  1  1  1  1
X.3      1 /A /A /A  A  A  A  1  1  1  1
X.4      1  1 /A  A  1 /A  A /A  A  1  1
X.5      1  A  1 /A /A  A  1 /A  A  1  1
X.6      1 /A  A  1  A  1 /A /A  A  1  1
X.7      1  1  A /A  1  A /A  A /A  1  1
X.8      1  A /A  1 /A  1  A  A /A  1  1
X.9      1 /A  1  A  A /A  1  A /A  1  1
X.10     3  .  .  .  .  .  .  .  .  B /B
X.11     3  .  .  .  .  .  .  .  . /B  B

A = E(3)
  = (-1+ER(-3))/2 = b3
B = 3*E(3)^2
  = (-3-3*ER(-3))/2 = -3-3b3
```

Via the online GAP user manual we easily find out how to read the output. □

Special purpose systems for commutative algebra and algebraic geometry allow us to manipulate ideals in polynomial rings (and more). They typically rely on Gröbner basis techniques, their "engine" is Buchberger's algorithm for computing Gröbner bases. The pioneering MACAULAY, and the more modern and complete MACAULAY2, SINGULAR, COCOA, and RISA/ASIR offer a variety of tools for experiments. FGB is a system just for the basic task of computing Gröbner bases. With BERGMAN one can compute Gröbner bases, Hilbert series and Poincaré series in both commutative and non-commutative graded algebras.

The examples in this article feature MACAULAY2 and SINGULAR, the two of the recent systems we know best, SCHUBERT, a MAPLE package for intersection theory, CASA, a MAPLE package offering a variety of tools (including plotting tools), and SURF, a system for visualizing curves and surfaces.

2. Gröbner basics

2.1. BUCHBERGER'S ALGORITHM

Hilbert (1890), in the first of his two landmark paper on invariant theory (algebraic geometry), introduced concepts which are crucial for the theory of Gröbner bases, too. Among others, he proved the basis theorem, the syzygy theorem, and the existence of the Hilbert polynomial. Gröbner bases, in particular monomial orders and initial ideals, were introduced by Gordan (1899) to give yet another proof of the basis theorem. Macaulay (1927) used monomial orders to characterize the Hilbert functions of graded ideals in polynomial rings. In fact, Macaulay proved that a graded ideal I in a polynomial ring R and its initial ideal $\mathrm{LT}(I)$ (with respect to any given monomial order) have the same Hilbert function. On his way he showed that the *standard monomials*, that is, the monomials not in $\mathrm{LT}(I)$, represent a vector space basis of R/I. This in turn motivated Gröbner to ask one of his students, Buchberger, to look for computational tools in R/I. Buchberger's solution (1965, 1970), his criterion and algorithm, was one of the starting points of computer algebra. Schreyer (1980) showed that Buchberger's algorithm allows to compute syzygies, too, and he gave a new proof of the syzygy theorem.

Buchberger's algorithm makes use of division with remainder. Generalizing the usual algorithm in one variable this requires to choose leading terms of polynomials in a way compatible with multiplication.

DEFINITION 2.1 (MONOMIAL ORDERS). *Let* $K[x] = K[x_1,\dots,x_n]$.

1. *A* monomial order *on* $K[x]$ *is a total order* $>$ *on the set of monomials* $\{x^\alpha \mid \alpha \in \mathbb{N}^n\}$ *which is multiplicative:*

$$x^\alpha > x^\beta \Rightarrow x^\alpha x^\gamma > x^\beta x^\gamma.$$

2. *Let* $>$ *be a monomial order on* $K[x]$. *By abuse of notation we write* $ax^\alpha > bx^\beta$ *(resp.* $ax^\alpha \geq bx^\beta$*) if* ax^α *and* bx^β *are non-zero terms in* $K[x]$ *with* $x^\alpha > x^\beta$ *(resp.* $x^\alpha \geq x^\beta$*). If* $f \in K[x]$ *is a non-zero polynomial, its largest term, denoted by* $\mathrm{LT}(f) = \mathrm{LT}_>(f)$, *is the* leading term *or* initial term *of* f. *We set* $\mathrm{LT}(0) = 0$.
3. *A monomial order* $>$ *on* $K[x]$ *is* global, *if* $x_i > 1$, $i = 1,\dots,n$, local, *if* $x_i < 1$, $i = 1,\dots,n$, *and* mixed, *otherwise.*
4. *If* F *is a free* $K[x]$*-module with a basis* $e_1,\dots,e_s$, *we speak of* monomials $x^\alpha e_i$ *and* terms $ax^\alpha e_i$ *in* F, *and define monomial orders on* F *and leading terms as above. We require in addition that*

$$x^\alpha e_i > x^\beta e_i \iff x^\alpha e_j > x^\beta e_j \quad \textit{for all} \quad i,j.$$

Then every monomial order on F induces a unique monomial order on $K[x]$, and notions like global and local carry over to monomial orders on free modules. □

Local and mixed orders are needed for computations in local rings (see Section 3.1). Note that $>$ is global iff it is a well-ordering.

THEOREM 2.2 (DIVISION WITH REMAINDER, DETERMINATE VERSION). *Let F be a free $K[x]$-module with a finite basis, let $>$ be a global monomial order on F, and let $f_1, \dots, f_r \in F \setminus \{0\}$. Then for every $f \in F$ there exist uniquely determined $g_1, \dots, g_r \in K[x]$, and a unique* remainder *$h \in F$ such that*

1. $f = g_1 f_1 + \dots + g_r f_r + h$,
2. *no term of $g_i \mathrm{LT}(f_i)$ is a multiple of $\mathrm{LT}(f_j)$ for some $j < i$, and*
3. *no term of h is a multiple of $\mathrm{LT}(f_i)$ for some i.* □

Here, for terms in F, notions like multiple or divisible are defined in the obvious way: $ax^{\alpha}e_i$ is *divisible* by $bx^{\beta}e_j \neq 0$ with *quotient* $a/b\, x^{\alpha-\beta} \in K[x]$ if $i = j$ and x^{α} is divisible by x^{β}. Hence condition 3 means that no term of h is in the monomial submodule of F generated by $\mathrm{LT}(f_1), \dots, \mathrm{LT}(f_r)$. Note that the theorem is obvious for monomial submodules. The general case follows as in the case of polynomials in one variable by induction on $\mathrm{LT}(f)$.

REMARK 2.3 (DIVISION WITH REMAINDER, INDETERMINATE VERSION). *If we only require*

1. $f = g_1 f_1 + \dots + g_r f_r + h$,

2'. $\mathrm{LT}(f) \geq \mathrm{LT}(g_i f_i)$, *whenever both sides are non-zero, and*

3'. *if h is non-zero,* $\mathrm{LT}(h) \notin (\mathrm{LT}(f_1), \dots, (\mathrm{LT}(f_r))$,

the expression in 1 is no longer uniquely determined. Any such expression is called a standard expression *of f in terms of the g_i with* remainder *h.* □

DEFINITION 2.4 (GRÖBNER BASES). *Let F be a free $K[x]$-module with a finite basis, let $>$ be a global monomial order on F, and let $I \subset F$ be a submodule. The* initial module *of I, denoted by $\mathrm{LT}(I)$, is the monomial submodule of F which is generated by the leading terms of the elements in I:*

$$\mathrm{LT}(I) = (\{\mathrm{LT}(f) \mid f \in I\}).$$

A Gröbner basis *of I is a set of elements $f_1, \dots, f_r \in I$ whose leading terms generate* $\mathrm{LT}(I)$:

$$(\mathrm{LT}(f_1), \dots, \mathrm{LT}(f_r)) = \mathrm{LT}(I).$$ □

REMARK 2.5. *Gröbner bases are well behaved under division with remainder:*

1. *$f_1,\dots,f_r \in F\setminus\{0\}$ form a Gröbner basis of $I=(f_1,\dots,f_r)$ iff for all $f\in I$ the remainder of f under determinate division by $f_1,\dots,f_r$ is zero. Indeed, if $f=\sum g_if_i \in I$, and if the g_i satisfy condition 2 of determinate division by $f_1,\dots,f_r$,*

$$\mathrm{LT}(f)=\max\{\mathrm{LT}(g_if_i)\mid i=1,\dots,r\}\in(\mathrm{LT}(f_1),\dots,\mathrm{LT}(f_r)).$$

2. *The remainder of $f\in F$ under determinate division by the elements of a Gröbner basis of I only depends on I and the monomial order $>$. It is a* normal form *of f* mod *I giving the expression for f* mod *I in terms of standard monomials.* □

The elements of a Gröbner basis of I generate I. Buchberger's criterion (1965, 1970), which we discuss next, allows to compute a Gröbner basis from any given finite set of generators of I.

NOTATION 2.6. *Let F be a free $K[x]$-module with basis $e_1,\dots,e_s$.*

1. *If $ax^\alpha e_i$, $bx^\beta e_i$ are two non-zero terms in F involving the same basis element, set*

$$\gcd(ax^\alpha e_i, bx^\beta e_i)=x_1^{\min(\alpha_1,\beta_1)}\cdot\ldots\cdot x_n^{\min(\alpha_n,\beta_n)}e_i.$$

2. *Let $>$ be a global monomial order on F, and let $f_1,\dots,f_r\in F\setminus\{0\}$. For each pair i,j such that $\mathrm{LT}(f_i)$ and $\mathrm{LT}(f_j)$ involve the same basis element, set*

$$m_{ij}=\mathrm{LT}(f_i)/\gcd(\mathrm{LT}(f_i),\mathrm{LT}(f_j)).$$ □

THEOREM 2.7 (BUCHBERGER'S CRITERION). *With notations as above, compute for each* "S-polynomial"

$$m_{ji}f_i-m_{ij}f_j$$

a standard expression with remainder h_{ij}. $f_1,\dots,f_r$ form a Gröbner basis of $I=(f_1,\dots,f_r)$ iff all h_{ij} are zero.

PROOF. By construction, $h_{ij}\in I$, thus $\mathrm{LT}(h_{ij})\in\mathrm{LT}(I)$. If $f_1,\dots,f_r$ is a Gröbner basis of I, all h_{ij} are zero by condition 3' of indeterminate division.

Conversely, suppose that all h_{ij} are zero. Then the corresponding standard expressions

$$m_{ji}f_i-m_{ij}f_j=\sum_\ell g_\ell^{ij}f_\ell$$

define relations (*syzygies*) G^{ij} on $f_1,\dots,f_r$. Formally, we consider a free $K[x]$-module F_1 with basis $\varepsilon_1,\dots,\varepsilon_r$, and view G^{ij} as an element of the kernel of the map

$$\varphi: F_1\to F,\ \varepsilon_i\mapsto f_i.$$

On F_1 we consider the *induced monomial order* defined as

$$x^\alpha\varepsilon_i > x^\beta\varepsilon_j \iff \begin{array}{l} x^\alpha \mathrm{LT}(f_i) > x^\beta \mathrm{LT}(f_j) \text{ or} \\ x^\alpha \mathrm{LT}(f_i) = x^\beta \mathrm{LT}(f_j) \text{ and } i > j. \end{array}$$

With respect to this monomial order $\mathrm{LT}(G^{ij}) = m_{ji}\varepsilon_i$ for $i > j$. Let $f = \sum a_i f_i \in I$. To show that the remainder of f under determinate division by $f_1, \ldots, f_r$ is zero, let $G = \sum g_i\varepsilon_i$ be the remainder of $A = \sum a_i\varepsilon_i \in F_1$ under determinate division by the G^{ij}, $i > j$, in some order. Then

$$f = \sum a_i f_i = \sum g_i f_i$$

since the G^{ij} are syzygies, and $G = 0$ or no term of G is a multiple of some $\mathrm{LT}(G^{ij}) = m_{ji}\varepsilon_i$, $i > j$, by condition 3 of determinate division in F_1. In F that means that the g_i satisfy condition 2 of determinate division of f by $f_1, \ldots, f_r$. Thus

$$\mathrm{LT}(f) = \max\{\mathrm{LT}(g_i f_i) \mid i = 1, \ldots, r\} \in (\mathrm{LT}(f_1), \ldots, \mathrm{LT}(f_r)),$$

and $f_1, \ldots, f_r$ is a Gröbner basis of I (see Remark 2.5). □

REMARK 2.8 (ALGORITHMS). *The criterion yields Buchberger's algorithm for computing Gröbner bases. In fact, the proof shows much more (*Schreyer, 1980*):*

1. *If $f_1, \ldots, f_r$ is a Gröbner basis of I, Buchberger's test gives syzygies G^{ij}, $i > j$. These syzygies generate all syzygies; in fact, they form a Gröbner basis of* $\ker\varphi$ *with respect to the induced monomial order on F_1. Indeed, if $G = \sum g_i\varepsilon_i$ is the remainder of $A = \sum a_i\varepsilon_i \in \ker\varphi$ under determinate division by the G^{ij}, then $G = 0$ since the g_i satisfy condition 2 for determinate division of $f = \sum g_i f_i$ by $f_1, \ldots, f_r$. This yields Schreyer's algorithm for computing a (not necessarily minimal) generating set of the syzygies on any finite set of generators of I: starting from the generators compute a Gröbner basis of I and the syzygies on the elements of the Gröbner basis, and substitute into the syzygies the expressions for the elements of the Gröbner basis in terms of the original generators.*
2. *In Buchberger's test it is enough to check the remainders of the S-polynomials corresponding to a minimal generating set of the syzygies on* $\mathrm{LT}(f_1), \ldots, \mathrm{LT}(f_r)$. □

COROLLARY 2.9 (HILBERT'S SYZYGY THEOREM, 1890). *Every finitely generated $R = K[x_1, \ldots, x_n]$-module M has a finite free resolution*

$$0 \to R^{r_n} \to \ldots \to R^{r_1} \to R^{r_0} \to M \to 0$$

of length at most n.

PROOF (Schreyer, 1980). We give a constructive proof, starting from a presentation

$$R^s \xrightarrow{\varphi} R^{r_0} \to M \to 0.$$

Compute a Gröbner basis, say $f_1, \dots, f_{r_1}$, of $\operatorname{im}\varphi$, and the syzygies G^{ij}, which form a Gröbner basis of the kernel of the corresponding map $\varphi_1 : R^{r_1} \to R^{r_0}$. Suppose we have sorted $f_1, \dots, f_{r_1}$ so that the exponent of x_n in $\mathrm{LT}(f_j)$ is smaller than that in $\mathrm{LT}(f_i)$ for $j < i$. Then none of the leading terms $\mathrm{LT}(G^{ij})$, $j < i$, involves x_n. Continuing in this way we see that after k steps no leading term of the Gröbner basis of $\ker\varphi_k$ involves the variables $x_{n-k+1}, \dots, x_n$. So our process stops after at most n steps. □

2.2. ELEMENTARY APPLICATIONS

EXAMPLE 2.10. *Consider the following computation in* MACAULAY2*:*

```
Macaulay 2, version 0.8.61
--Copyright 1993-2000, all rights reserved, D. R. Grayson and M. E. Stillman
--Factory 1.2c from Singular, copyright 1993-1997, G.-M. Greuel, R. Stobbe
--Factorization and characteristic sets 0.3.1, copyright 1996, M. Messollen
--GC 6.0 alpha 2, copyright, H-J. Boehm, A. Demers, Xerox, Silicon Graphics
--GNU libc and libg++, copyright, Free Software Foundation
--GNU MP 3.1, copyright, Free Software Foundation
```

We choose as field of definition $K = \mathbb{Q}$. However, the questions treated in this section are not of arithmetic nature, our computations concern schemes defined over $\overline{\mathbb{Q}}$.

```
i1 : R=QQ[x_0..x_3]
o1 = R
o1 : PolynomialRing
```

`i1` *stands for the first input,* `o1` *for the first output of our* MACAULAY2 *session.*

```
i2 : a=matrix{{x_0,x_1,0_R,0_R}}
o2 = | x_0 x_1 0 0 |
             1      4
o2 : Matrix R  <--- R
```

We randomly choose a graded 2×4 matrix with linear entries in R, that is, with homogeneous entries in R of degree 1. The ; *tells* MACAULAY2 *to suppress part of the output.*

```
i3 : b=random(R^2,R^{4:-1});
             2      4
o3 : Matrix R  <--- R
i4 : m=transpose(b||a)
```

```
o4 = {-1} | 5/2x_1+x_2                          -8/9x_0-2/5x_1-2/3x_2              x_0 |
     {-1} | -9/5x_0-3/2x_1-4/3x_2-7/4x_3 1/3x_0-4/7x_1+1/3x_2-10x_3      x_1 |
     {-1} | 1/2x_0+2/3x_1+1/6x_2-5/2x_3  4/5x_0-3/7x_1-4x_2-3/2x_3       0   |
     {-1} | -2x_0+2/5x_1+2/3x_2-5/4x_3   -9/5x_0+3/10x_1-1/2x_2+7/5x_3 0   |
              4       3
o4 : Matrix R  <--- R
i5 : I=minors(3,m);
o5 : Ideal of R
```

The projective scheme $V(I) \subset \mathbb{P}^3$ *defined by the determinantal ideal* I *is unmixed since it has the expected codimension 2:*

```
i6 : codim I
o6 = 2
i7 : degree I
o7 = 6
```

The computation of the dimension and the degree of $V(I)$ *is actually a byproduct of the computation of the Hilbert polynomial of* R/I*, which can, in principle, be read off the minimal free resolution of* R/I*.*

```
i8 : betti res I
o8 = total: 1 4 3
         0: 1 . .
         1: . . .
         2: . 4 3
o8 : Net
```

A generally much faster method is to apply Macaulay's result and compute the Hilbert polynomial of $R/\mathrm{LT}(I)$ *(for monomial ideals this computation is of purely combinatorial nature).*

```
i9 : hilbertPolynomial coker gens I
o9 = - 8*P  + 6*P
          0      1
o9 : ProjectiveHilbertPolynomial
```

So the Hilbert polynomial is

$$P_{R/I}(t) = 6(t+1) - 8 = 6t + 1 - 3.$$

Hence $V(I)$ *is a curve of arithmetic genus 3. We compute its singular locus via the implicit function theorem (recall that* $V(I)$ *is unmixed):*

```
i10 : singI=minors(codim I,jacobian I)+I;
o10 : Ideal of R
i11 : codim singI
o11 = 3
i12 : degree singI
```

```
o12 = 2
i13 : betti singI
o13 = generators: total: 1 39
                      0: 1  .
                      1: .  .
                      2: .  4
                      3: . 35

o13 : Net
```

The formula `i10` *actually gives the singular locus of the affine cone over the projective scheme* $V(I)$. *Hence the result usually comes with a big embedded component at the irrelevant maximal ideal* $\mathfrak{m}$ *of* R. *The homogeneous ideal of the singular locus of* $V(I)$ *itself can be obtained via saturation: if* $\mathfrak{a} \subset R = K[x_0,\dots,x_n]$ *is a graded ideal, then its* saturation *(with respect to* $\mathfrak{m}$*),*

$$\mathfrak{a}^{sat} = \{r \in R \mid \exists i \in \mathbb{N} : r\mathfrak{m}^i \subset \mathfrak{a}\},$$

is the homogeneous ideal of the scheme defined by $\mathfrak{a}$ *in* $\mathbb{P}^n$.

We describe one algorithm for computing the saturation. Let $\mathfrak{a}, \mathfrak{b} \subset R$ *be two ideals. Then the* ideal quotient *of* $\mathfrak{a}$ *by* $\mathfrak{b}$ *is defined to be*

$$(\mathfrak{a} : \mathfrak{b}) = \{r \in R \mid r\mathfrak{b} \subset \mathfrak{a}\}.$$

It can be computed via syzygies as follows. Suppose that $\mathfrak{a} = (a_1,\dots,a_r)$ *and* $\mathfrak{b} = (b_1,\dots,b_s)$ *are given by generators. Then* $(\mathfrak{a} : \mathfrak{b})$ *is generated by the entries of the first row of a matrix whose columns generate the syzygies on the elements of* R^s *defined by the columns of the matrix*

$$\begin{pmatrix} b_1 & a_1 & \dots & a_r & 0 & \dots & & & & \dots & 0 \\ b_2 & 0 & \dots & 0 & a_1 & \dots & a_r & 0 & & \dots & 0 \\ \vdots & & & & & & \ddots & & & & \\ b_s & 0 & \dots & & & \dots & 0 & a_1 & & \dots & a_r \end{pmatrix}.$$

Note that the intersection $\mathfrak{a} \cap \mathfrak{b}$ *can be computed in a similar way. The* saturation

$$(\mathfrak{a} : \mathfrak{b}^\infty) = \{r \in R \mid \exists i \in \mathbb{N} : r\mathfrak{b}^i \subset \mathfrak{a}\} = \bigcup_{i=0}^{\infty} (\mathfrak{a} : \mathfrak{b}^i)$$

of $\mathfrak{a}$ *with respect to* $\mathfrak{b}$ *is obtained by iterating* $(\dots((\mathfrak{a} : \mathfrak{b}) : \mathfrak{b})\dots : \mathfrak{b})$ *until the result stabilizes.*

Let us resume our MACAULAY2 *session:*

```
i14 : time singIsat=saturate singI;
     -- used 0.17 seconds
o14 : Ideal of R
```

```
i15 : betti  singIsat
o15 = generators: total: 1 3
                      0: 1 2
                      1: . 1

o15 : Net
i16 : singIsat
                      2   151        129  2
o16 = ideal (x , x , x  - ---*x x  - ---*x )
              1   0   2   155  2 3    62  3

o16 : Ideal of R
```

In the singular locus we find, as expected from our construction, the equations of the line $L = \mathrm{V}(x_0, x_1)$. In fact, L turns out to be a component. The residual is obtained by an ideal quotient computation.

```
i17 : L=ideal ((gens singIsat)_{0,1})
o17 = ideal (x , x )
              1   0

o17 : Ideal of R
i18 : J=I:L;
o18 : Ideal of R
i19 : hilbertPolynomial(coker gens J)
o19 = - 6*P  + 5*P
           0      1

o19 : ProjectiveHilbertPolynomial
i20 : betti res J
o20 = total: 1 3 2
          0: 1 . .
          1: . 1 .
          2: . 2 2

o20 : Net
i21 : degree (L+J) == degree singI
o21 = true
i22 : codim (minors(codim J,jacobian J)+J)
o22 = 4
```

Thus $\mathrm{V}(I)$ is the union of a smooth curve of genus 2 and degree 5 with a secant line. □

Theoretically, for the type of computations done so far, one can choose any global monomial order. In practice, the degree reverse lexicographic order, which is the default order of MACAULAY2, is preferable (see Bayer and Stillman, 1987).

DEFINITION 2.11 ($>_{\mathrm{drlex}}$). *The* degree reverse lexicographic order *on $K[x]$ is defined by*

$$x^\alpha >_{\mathrm{drlex}} x^\beta \iff \deg x^\alpha > \deg x^\beta \text{ or } \deg x^\alpha = \deg x^\beta \text{ and the last non-zero entry of } \alpha - \beta \in \mathbb{Z}^n \text{ is negative.}$$ □

REMARK 2.12. *Let F be a free $K[x]$-module with basis $e_1, \dots, e_s$. Any monomial order $>$ on $K[x]$ can be extended to a monomial order $>$ on F in at least the following two ways:*

$$x^\alpha e_i > x^\beta e_j \Leftrightarrow i > j \text{ or } i = j \text{ and } x^\alpha > x^\beta,$$

giving priority to the components, and

$$x^\alpha e_i > x^\beta e_j \Leftrightarrow x^\alpha > x^\beta \text{ or } x^\alpha = x^\beta \text{ and } i > j,$$

giving priority to the monomials in $K[x]$. □

$>_{\text{drlex}}$ is used for the following basic tasks:

- *Solvability.* Given an ideal $I \subset K[x]$, decide whether $1 \in I$. More generally, Remark 2.5 allows one to settle:
- *Module membership.* Given a free $K[x]$-module F with a finite basis, $f \in F$ and a submodule $I \subset F$, decide whether $f \in I$.
- Compute *Hilbert polynomials* and *Hilbert functions* (*Hilbert series*). In particular, compute the dimension and the degree of a variety.
- Compute *syzygies*.
- Compute *ideal intersections* $I \cap J$.
- Compute *ideal quotients* $(I : J)$.

Many problems in constructive module theory, where one manipulates $K[x]$-modules given by generators and relations, can be reduced to solving systems of linear equations over $K[x]$, that is, to syzygy computations over $K[x]$.

EXAMPLE 2.13 (HOM). *Given free presentations of two $K[x]$-modules M and N, we want to compute a free presentation of* $\mathrm{Hom}_{K[x]}(M,N)$. *Any homomorphism from M to N lifts to a homomorphism of free presentations:*

$$\begin{array}{ccccccc} 0 \leftarrow & M & \leftarrow & F_0 & \leftarrow & F_1 \\ & \downarrow \varphi & & \downarrow \varphi_0 & & \downarrow \varphi_1 \\ 0 \leftarrow & N & \leftarrow & G_0 & \leftarrow & G_1 \end{array}$$

The coefficients of the matrices φ_0, φ_1 are the solutions of a linear system of equations over $K[x]$, that is, they are syzygies. The diagrams corresponding to the zero in $\mathrm{Hom}_{K[x]}(M,N)$ *are those with φ_0 factoring through G_1. Thus computing* Hom *reduces to two syzygy calculations.* □

For some types of computations one needs special monomial orders. An important example is *elimination:* Let I be an ideal in the polynomial ring

$$K[x,y] = K[x_1, \dots, x_n, y_1, \dots y_m].$$

We explain how to eliminate the variables $x_1, \dots, x_n$ from I, that is, how to compute $I \cap K[y]$.

DEFINITION 2.14. *A monomial order $>$ on $K[x,y]$ is an* elimination order *with respect to $x_1,\ldots,x_n$ if the following holds: if $f \in K[x,y]$ with $\mathrm{LT}(f) \in K[y]$, then $f \in K[y]$.* □

EXAMPLE 2.15. *Given a global monomial order $>_1$ on $K[x]$ and a monomial order $>_2$ on $K[y]$, the* product order *or* block order $> = (>_1,>_2)$ *on $K[x,y]$, defined by*

$$x^\alpha y^\beta > x^\gamma y^\delta \Leftrightarrow x^\alpha >_1 x^\gamma \text{ or } x^\alpha = x^\gamma \text{ and } y^\beta >_2 y^\delta,$$

is an elimination order with respect to $x_1,\ldots,x_n$. □

PROPOSITION 2.16 (ELIMINATION). *Let $>$ be a global elimination order on $K[x,y]$ with respect to $x_1,\ldots,x_n$, and let $\mathcal{G}$ be a Gröbner basis of I with respect to $>$. Then $\mathcal{G} \cap K[y]$ is a Gröbner basis of $I \cap K[y]$ with respect to the restriction of $>$ to $K[y]$.* □

REMARK 2.17. *Computing a Gröbner basis with respect to an elimination order is costly. In some cases a better performance is achieved as follows. First compute a Gröbner basis of I with respect to $>_{\mathrm{drlex}}$. Then compute the Hilbert function of I from this Gröbner basis. Finally compute a Gröbner basis with respect to an elimination order by using a Hilbert driven version of Buchberger's algorithm* (Traverso, 1996).

Elimination can be used to compute the kernel of a morphism

$$\phi : K[y_1,\ldots,y_m] \to K[x_1,\ldots,x_n]/I, \quad y_i \mapsto f_i:$$

PROPOSITION 2.18 (ALGEBRA RELATIONS). *Pick polynomial representatives F_i of f_i, and let J be the ideal*

$$J = IK[x,y] + (F_1 - y_1,\ldots,F_m - y_m)_{K[x,y]}.$$

Then

$$\ker\phi = J \cap K[y].$$ □

Geometrically, this means to compute the closure of the image of a variety under a morphism.

EXAMPLE 2.19. *Consider the map*

$$S^2 \to \mathbb{R}^3, \ (x_1,x_2,x_3) \mapsto (x_1x_2, x_1x_3, x_2x_3),$$

from the real 2-sphere to the real 3-space. Its image is the famous Steiner Roman surface whose equation can be computed with MACAULAY2 *as follows:*

```
i1 : ringS2=QQ[x_1..x_3]/(x_1^2+x_2^2+x_3^2-1)
o1 = ringS2
o1 : QuotientRing
i2 : ringR3=QQ[y_1..y_3]
o2 = ringR3
o2 : PolynomialRing
i3 : parametrization=map(ringS2,ringR3,{x_1*x_2,x_1*x_3,x_2*x_3});
o3 : RingMap ringS2 <--- ringR3
i4 : steiner= kernel parametrization
             2 2    2 2    2 2
o4 = ideal(y y  + y y  + y y  - y y y )
            1 2    1 3    2 3    1 2 3
o4 : Ideal of ringR3
```

To visualize the Steiner Roman surface one can use packages such as SURF *(see Section 9).*

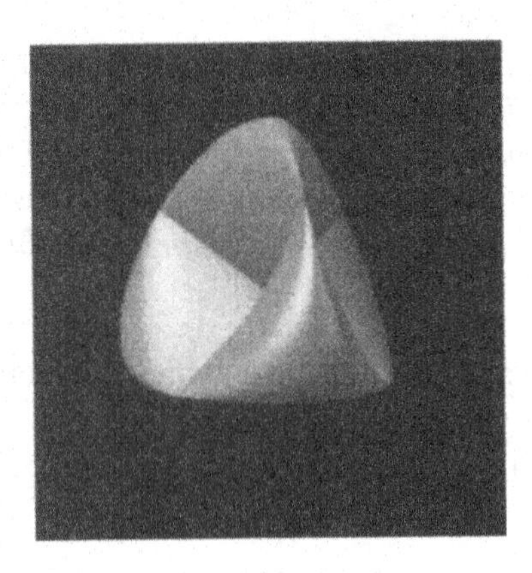

□

The computation of the projective closure of an affine variety $V(I)$ requires special monomial orders, too:

PROPOSITION 2.20 (HOMOGENIZATION). *Let $I \subset K[x_1,\dots,x_n]$ be an ideal. Pick a global monomial order $>$ on $K[x_1,\dots,x_n]$ which respects degrees, that is, which satisfies*

$$\deg x^\alpha > \deg x^\beta \Rightarrow x^\alpha > x^\beta,$$

and set

$$x^\alpha x_0^d >_h x^\beta x_0^e \Leftrightarrow x^\alpha > x^\beta \text{ or } (x^\alpha = x^\beta \text{ and } d > e).$$

Then $>_h$ is a global monomial order on $K[x_0,x_1,\dots,x_n]$ and the following holds if we homogenize with respect to x_0: if $f_1,\dots,f_r$ form a Gröbner basis of I with respect to $>$, the homogenized polynomials $f_1^h,\dots,f_r^h$ form a Gröbner basis of the homogenized ideal I^h. □

3. Gröbner bases in more general rings

3.1. LOCAL RINGS

We may think of the local ring $\mathcal{O}_p$ of a point $p \in \mathbb{A}^n$ as a subring of the ring of formal power series $K[[x_1,\ldots,x_n]]$. If we choose a local monomial order, then every power series has a leading term and the concept of division with remainder and Gröbner bases, here introduced by Hironaka (1964) and Grauert (1972) under the name of standard bases, carries over. This gives not yet an algorithm, however, because in the division theorem we may get a series which only formally converges. The missing step towards a true algorithm was realized by Mora (1982).

THEOREM 3.1 (MORA DIVISION WITH REMAINDER). *Let $>$ be any monomial order on $K[x] = K[x_1,\ldots,x_n]$, and let $f_1,\ldots,f_r \in K[x]\setminus\{0\}$ be polynomials. Then for every $g \in K[x]$ there exist $g_1,\ldots,g_r,h \in K[x]$ and a polynomial $u \in K[x]$ with $u(p) = 1$, that is, u is a unit in $\mathcal{O}_p$, such that*

1. $ug = g_1 f_1 + \ldots + g_r f_r + h$,
2'. $\mathrm{LT}(ug) \geq \mathrm{LT}(g_i f_i)$, *whenever both sides are non-zero, and*
3'. *if h is non-zero,* $\mathrm{LT}(h) \notin (\mathrm{LT}(f_1),\ldots,\mathrm{LT}(f_r))$. □

So for an element $g/v \in \mathcal{O}_p \subset K(x_1,\ldots,x_n)$, $v(p) \neq 0$, we have $g/v \equiv h/(uv)$ mod $(f_1,\ldots,f_r)\mathcal{O}_p$. To prove the theorem we give Mora's division algorithm. The key new ingredient is defined as follows:

DEFINITION 3.2. *Let $>$ be a monomial order on $K[x]$, and let f be a non-zero polynomial in $K[x]$. The* ecart *of f is*

$$\mathrm{ecart}(f) := \deg f - \deg \mathrm{LT}(f). \qquad \square$$

ALGORITHM 3.3 (MORA'S DIVISION ALGORITHM). *Given a monomial order $>$ on $K[x]$ and polynomials $f_1,\ldots,f_r \in K[x]\setminus\{0\}$, $g \in K[x]$, compute a remainder h as above.*

1. *Set $h = f$ and $T = \{f_1,..,f_r\}$.*
2. *while $(h \neq 0)$ and $T_h := \{f \in T \mid \mathrm{LT}(f)$ divides $\mathrm{LT}(h)\} \neq \emptyset$ do*
   2.1. *Choose $f \in T_h$ with* $\mathrm{ecart}(f)$ *minimal;*
   2.2. *If* $(\mathrm{ecart}(f) > \mathrm{ecart}(h))$ *then $T = T \cup \{h\}$;*
   2.3. *$h = h - mf$, where m is the term such that* $\mathrm{LT}(h) - m\mathrm{LT}(f) = 0$;
3. *return h.* □

That Mora's division algorithm terminates is most easily seen by homogenization (see for example Greuel and Pfister, 1996). If the input is homogeneous then the ecart is always zero and we obtain the indeterminate division algorithm as in Section 2.

With Mora division Buchberger's algorithm works for not necessarily global orders, too (besides local orders we also need mixed orders, namely for elimination; see Greuel and Pfister, 1996). This algorithm is implemented in `SINGULAR`.

EXAMPLE 3.4. *We start a* SINGULAR *session by defining the polynomial ring* $\mathbb{F}_{32003}[x,y,z]$ *together with the* negative degree reverse lexicographic order, *which is denoted by* ds *in* SINGULAR.

```
                         SINGULAR                              /
  A Computer Algebra System for Polynomial Computations       /   version 2-0-0
                                                             0<
     by: G.-M. Greuel, G. Pfister, H. Schoenemann             \   February 2001
FB Mathematik der Universitaet, D-67653 Kaiserslautern         \
> ring r = 32003,(x,y,z),ds;
> int a,b,c=37,27,6;
> int t=0;
> poly f0 = x^a+y^b+z^(3*c)+x^(c+2)*y^(c-1)+x^(c-1)*y^(c-1)*z3+
.                     x^(c-2)*y^c*(y^2+t*x)^2;
> f0;
x8y5+x5y5z3+x4y10+z18+y27+x37
> int aa=timer;
```

timer *gives the computing time in full seconds. We compute the Milnor number*

$$\dim \mathcal{O}_0/(\partial f_0/\partial x, \partial f_0/\partial y, \partial f_0/\partial z)$$

(see Greuel and Pfister, 1996, Corollary 3.4).

```
> ideal i0=jacob(f0);
> ideal j0=std(i0);
> vdim(j0);
4840
> timer-aa;
0
> t=1;
> poly f1=x^a+y^b+z^(3*c)+x^(c+2)*y^(c-1)+x^(c-1)*y^(c-1)*z3
.             +x^(c-2)*y^c*(y^2+t*x)^2;
> f1;
x6y6+x8y5+2x5y8+x5y5z3+x4y10+z18+y27+x37
> ideal i1=jacob(f1);
> ideal j1=std(i1);
> vdim(j1);
4834
> timer-aa;
9
```

□

3.2. *D*-MODULES

Consider

$$D = D_n = K\langle x_1, \ldots, x_n, \partial_1, \ldots, \partial_n \rangle,$$

the Weyl algebra of partial differential operators with polynomial coefficients. Here $\partial_i = \frac{\partial}{\partial x_i}$, and we have the relations

$$x_i x_j = x_j x_i, \ \partial_i \partial_j = \partial_j \partial_i,$$

and, the product rule,

$$\partial_i x_j = x_j \partial_i + \delta_{ij}.$$

The notions of monomials and Gröbner basis carry over to left ideals and finitely presented left D-modules (see for example Saito, Sturmfels and Takayama, 1999). Every non-zero D-module has dimension at least n, and those of dimension n are called holonomic. The most important holonomic module is $R_n = K[x_1, \ldots, x_n]$. Others include $R_n[f^{-1}]$, the localization of R_n by a polynomial f.

Since Hom_D and Ext_D can be computed using Gröbner bases (see Section 4.1 below) we can for instance compute polynomial solutions to a system of partial differential equations $P_1, \ldots, P_r \in D_n$ by computing $\mathrm{Hom}_{D_n}(D_n/I, R_n)$, where $I = D_n(P_1, \ldots, P_r)$.

In (Walter, 2001) the algorithm to compute for example the Bernstein-Sato polynomial $b_f(s) \in K[s]$ of a polynomial f is explained and an implementation in MACAULAY2 presented. Other examples include the computation of the de Rham cohomology of the complement of an affine hypersurface. In principle, the de Rham cohomology of a projective variety can be computed using D-modules (Walter, 2000), but, to our knowledge, there is no implementation yet.

3.3. OTHERS

The method of Gröbner bases has been extended to various classes of non-commutative algebras over fields and other rings including enveloping algebras of Lie algebras, algebras of solvable type, Grassman algebras, and Clifford algebras (see Madlener and Reinert, 1998 for an overview). Gröbner bases in exterior algebras are implemented in MACAULAY2 and SINGULAR. BERGMANN can handle Gröbner bases computations in commutative and non-commutative graded algebras.

4. Homological methods

4.1. EXT AND TOR

Given finitely generated, graded modules M and N over the polynomial ring $S = K[x_0, \ldots, x_n]$, one can compute the modules $\mathrm{Ext}^i_S(M,N)$ and $\mathrm{Tor}^S_i(M,N)$ from a free resolution of M over S. The situation is more difficult over quotient rings $R = S/I$ since in this case $\mathrm{Ext}^i_R(M,N)$ and $\mathrm{Tor}^R_i(M,N)$ might be non-zero for arbitrary large i. Any finite piece of a free resolution of M, however, can be computed.

EXAMPLE 4.1. *We consider a* MACAULAY2 *code, which computes the resolution of the base field K over the homogenous coordinate ring R of the rational normal curve in* $\mathbb{P}^3$. *The ith betti number is the dimension of* $\mathrm{Ext}^i_R(K,K)$.

```
i1 : K=ZZ/101;
i2 : S=K[x_0..x_3];
i3 : m=matrix(apply(3,i->apply(2,j->x_(i+j))))
o3 = | x_0 x_1 |
     | x_1 x_2 |
     | x_2 x_3 |
             3       2
o3 : Matrix S  <--- S
i4 : R=S/minors(2,m);
i5 : time betti res(coker vars R,LengthLimit=>13)
     -- used 9.85 seconds
o5 = total: 1 4 9 18 36 72 144 288 576 1152 2304 4608 9216 18432
         0: 1 4 9 18 36 72 144 288 576 1152 2304 4608 9216 18432
o5 : Net
```

□

The Poincaré series

$$P^R_M(t,u) = \sum_{i\in\mathbb{N},s\in\mathbb{Z}} \dim_K \mathrm{Ext}^i_R(M,K)_s u^{-s} t^i \in \mathbb{Z}[u,u^{-1}][[t]]$$

summarizes part of the numerical data of the minimal free resolution of M over R. In the example above we may guess that

$$P^R_K(t,u) = 1 + 4tu + 9tu\sum_{i=0}^{\infty}(2tu)^i = \frac{1+2tu+(tu)^2}{1-2tu}.$$

In general, however, such a simple expression does not necessarily exist: examples of graded rings R are known, where $P^R_K(t,1)$ is not a rational function, that is, $P^R_K(t,1) \notin \mathbb{Z}[[t]] \cap \mathbb{C}(t)$ (Anick, 1980). Moreover, Roos (1993) gave examples of rings $R = R(r)$ for each $r \geq 2$ such that $\mathrm{Ext}^i_R(K,K)_s = 0$ for all $0 \leq i < r$ and $s \neq i$, and $\mathrm{Ext}^r_R(K,K)_{r+1} \neq 0$. So our guess in the example above, that the free resolution of the base field might be linear for ever, could be wrong (in fact, it is not wrong in this case). BERGMAN provides excellent software to investigate these questions for quadratic not necessarily commutative algebras.

By work of Avramov and others it is known that $\dim_K \mathrm{Ext}^i_R(K,K)$ has polynomial growth iff R is a complete intersection (see Avramov, 1996). In the case of a complete intersection R the irregular behavior of the free resolutions disappears. Avramov and Grayson (2001) present a MACAULAY2 code which is able to compute $\mathrm{Ext}^i_R(M,N)$ simultaneously for all i in this case. It is a big challenge to develop efficient algorithms to compute $\mathrm{Ext}^i_R(M,N)$ for i large, say $i = 100$, for the case that R is not a complete intersection.

In our own work one of the most frequent applications of Ext is the computation of the dualizing sheaf of a variety.

EXAMPLE 4.2. *We consider a random curve of genus* 14, *which we realize as a degree 14 space curve. The* MACAULAY2 *script which produces the ideal of the curve will be discussed in Section 6.3.*

```
i7 : R=ZZ/5[x_0..x_3]
o7 = R
o7 : PolynomialRing
i8 : time I=randomGenus14Curve(R);
     -- 154 modules tested
     -- good modules tested:1
     -- 6.03 seconds used to check smoothness once
     -- curves tested:1
     -- used 16.22 seconds
o8 : Ideal of R
i9 : hilbertPolynomial(R^1/I)
o9 = - 27*P  + 14*P
           0       1
o9 : ProjectiveHilbertPolynomial
i10 : betti res I
o10 = total: 1 13 17 5
          0: 1 .  .  .
          1: . .  .  .
          2: . .  .  .
          3: . .  .  .
          4: . .  .  .
          5: . 13 17 4
          6: . .  .  1
o10 : Net
i11 : omega=Ext^2(R^1/I,R^{-4});
i12 : betti res omega
o12 = total: 5 11 6
         -1: 3 .  .
          0: 2 11 5
          1: . .  .
          2: . .  1
          3: . .  .
          4: . .  .
o12 : Net
```

Thus $h^1(\mathcal{O}_C(1)) = h^0(\omega_C(-1)) = 3$ *and* $h^0(\omega_C) = 3*4+2 = 14$. □

4.2. COHOMOLOGY OF COHERENT SHEAVES

Computing the cohomology of a coherent sheaf $\mathcal{F}$ on $\mathbb{P}^n$ might mean, for example, to compute one of the dimensions $\mathrm{h}^j\,\mathcal{F}(i) := \dim_K \mathrm{H}^j\,\mathcal{F}(i)$, or to compute these dimensions in a certain range of twists, or to compute the graded modules

$$\mathrm{H}^j_*\,\mathcal{F} := \bigoplus_{i\in\mathbb{Z}} \mathrm{H}^j\,\mathcal{F}(i)$$

over the homogeneous coordinate ring of $\mathbb{P}^n$. To fix our notation in this context, let V be a vector space of dimension $n+1$ over K, $W = V^*$ its dual space, $\mathbb{P}^n = \mathbb{P}(W)$ the projective space of 1-quotients of W, and $S = \mathrm{Sym}_K(W)$ its homogeneous coordinate ring. Serre's sheafification functor $M \mapsto \widetilde{M}$ allows us to consider the coherent sheaf $\mathcal{F}$ as an equivalence class of finitely generated graded S-modules, where we identify two such modules M and M' if, for some r, the truncated modules $M_{\geq r}$ and $M'_{\geq r}$ are isomorphic.

We discuss two of the methods for computing sheaf cohomology, one relying on the ability to compute free resolutions over the symmetric algebra S, and one which asks for syzygy computations over the exterior algebra E on V.

The first method, described by Eisenbud (1998) (see also Smith, 2000), is based on local duality:

THEOREM 4.3. *Let M be a finitely generated graded S-module, and let $\mathcal{F} = \widetilde{M}$ be the associated coherent sheaf. For all $j \geq 1$ we have*

$$\mathrm{H}^j_*\,\mathcal{F} \cong \mathrm{Ext}_S^{n-j}(M,\omega)^{\vee},$$

where $\omega = S(-n-1)$, and where $N^{\vee}$ denotes the graded vector space dual of N as an S-module.

The resulting algorithm is implemented in MACAULAY2 as the function HH. If F is a sheaf object in MACAULAY2, then

```
HH^j(F(>=d))
```

returns the truncated jth cohomology module $\mathrm{H}^j_{\geq d}\,\mathcal{F}$.

The second method is due Eisenbud, Fløystad, and Schreyer (2001), with explicit code given in (Decker and Eisenbud, 2001). It makes use of a new, constructive version of the Bernstein-Gel'fand-Gel'fand (BGG) correspondence, originally described in (BGG, 1978). The BGG correspondence consists of a pair of adjoint functors **R** and **L** which define an equivalence between the derived category of bounded complexes of finitely generated graded S-modules and the derived category of bounded complexes of finitely generated graded E-modules.

Let us explain how $\mathbf{R}(M)$ is defined for a finitely generated graded S-module $M = \bigoplus_i M_i$, considered as a complex concentrated in cohomological degree 0. We grade S and E by taking elements of W to have degree 1, and elements of V to have degree -1. $\mathbf{R}(M)$ is the sequence of free E-modules and maps

$$\mathbf{R}(M): \cdots \longrightarrow F^{j-1} \xrightarrow{\phi_{j-1}} F^j \xrightarrow{\phi_j} F^{j+1} \longrightarrow \cdots$$

given as follows. We set

$$\omega_E = \mathrm{Hom}_K(E,K) = E \otimes \bigwedge^{n+1} W \cong E(-n-1),$$

$$F^j = \mathrm{Hom}_K(E,M_j) = M_j \otimes_K \omega_E,$$

where M_j is considered as a vector space concentrated in degree j, and let $\phi_j : F^j \to F^{j+1}$ be the map taking α to

$$[e \mapsto \sum_i x_i \alpha(e_i \wedge e)],$$

where $\{x_i\}$ and $\{e_i\}$ are dual bases of W and V respectively. One checks that $\mathbf{R}(M)$ is indeed a complex. In fact, it is a linear complex, and it is eventually exact:

THEOREM 4.4 (EISENBUD, FLØYSTAD, AND SCHREYER, 2001). *Let M be a finitely generated graded S-module, and let r be its Castelnuovo-Mumford regularity r. The complex $\mathbf{R}(M)$ is exact at $\mathrm{Hom}_K(E,M_i)$ for all $i \geq s$ iff $s > r$.* □

Starting from $\mathbf{T}^{>r}(M) := \mathbf{R}(M_{>r})$ we obtain a doubly infinite, exact, E-free complex $\mathbf{T}(M)$, the *Tate resolution* of M, by adjoining a minimal free resolution of the kernel of $\mathrm{Hom}_K(E,M_{r+1}) \to \mathrm{Hom}_K(E,M_{r+2})$. In fact, we may construct $\mathbf{T}(M)$ starting from any truncation $\mathbf{R}(M_{>s})$, $s \geq r$. Whence $\mathbf{T}(M)$ only depends on the sheaf $\mathcal{F} = \widetilde{M}$ associated to M. $\mathbf{T}(\mathcal{F}) := \mathbf{T}(M)$ is also called the *Tate resolution* of $\mathcal{F}$.

THEOREM 4.5 (EISENBUD, FLØYSTAD, AND SCHREYER, 2001). *Let M be a finitely generated graded S-module, and let $\mathcal{F} = \widetilde{M}$ be the associated coherent sheaf. The term of the complex $\mathbf{T}(\mathcal{F})$ with cohomological degree i is*

$$\bigoplus_j H^j \mathcal{F}(i-j) \otimes \omega_E,$$

where $H^j \mathcal{F}(i-j)$ is regarded as a vector space concentrated in degree $i-j$. □

Thus each cohomology group of each twist of the sheaf $\mathcal{F}$ occurs exactly once in a term of $\mathbf{T}(\mathcal{F})$. We can thus compute part of the cohomology of $\mathcal{F}$ by computing part of the Tate resolution $\mathbf{T}(\mathcal{F})$. A corresponding MACAULAY2 function, `sheafCohomology`, is described in (Decker and Eisenbud, 2001). If m is a presentation matrix of M, and if $l \leq h$ are integers, then

```
betti sheafCohomology(m,E,l,h)
```

prints a cohomology table for $\mathcal{F}$ of the form

$$\begin{array}{ccc} h^0\,\mathcal{F}(h) & \dots & h^0\,\mathcal{F}(l) \\ h^1\,\mathcal{F}(h-1) & \dots & h^1\,\mathcal{F}(l-1) \\ \vdots & & \vdots \\ h^n\,\mathcal{F}(h-n) & \dots & h^n\,\mathcal{F}(l-n)\,. \end{array}$$

We refer to Section 5.1 below for an example.

4.3. BEILINSON MONADS AND RESULTANT FORMULAS

The technique of monads provides powerful tools for problems such as the construction and classification of coherent sheaves with prescribed invariants (see for example Okonek, Schneider, and Spindler, 1980). The basic idea is to represent arbitrary coherent sheaves in terms of simpler sheaves such as line bundles or bundles of differentials, and in terms of homomorphisms between these simpler sheaves.

DEFINITION 4.6. *A* monad *on* $\mathbb{P}^n$ *with* homology $\mathcal{F}$ *is a bounded complex*

$$\cdots \longrightarrow \mathcal{K}^{-1} \longrightarrow \mathcal{K}^0 \longrightarrow \mathcal{K}^1 \longrightarrow \cdots$$

of coherent sheaves on $\mathbb{P}^n$ *with homology* $\mathcal{F}$ *at* $\mathcal{K}^0$, *and with no homology otherwise.* □

If M is a finitely generated graded S-module, with associated sheaf $\mathcal{F} = \widetilde{M}$, then the sheafification of the minimal free resolution of M is a monad for $\mathcal{F}$ which involves direct sums of line bundles and thus homogeneous matrices over S. The Beilinson monad for $\mathcal{F}$ involves direct sums of twisted bundles of differentials, and thus, as we will see, homogeneous matrices over E. It is much more directly connected with cohomology than the free resolution. Eisenbud, Fløystad, and Schreyer (2001) construct the Beilinson monad from the Tate resolution explicitly.

To explain the construction we recall that the ith bundle of differentials $\Omega^i_{\mathbb{P}(W)} = \bigwedge^i T^*_{\mathbb{P}(W)}$ fits as $\ker(\bigwedge^i W \otimes \mathcal{O}(-i) \to \bigwedge^{i-1} W \otimes \mathcal{O}(-i+1))$ into the exact sequence

$$0 \longrightarrow \textstyle\bigwedge^{n+1} W \otimes \mathcal{O}(-n-1) \longrightarrow \cdots \longrightarrow \bigwedge^0 W \otimes \mathcal{O} \longrightarrow 0$$

which is obtained by sheafifying the Koszul complex resolving the "trivial" graded S-module $K = S/(W)$. By lifting homomorphisms between the bundles to homomorphisms between the Koszul resolutions one shows (see for example Eisenbud and Decker, 2001):

LEMMA 4.7. *There are canonical isomorphisms*

$$\textstyle\bigwedge^{i-j} V \to \mathrm{Hom}(\Omega^i(i), \Omega^j(j)), \quad 0 \le i, j \le n.$$

Under these isomorphisms an element $e \in \bigwedge^{i-j} V$ *acts by contraction on the fibers of the* $\Omega^i(i)$. □

The construction of the Beilinson monad can now be easily described. Given $\mathbf{T}(\mathcal{F})$ we define $\Omega(\mathcal{F})$ to be the complex of vector bundles on $\mathbb{P}^n$ obtained by replacing each summand $\omega_E(i)$ by the sheaf $\Omega^i(i)$, and by using the isomorphisms

$$\mathrm{Hom}_E(\omega_E(i),\omega_E(j)) \cong \wedge^{i-j}V \cong \mathrm{Hom}(\Omega^i(i),\Omega^j(j))$$

to provide the maps.

THEOREM 4.8 (EISENBUD, FLØYSTAD, AND SCHREYER, 2001). *Let $\mathcal{F}$ be a coherent sheaf on $\mathbb{P}^n$. Then $\mathcal{F}$ is the homology of $\Omega(\mathcal{F})$ in cohomological degree 0, and $\Omega(\mathcal{F})$ has no homology otherwise. $\Omega(\mathcal{F})$ is called the* Beilinson monad *for $\mathcal{F}$.* □

EXAMPLE 4.9. *There are only a few vector bundles known on $\mathbb{P}^n$ which are indecomposable and of low rank, say of rank $\leq n-2$. One example is the Horrocks-bundle $\mathcal{F}_H$ on $\mathbb{P}^5$* (Horrocks, 1978) *which is indecomposable if char $K \neq 2$* (Decker, Manolache, and Schreyer, 1992)*. We construct this bundle as the homology of its Beilinson monad* (Decker, Manolache, and Schreyer, 1992) *by using the* `Macaulay2` *function* `beilinson` *given in* (Eisenbud and Decker, 2001)*. The* `beilinson` *function, the* `sheafCohomology` *function, which we will need in Section 5.1 below, and some functions used by* `beilinson` *or* `sheafCohomology` *are not yet part of* MACAULAY2*. They therefore have to be defined in the first part of our* MACAULAY2 *session (which we omit for lack of space).*

```
i8 : S = ZZ/101[x_0..x_5];
i9 : E=ZZ/101[e_0..e_5,SkewCommutative=>true];
i10 : beta=matrix({{e_0,e_4*e_5},{e_1,e_5*e_3},{e_2,e_3*e_4},
                {e_3, e_1*e_2}, {e_4, e_2*e_0}, {e_5,e_0*e_1}});
                 6       2
o10 : Matrix E  <--- E
i11 : beta=map(E^6,E^{-1,-2},beta)
o11 = | e_0 e_4e_5  |
      | e_1 -e_3e_5 |
      | e_2 e_3e_4  |
      | e_3 e_1e_2  |
      | e_4 -e_0e_2 |
      | e_5 e_0e_1  |
          6       2
o11 : Matrix E  <--- E
i12 : alpha=syz beta
o12 = {1} | e_0e_1e_2-e_3e_4e_5   e_0e_1e_3e_4+e_0e_2e_3e_5+e_1e_2e_4e_5 |
      {2} | -e_0e_3-e_1e_4-e_2e_5 -e_0e_1e_2-e_3e_4e_5                   |
               2       2
o12 : Matrix E  <--- E
i13 : beta=beilinson(beta,S);
o13 : Matrix
i14 : alpha=beilinson(alpha,S);
```

```
o14 : Matrix
i15 : time FH = prune homology(beta,alpha);
     -- used 0.36 seconds
i16 : time betti res FH
     -- used 0.15 seconds
o16 = total: 35 84 86 42 8
          2: 8  6  1  .  .
          3: 27 78 85 42 8
          4: .  .  .  .  .
          5: .  .  .  .  .
o16 : Net
```

The last diagram shows the graded betti numbers of $\mathcal{F}_H$. We see in particular that $\mathcal{F}_H$ is generated by 8 sections in twist 2 and 27 sections in twist 3. We will analyze the zero scheme of a general section of $\mathcal{F}_H(2)$ in Section 5.1 below. □

More generally, starting again from the Tate resolution $\mathbf{T}(\mathcal{F})$, Eisenbud and Schreyer (2001) explicitly determine a complex $\mathbf{U}(\mathcal{F})$ on the Grassman variety $\mathbb{G}(k,n)$ of codimension k planes in $\mathbb{P}^n$ which represents the sheaf $R\pi_2(\pi_1^* \mathcal{F})$ defined via the incidence diagram

$$\begin{array}{ccc} \mathbb{F} = \{(p,L) \mid p \in L\} & \xrightarrow{\pi_2} & \mathbb{G}(k,n) \\ \downarrow \pi_1 & & \\ \mathbb{P}^n & & \end{array} .$$

In this way they obtain new formulas of Chow forms and resultants in terms of the Plücker coordinates. For instance, the resultant of 3 quadrics in 3 variables is given by the Pfaffian of the matrix

$$\begin{pmatrix} 0 & [245] & [345] & [135] & [045] & [035] & [145] & [235] \\ -[245] & 0 & -[235] & [035] & [025] & [015] & [125] & -[125]+[045] \\ -[345] & [235] & 0 & [134] & [035] & [034] & [135] & [234] \\ -[135] & -[035] & -[134] & 0 & [023] & [013] & [123]-[034] & -[123] \\ -[045] & -[025] & -[035] & -[023] & 0 & [012] & -[015] & -[024]+[015] \\ -[035] & -[015] & -[034] & -[013] & -[012] & 0 & [023]-[014] & -[023] \\ -[145] & -[125] & -[135] & -[123]+[034] & [015] & -[023]+[014] & 0 & -[124]+[035] \\ -[235] & [125]-[045] & -[234] & [123] & [024]-[015] & [023] & [124]-[035] & 0 \end{pmatrix}$$

Here the monomials in the three variables a,b,c are ordered as $a^2, ab, ac, b^2, bc, c^2$, and the brackets $[ijk]$ denote the corresponding Plücker coordinates of the net of quadrics.

5. Modifying algebraic sets

5.1. PRIMARY DECOMPOSITION

Depending on the application we might be interested in computing, for instance, the radical of a given ideal, its equidimensional parts, its associated primes, or a complete primary decomposition. The first algorithms for these tasks were given by Grete Hermann (1926), a student of Emmy Noether. Hermann used generic projections, that is elimination, based on resultant techniques, to reduce to the case of principal ideals. Thus computing radicals reduces to the computation of the square-free part of polynomials, and computing primary decompositions reduces to factorizing polynomials. In contrast to the algorithms discussed in Section 2, algorithms for polynomial factorization highly depend on the nature of the underlying field. We refer to the survey papers by Kaltofen (1982, 1990, 1992) for part of the history of univariate and multivariate polynomial factorization over various coefficient domains.

The problem to provide efficient algorithms for the tasks mentioned above is difficult and still one of the big challenges for computer algebra. A survey on some of the modern algorithms can be found in (Decker, Greuel, and Pfister, 1998). Most of the algorithms follow the basic strategy of Hermann. They use elimination, now based on Gröbner basis techniques, to reduce to square-free decomposition or polynomial factorization. A different approach is taken by Eisenbud, Huneke, and Vasconcelos (1992), who avoid generic projections, relying on syzygy methods instead.

EXAMPLE 5.1. *We continue the* MACAULAY2 *session started in Example 4.10 and compute the zero scheme X of a general section of the twisted Horrocks bundle* $\mathcal{F}_H(2)$.

```
i17 : betti(allSections=syz transpose presentation FH)
o17 = total: 35 35
          -3: 27  .
          -2:  8  .
          -1:  .  .
           0:  .  .
           1:  .  8
           2:  . 27
o17 : Net
```

To pick a "general" section of $\mathcal{F}_H(2)$ *would mean to make a random choice. To ease our computations we pick a section which, as it will turn out, is just general enough.*

```
i18 : pickSection=map(S^1,S^27,0) | map(S^1, S^8,matrix{{4:1_S,4:0}});
                     1        35
o18 : Matrix S  <--- S
```

```
i19 : X=ideal(pickSection*allSections);
o19 : Ideal of S
i20 : codim X
o20 = 3
i21 : degree X
o21 = 14
i22 : time componentsX=decompose X;
     -- used 113.97 seconds
i23 : apply(componentsX,i->degree i)
o23 = {1, 2, 2, 2, 1, 6}
o23 : List
i24 : apply(componentsX,i->codim(minors(3,jacobian gens i)+i))
o24 = {infinity, 6, 6, 6, infinity, 6}
o24 : List
```

We see that X has 6 smooth components: 2 planes, 3 quadrics, and a surface of degree 6. What is the degree 6 surface? We compute information on the cohomology of the ideal sheaf of the surface by using the function sheafCohomology *discussed in Section 4.2.*

```
i25 : Y=componentsX_5;
i26 : time betti sheafCohomology(gens Y,E,-2,2)
     -- used 0.69 seconds
o26 = total: 19 8 8 19 37
         -2: 19 7 1  .  .
         -1:  . . .  .  .
          0:  . 1 7 19 37
o26 : Net
```

From the betti diagram we see that Y has geometric genus and irregularity $p_g = q = 0$, that $\mathrm{h}^0\mathcal{O}_X(H) = 7$, so Y is projected from $\mathbb{P}^6$, and that $\mathrm{h}^0\mathcal{O}_X(K+H) = 1$. By adjunction theory, Y is a del Pezzo surface. Decker, Manolache and Schreyer (1992) *showed that the zero scheme of a general section of $\mathcal{F}_H(2)$ decomposes in this way.* □

REMARK 5.2. *1.* MACAULAY2 *can compute radicals and top-dimensional parts. But currently it cannot handle primary decompositions in full generality. The* decompose *command computes the minimal associated primes only, by an algorithm which combines characteristic set methods with elimination to reduce to polynomial factorization (see* Wang, 1989*). For computing characteristic sets and factorization* MACAULAY2 *refers to a sub-package of* SINGULAR. *In* SINGULAR, *all algorithms discussed in (*Decker, Greuel, and Pfister, 1998*) are implemented (including, for primary decomposition, the algorithm of* Gianni, Trager, and Zacharias, 1988, *and the algorithm of* Shimoyama and Yokoyama, 1996 *combined with that of* Wang, 1989*). No "generally best" algorithm is known. We recommend to compare the performance of the different algorithms for the examples* *one has in mind. An example involving the* `SINGULAR` *library* `primdec.lib` *will be given in Section 6.1.*

2. On the computer we work over a non-algebraically closed field K such as a prime field or a number field. The known algorithms decompose a given ideal over its field of definition and not over the extension field where the decomposition into the absolutely irreducible components occurs. □

5.2. NORMALIZATION

Algorithms for normalization were given by several authors (Stolzenberg, 1968; Seidenberg, 1975; Vasconcelos, 1991; Gianni and Trager, 1997; de Jong, 1998). The algorithm proposed by de Jong is based on a criterion for normality due to Grauert and Remmert (1971):

THEOREM 5.3. *Let R be a Noetherian, reduced ring. Let J be a radical ideal containing a non–zero divisor such that the zero set of J, $\mathrm{V}(J)$, contains the non-normal locus of* $\mathrm{Spec}(R)$. *Then R is normal if and only if $R = \mathrm{Hom}_R(J,J)$.* □

The algorithm is implemented in `SINGULAR` (see Decker, Greuel, de Jong, and Pfister, 1999). It takes as input a radical ideal I in a polynomial ring R, and computes the normalization of R/I, that is, the integral closure of R in its total quotient ring $Q(R)$ (if I is not radical, first compute `I=radical(I);`). The result describes the normalization as a product of affine rings together with the normalization map.

EXAMPLE 5.4. *We start a* `SINGULAR` *session by loading the normalization library.*

```
> LIB "normal.lib";
> ring R = 0,(x(1..3)),dp;
> ideal I = x(1)^3*x(2)^2+x(1)^3*x(3)^2+x(1)*x(2)^2*x(3)^2-x(1)^2*x(2)*x(3);
> int t=timer;
> list NOR=normal(I);
// 'normal' created a list of 2 ring(s).
// To see the rings, type (if the name of your list is nor):
     show( nor);
// To access the 1-st ring and map (similair for the others), type:
     def R = nor[1]; setring R;  norid; normap;
// R/norid is the 1-st ring of the normalization and
// normap the map from the original basering to R/norid
> timer-t;
1
> show(NOR);
```

```
// list, 2 element(s):
[1]:
   // ring: (0),(T(1),T(2),T(3),T(4),T(5)),(dp(5),C);
   // minpoly = 0
// objects belonging to this ring:
// normap               [0]  ideal, 3 generator(s)
// norid                [0]  ideal, 6 generator(s)
[2]:
   // ring: (0),(T(1),T(2)),(dp(2),C);
   // minpoly = 0
// objects belonging to this ring:
// normap               [0]  ideal, 3 generator(s)
// norid                [0]  ideal, 1 generator(s)
> def R1=NOR[1];
> setring R1;
> norid;
norid[1]=2T(3)2-T(4)T(5)+T(4)
norid[2]=T(2)T(3)+T(1)T(4)
norid[3]=2T(1)T(3)+T(2)T(5)-T(2)
norid[4]=2T(2)2+2T(4)2+T(4)T(5)+T(4)
norid[5]=2T(1)T(2)-2T(3)T(4)-T(3)T(5)-T(3)
norid[6]=4T(1)2+2T(4)T(5)+T(5)2-2T(4)-1
> normap;
normap[1]=T(1)
normap[2]=T(2)
normap[3]=T(3)
> def R2=NOR[2];
> setring R2;
> norid;
norid[1]=0
> normap;
normap[1]=0
normap[2]=T(1)
normap[3]=T(2)
```

Thus the normalization has two components, an affine part of the Veronese surface, and a plane. This is no big surprise since we started with the equation of the Steiner Roman surface multiplied by x_1. □

5.3. PUISEUX EXPANSION

Given a germ of a plane curve singularity

$$f(x,y) = 0\,,$$

SINGULAR can compute the Puiseux expansion of a branch of f,

$$x(t) = t^d$$
$$y(t) = a_0 t^{\alpha_0} + a_1 t^{\alpha_1} + \ldots$$

To cover the case of positive characteristic, SINGULAR actually computes the Hamburger-Noether expansion, where also $x(t)$ is allowed to be a power series (see Campillo, 1980).

EXAMPLE 5.5. *We start now a* SINGULAR *session by loading the Hamburger-Noether library.*

```
> LIB "hnoether.lib";
> ring r = 0,(x,y),ds;
> poly f = (y2+x3)^2+x5y;
> list hne = HNdevelop(f);
HNE of one branch found
// result: 1 branch(es) successfully computed,
//          basering has changed to HNEring
> displayHNE(hne);
// Hamburger-Noether development of branch nr.1:
HNE[1]=-y+z(0)*z(1)
HNE[2]=-x-z(1)^2+z(1)^2*z(2)
HNE[3]=-z(2)^2-2*z(2)^3
> displayInvariants(hne[1]);
 characteristic exponents   : 4,6,7
 generators of semigroup    : 4,6,13
 Puiseux pairs              : (3,2)(7,2)
 degree of the conductor    : 16
 delta invariant            : 8
 sequence of multiplicities: 4,2,2,1
> list L = param(extdevelop(hne[1],7));
a(2,3) = -2
a(2,4) = -3
a(2,5) = -4
a(2,6) = -5
a(2,7) = -6
// ** Warning: result is exact up to order 9 in x and 11 in y !
> L;
[1]:
   _[1]=-x4-3x5-6x6-10x7-15x8-21x9-14x10-6x11+3x12+13x13+24x14+36x15
   _[2]=x6+5x7+15x8+35x9+70x10+126x11+189x12+243x13+270x14+250x15+161x16-21x17
-175x18-287x19-342x20-324x21-216x22
>
```

Thus f is a uni-branched singularity with two Puiseux pairs. At present the implementation handles uni-branched singularities over $\mathbb{Q}$ *and several branches over finite fields.* □

5.4. RATIONAL PARAMETRIZATION

Given a plane curve $C \subset \mathbb{P}^2$ of genus 0 defined over $\mathbb{Q}$, the MAPLE package CASA allows one to compute a rational parametrization of C defined over $\mathbb{Q}$ or over a quadratic extension of $\mathbb{Q}$ (to extend $\mathbb{Q}$ might be necessary for curves of even degree; see Winkler, 1996 and the references there for information on the algorithms).

EXAMPLE 5.6. *After having installed* CASA *we can load it into* MAPLE.

```
> with(casa);

           |__|
           |  |          Welcome to CASA 2.5 for Maple V.5
     |   /\|  |/\
    /=\__| []  |
   /            \_       Copyright (C) 1990-2000 by Research Institute
   |              \      for Symbolic Computation (RISC-Linz), the
   \   CASA 2.5   |      University of Linz, A-4040 Linz, Austria.
     |            |
   _|      |||    |      For help type '?casa' or '?casa,<topic>'.
__/        |||    |_

  [BCH2, BCHDecode, CyclicEncode, DivBasisL, GWalk, GoppaDecode,
        GoppaEncode, GoppaPrepareDu, GoppaPrepareSV, GoppaPrepareSa,
        GoppaPrimary, Groebnerbasis, InPolynomial, NormalPolynomial,
        OutPolynomial, PolynomialRoots, RPHcurve, SakataDecode,
        SubsPolynomial, _casaAlgebraicSet, adjointCurve, algset,
        casaAttributes, casaVariable, computeRadical, conic,
        decompose, delete, dimension, equalBaseSpaces,
        equalProjectivePoints, finiteCurve, finiteField, generators,
        genus, homogeneousForm, homogeneousPolynomial, homogenize,
        implDifference, implEmpty, implEqual, implIdealQuo,
        implIntersect, implOffset, implSubSet, implUnion,
        implUnionLCM, imult, independentVariables, init, isProjective,
        leadingForm, makeDivisor, mapOutPolynomial, mapSubsPolynomial,
        mgbasis, mgbasisx, mkAlgSet, mkImplAlgSet, mkParaAlgSet,
        mkPlacAlgSet, mkProjAlgSet, mnormalf, msolveGB, msolveSP,
        mvresultant, neighbGraph, neighborhoodTree, numberOfTerms,
        pacPlot, paraOffset, parameterList, passGenCurve, planecurve,
        plotAlgSet, pointInAlgSet, projPoint, properParametrization,
        properties, rationalPoint, realroot_a, realroot_sb,
        setPuiseuxExpansion, setRandomParameters, singLocus,
        singularities, ssiPlot, subresultantChain, tangSpace,
        toAffine, toImpl, toPara, toPlac, toProj, toProjective,
        tsolve, variableDifferentFrom, variableList]
```

```
> f:=11*y^7+7*y^6*x+8*y^5*x^2-3*y^4*x^3-10*y^3*x^4-10*y^2*x^5-x^7-33*y^6
> -29*y^5*x-13*y^4*x^2+26*y^3*x^3+30*y^2*x^4+10*y*x^5+3*x^6+33*y^5+37*y^4*x
> -8*y^3*x^2-33*y^2*x^3-20*y*x^4-3*x^5-11*y^4-15*y^3*x+13*y^2*x^2+10*y*x^3+x^4;
              6         5  2       4  3        3  4        2  5        5
  f := 7 y  x + 8 y  x  - 3 y  x  - 10 y  x  - 10 y  x  - 29 y  x
               4  2        3  3        2  4          5        4
        - 13 y  x  + 26 y  x  + 30 y  x  + 10 y x  + 37 y  x
             3  2        2  3           4         3          2  2
        - 8 y  x  - 33 y  x  - 20 y x  - 15 y  x + 13 y  x
                3        7         6         5        4    7       6       5    4
        + 10 y x  + 11 y  - 33 y  + 33 y  - 11 y  - x  + 3 x  - 3 x  + x
> A:=mkImplAlgSet([f],[x,y]):
> Ap:=plotAlgSet(A,x=-1..3,y=-1..2,numpoints=200,axes=boxed,tickmarks=[5,4],
  color=black,scaling=constrained):plotsetup(x11):with(plots):display(Ap);
           Time for isolating critical points: , 31.800
         Time for finding intermediate points: ,  8.040
                       Time for others: , .091
bytes used=63101484, alloc=3603820, time=40.57
```

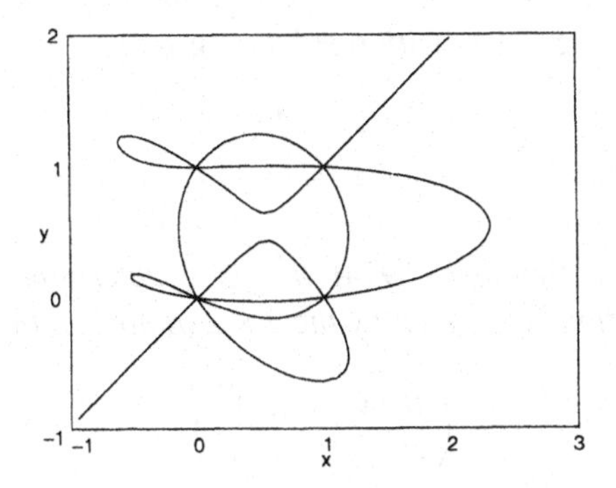


```
> toPara(A,t);
bytes used=127850248, alloc=4324584, time=55.10
Parametric_Algebraic_Set([

                  5        6       7          2        3                4
   -121 - 375 t  - 63 t  + 56 t  - 890 t  + 139 t  + 594 t + 660 t
   ----------------------------------------------------------------,
                          5        3         4        7       6         2
   -121 + 720 t - 62 t  + 1789 t  - 725 t  + 50 t  + 15 t  - 1665 t

      7         5        6                      4        3        2
    8 t  + 473 t  - 153 t  + 77 - 467 t - 299 t  - 577 t  + 939 t
   ----------------------------------------------------------------], [t])
                          5        3         4        7       6         2
   -121 + 720 t - 62 t  + 1789 t  - 725 t  + 50 t  + 15 t  - 1665 t
```

□

We refer to (Decker, Ein, and Schreyer, 1993) and (Decker and Schreyer, 2000) for a method to compute rational parametrizations of some rational surfaces via adjunction theory.

6. Parameter spaces

6.1. DEFORMATIONS

Given an isolated singularity $X_0 = (X,x)$, SINGULAR can compute the miniversal deformation of X along the line of Schlessinger's algorithm (see Martin, 1999 for information on the implementation) . So first $T^1(X_0)$ and $T^2(X_0)$ are computed, and then successive liftings give the equations of the base space of the deformation and of the versal family up to a certain order.

In lucky cases, for example, if the singularity is quasi-homogenous and all elements of $T^1(X_0)$ have negative degrees, the computation stops after finitely many steps, and one can compute the versal deformation completely.

EXAMPLE 6.1. *Here is the* SINGULAR *computation for the versal deformation of a rational surface singularity with resolution graph*

$$\begin{array}{cccc} -2 & -5 & -2 & -2 \\ \circ\!\!-\!\!-\!\!- & \circ\!\!-\!\!-\!\!- & \circ\!\!-\!\!-\!\!- & \circ \end{array}$$

*According to (*Riemenschneider, 1974*) this is a determinantal cyclic quotient singularity. The equations are given by the* 2×2 *minors of the matrix*

$$\begin{pmatrix} u & v & w & x & y^3 \\ v^2 & w & x & y & z \end{pmatrix}$$

in a $\mathbb{C}^6$ *with coordinates* $(u, v, \ldots, z)$.

```
> LIB "deform.lib";
> printlevel     = 1;
> ring Ro        = 0,(u,v,w,x,y,z),ds;
> matrix m[2][5] = u,v,w,x,y^2,v^3,w,x,y,z;
> ideal Fo       = minor(m,2);
> print(Fo,"%s");
xz-y3,wz-xy2,vz-wy2,uz-v3y2,x2-wy,wx-vy,-uy+v3x,-w2+vx,ux-v3w,-uw+v4
> int t=timer;
> versal(Fo);
// dim T_1 = 9
// dim T_2 = 8
// ready: T_1 and T_2
// start computation in degree 2.
// start computation in degree 3.
```

```
// start computation in degree 4.
// no obstruction
// start computation in degree 5.
// finished
// Result belongs to ring Px.
// Equations of total space of miniversal deformation are
// given by Fs, equations of miniversal base space by Js.
// Make Px the basering and list objects defined in Px by typing:
   setring Px; show(Px);
   listvar(matrix);
// NOTE: rings Qx, Px, So are alive!
// (use 'kill_rings("");' to remove)
> timer-t;
7
> setring(Px);
> show(Px);
// ring: (0),(A,B,C,D,E,F,G,H,I,u,v,w,x,y,z),(ds(9),ds(6),C);
// minpoly = 0
// objects belonging to this ring:
// Rs                    [0]  matrix 10 x 20
// Fs                    [0]  matrix 1 x 10
// Js                    [0]  matrix 1 x 8
> print(Js,"%s");
2AG+GH+B2F-FI2,-BE-EI+4A2D-DH2,-BG-GI,-2AB-BH,
2AE+EH,2AH+H2,-BI-I2,2AB+2BH+2AI+2HI
```

Thus the base space is defined by 8 equations in a $\mathbb{C}^9$ with variables $A,\ldots,I$. Just from the equations, it is hard to see what the base space looks like. Primary decomposition can help:

```
> LIB "primdec.lib";
> list pr = primdecGTZ(Js);
> size(pr);
4
```

So there are 4 primary components.

```
> print(pr[1][1],"%s");
B+I,2A+H
> print(pr[2][1],"%s");
I,H,G,F,E,A
> print(pr[3][1],"%s");
I,H,G,E,D,B
> print(pr[4][1],"%s");
I2,HI,H2,G2,EG,E2,BI,BG+GI,BEH-EHI,B2H,B2E,B3,BH+2AI,AH,2AG+GH+B2F,
2AE+EH,AB-AI,-BE-EI+4A2D,A3
> print(pr[4][2],"%s");
I,H,G,E,B,A
```

Hence we have three linear components of dimension 7, 3 and 3 respectively, and another non-reduced component, whose reduction is again a linear space of dimension 3. The 7-dimensional component is the so called Artin component *(see* Wahl, 1979*).* □

The mini-versal deformation of a module M over a fixed affine ring R can also be computed in SINGULAR. The tangent space of the base is $\mathrm{Ext}^1_R(M,M)$, the obstructions lie in $\mathrm{Ext}^2_R(M,M)$.

EXAMPLE 6.2. *We compute the deformation up to order 7 of the module of global sections of a degree 1 line bundle on a plane elliptic curve.*

```
> LIB "deform.lib";
> printlevel = 1;
> ring  Ro = 0,(x,y,z),ds;
> ideal Io = x3+y3-z^3;
> matrix Mo[2][2] = x,z-y,y2+yz+z2,x2;
> print(Mo);
x,        -y+z,
y2+yz+z2,x2
> mod_versal(Mo,Io,7);
// vdim (Ext^2) = 2
// vdim (Ext^1) = 2
// ready: Ext1 and Ext2
// Ext1 is quasi-homogeneous represented: 0,1
// infinitesimal extension
x+Az+B,  -y+z,
y2+yz+z2,x2-Axz-Bx
// start deg = 2
// start deg = 3
// start deg = 4
// start deg = 5
// start deg = 6
// start deg = 7
// quasi-homogeneous weights of miniversal base
0,1
// Result belongs to qring Qx
// Equations of total space of miniversal deformation are in Js
3AB2-3A4B2,B3
// Matrix of the deformed module is Ms and lifted syzygies are Ls.
// Make Qx the basering and list objects defined in Qx by typing:
   listvar(ring);setring Qx; show(Qx);listvar(ideal);listvar(matrix);
// NOTE: rings Qx, Ox, So are still alive!
// (use: 'kill_rings();' to remove them)
```

So the tangent space has dimension 2. From the equations, which are correct up to order 6, we can guess that the variable B *corresponds to an obstructed direction.*

```
> setring(Qx);
> Ls;
Ls[1,1]=y-z+1/3A2x-2/3A2B+1/9A6z+2/3A5B
Ls[1,2]=x2-Axz-Bx+2ABz+B2+1/9A4xy+2/9A4xz-1/9A5yz-2/9A5z2-5/9A4By-10/9A4Bz
Ls[2,1]=x+Az+B
Ls[2,2]=-y2-yz-z2+1/3A2xy+2/3A2xz-2/3A2By-4/3A2Bz+1/9A6yz+1/9A6z2
+2/3A5By+2/3A5Bz
```

Notice that the deformed presentation matrix contains higher and higher powers of `A`. *To understand why, we note that the deformation parameter* `A` *corresponds to deforming the line bundle of degree* 1 *on the elliptic curve, which means to move the point defined by* `Ls[1,1]=Ls[2,1]=B=0` *on the curve. Since elliptic curves have no rational parametrization the presentation matrix* `Ls` *will contain arbitrary high powers of* `A`. □

6.2. INVARIANT RINGS

Let G be a group, and let $\rho : G \to \mathrm{GL}(V)$ be a representation of G on a K-vector space V of finite dimension n. Then G acts linearly on the K-algebra $K[V] \cong K[x_1,\dots,x_n]$ of polynomial functions on V. The corresponding *invariant ring* is the sub-algebra

$$K[V]^G \subset K[V]$$

of invariant functions. The first fundamental problem of invariant theory is to decide, whether $K[V]^G$ is finitely generated as a K-algebra, and, if so, to compute a finite set of generators. The second fundamental problem of invariant theory is to compute the algebra relations on the generators. Elimination is a (costly) solution to the second problem (see Proposition 2.17). In what follows we focus on the first problem.

Suppose that G is linearly reductive, and that ρ is a rational representation of G. Then there is a K-linear map $R : K[V] \to K[V]$, the *Reynolds operator*, which projects $K[V]$ onto $K[V]^G$, and which is a $K[V]^G$-module homomorphism. Hilbert (1890), in his first landmark paper on invariant theory, showed that $K[V]^G$ is finitely generated in this case. To begin with he concluded from the basis theorem, which he proved for this purpose, that the ideal $I_{\mathcal{N}}$ generated by all homogeneous invariants of positive degree is in fact generated by finitely many of them, say by $f_1,\dots,f_s$. Then he used the Reynolds operator and induction to show that $f_1,\dots,f_s$ generate $K[V]^G$ as a K-algebra. As it stands, this proof is non-constructive, since no recipe is given for computing the ideal generators $f_1,\dots,f_s$. Only recently Derksen (1999) refined Hilbert's ideas and turned them into an algorithm, which is easy to implement. Derksen first showed that if $f_1,\dots,f_s$ is any set of homogeneous generators of $I_{\mathcal{N}}$ (not necessarily invariants), then $K[V]^G$ is generated by $R(f_1),\ldots,R(f_s)$ (as usual, the abstract Reynolds operator can be replaced by any other means of generating the invariants of a given degree, for example, by Cayley's Ω-process). To find generators of $I_{\mathcal{N}}$, Derksen considered the map

$$\psi : G \times V \to V \times V, \quad (\pi, v) \to (v, \pi v),$$

and the ideal $J \subset K[x_1,\ldots,x_n,y_1,\ldots,y_n]$ defining the Zariski closure of the image of ψ. He proved that if $f_1(x,y),\ldots,f_s(x,y)$ are generators of J, then $f_1(x,0),\ldots,$ $f_s(x,0)$ generate $I_{\mathcal{N}}$. Generators of J, in turn, can be computed via elimination (see Proposition 2.17). Hilbert (1893) himself, in his more constructive second landmark paper on invariant theory, presented two ways of reducing the computation of invariants to normalization (see also Sturmfels, 1993 and Decker and de Jong, 1998). For this purpose he proved the Hilbert-Mumford criterion (which, in practical terms, requires elimination), the Nullstellensatz, and the Noether normalization theorem (in the graded case). Except for rather small examples both, Derksen's and Hilbert's algorithm, are not of practical interest (requiring a Gröbner basis computation with respect to an elimination order in a large number of variables).

The situation is better for finite groups. Emmy Noether (1926) showed that in this case $K[V]$ is integral over $K[V]^G$, in particular $\operatorname{trdeg}_K K[V]^G = \operatorname{trdeg}_K K[V]$ $= n$, and deduced, with a non-constructive proof, that $K[V]^G$ is finitely generated. How to compute generators? Since we already know that $K[V]^G$ has dimension n we may proceed in two steps. First we compute *primary invariants* $p_1,\ldots,p_n$, that is, we compute a graded Noether normalization $K[p_1,\ldots,p_n]$ of $K[V]^G$. Then we compute *secondary invariants*, that is, a minimal system of homogeneous invariants $s_0 = 1, s_1,\ldots,s_m$ generating $K[V]^G$ as a $K[p_1,\ldots,p_n]$-module. To describe the basic ideas in more detail we suppose that G is explicitly given as a matrix group $G \subset \mathrm{GL}(V) \cong \mathrm{GL}_n(K)$ by a set of generating matrices. We distinguish two cases.

If the group order $|G|$ is invertible in K (the *non-modular case*), G is linearly reductive. In fact, there is the explicit Reynold's operator

$$R : K[V] \to K[V], \quad f(x) \mapsto \frac{1}{|G|} \sum_{\pi \in G} (\pi f)(x).$$

Moreover, in this case, we can precompute the Hilbert series of $K[V]^G$ by Molien's theorem (1897) so that for each given degree d the dimension of $K[V]_d$ is known to us a priori. Thus we can compute a K-basis of $K[V]_d^G$ by applying R successively to the monomials in $K[V]_d$ until the correct number of linearly independent invariants has been found. One way of computing primary invariants is to proceed degree by degree, and, if $p_1,\ldots,p_{i-1}$ are already constructed, to search for a homogeneous invariant p_i satisfying $\dim(p_1,\ldots,p_i) = n - i$ (see Decker, Heydtmann and Schreyer, 1997; an alternative method is due to Kemper, 1999). The computation of secondary invariants is based on the fact that $K[V]^G$ is Cohen-Macaulay. This is true more generally for any linearly reductive group acting rationally on V by a result of Hochster and Roberts (1974) (see for example Stanley, 1979 or Decker and de Jong, 1998 for a short proof in the case considered here). That $K[V]^G$ is Cohen-Macaulay means that $K[V]^G$ is a free $K[p_1,\dots,p_n]$-module. This gives information on the Hilbert series of $K[V]^G$ (depending on the degrees of the primary invariants). By comparing with Molien's series we obtain a polynomial from which the degrees of the secondary invariants and their number in each degree can be read off (the total number of secondary invariants equals $\prod_{i=1}^n \deg(p_i)/|G|$). This yields a straightforward algorithm for computing secondary invariants, (see Sturmfels,1993; the version given by Kemper and Steel, 1999 singles out the secondary invariants which are not power products of secondary invariants of lower degree).

EXAMPLE 6.3. *We use* SINGULAR *to compute the invariant ring of the Heisenberg group H_3 in its Schrödinger representation. That is, H_3 is the subgroup of* $\mathrm{GL}(3,\mathbf{Q}(\xi))$, $\xi = e^{\frac{2\pi i}{3}}$, *generated by*

$$\sigma = \begin{pmatrix} 0 & 0 & 1 \\ 1 & 0 & 0 \\ 0 & 1 & 0 \end{pmatrix} \quad \text{and} \quad \tau = \begin{pmatrix} 1 & 0 & 0 \\ 0 & \xi & 0 \\ 0 & 0 & \xi^2 \end{pmatrix}.$$

```
> LIB"finvar.lib";
> ring R=(0,a),(x,y,z),dp;
> minpoly=a2+a+1;
> " need primitive third roots of unity";
> matrix A[3][3]=0,0,1,1,0,0,0,1,0;
> matrix B[3][3]=a,0,0,0,a2,0,0,0,a3;
> int t=timer;
> matrix P,S,IS=invariant_ring(A,B,intvec(0,0,1));

  Generating the entire matrix group. Whenever a new group element is found,
  the coressponding ring homomorphism of the Reynolds operator and the
  corresponding term of the Molien series is generated.

  Group element 3 has been found.
               ...
  Group element 27 has been found.

  Now we are done calculating Molien series and Reynolds operator.

  We can start looking for primary invariants...

  Computing primary invariants in degree 3:
  We find: xyz
  We find: x3+y3+z3
```

```
  Computing primary invariants in degree 6:
  We find: x3y3+x3z3+y3z3

  We found all primary invariants.

  Polynomial telling us where to look for secondary invariants:
   x9+1

  In degree 0 we have: 1

  Searching in degree 9, we need to find 1 invariant(s)...
  We find: x3y6+x6z3+y3z6

  We're done!

> timer-t;
1
```

□

In any case, we can compute a K-basis of $K[V]_d^G$ for each given degree d by solving a linear system of equations depending on the (finitely many) given generators of G. Indeed, to be invariant under just one element of G imposes a linear condition on the polynomials of degree d. Thus we can compute primary invariants also in the modular case by dimension checks as above. The computation of secondary invariants is reduced to that in the non-modular case by a trick of Kemper (1996), who considers a subgroup H of G whose order is invertible in K (for example, the trivial subgroup will do).

EXAMPLE 6.4. *We consider the cyclic group of order 4 acting on* $\mathbb{F}_2^4$.

```
> ring R=2,(x(1..4)),dp;
> matrix A[4][4];
> A[1,4]=1; A[2,1]=1; A[3,2]=1; A[4,3]=1;
> print(A);
0,0,0,1,
1,0,0,0,
0,1,0,0,
0,0,1,0
> matrix P,S=invariant_ring(A);
> P;
P[1,1]=x(1)+x(2)+x(3)+x(4)
P[1,2]=x(1)*x(3)+x(2)*x(4)
P[1,3]=x(1)*x(2)+x(2)*x(3)+x(1)*x(4)+x(3)*x(4)
P[1,4]=x(1)*x(2)*x(3)*x(4)
> size(S);
5
```

We hence need more than $\prod_{i=1}^{4} \deg(p_i)/|G| = 4$ *secondary invariants. So* $K[V]^G$ *is not Cohen-Macaulay. A check on the secondary invariants shows that the invariants of degree* $\leq |G| = 4$ *do not generate* $K[V]^G$*, that is,* Noether's degree bound (1916) *does not hold in the modular case (see* Richman, 1990*) for examples involving generators of arbitrary high degree and fixed group order).* □

Excellent software for computing invariant rings of finite groups is also provided by MAGMA.

6.3. CONSTRUCTION OF SPECIAL VARIETIES

The classification of projective varieties of low degree has a long history. The interest in surfaces in $\mathbb{P}^4$, for example, goes back to the Italian school of algebraic geometry around 1900. Recently, the following celebrated result has stimulated a lot of research:

THEOREM 6.5 (ELLINGSRUD AND PESKINE, 1989). *There exists a constant* d_0 *such that every smooth surface in* $\mathbb{P}^4$ *of degree* $d > d_0$ *is of general type.* □

In particular, only finitely many components of the Hilbert scheme contain surfaces of Kodaira dimension ≤ 1. Examples were constructed up to degree $d = 15$ (see Decker, Ein, and Schreyer, 1989, Decker and Popescu, 1995, and Decker and Schreyer, 2000 for a survey). On the other hand, the best bound known today is $d_0 = 52$ (Decker and Schreyer, 2000).

A large part of the about 60 components known so far was constructed with computer algebra first. In fact, with a few exceptions, the known examples can be constructed in a systematic way with computer algebra as an application of the method of Decker, Ein, and Schreyer (1993). The exceptions are the surfaces in which the construction relies most substantially on computer algebra, namely 3 families of rational surfaces and a family of Enriques surfaces, all of degree 11 and sectional genus 10, which were constructed by using small finite fields and random trials (Schreyer, 1996). It is not known whether these 4 families are uni-rational (for most of the other families this follows as a byproduct from the method of Decker, Ein, and Schreyer, 1993).

The same idea of guessing solutions over finite fields is applied in (Schreyer and Tonoli, 2001) to give a computer algebra code which picks curves of genus 14 at random over very small finite fields.

EXAMPLE 6.6. *The following* MACAULAY2 *function produces smooth curves* C *of genus 14 and degree 14 in* $\mathbb{P}^3$*. The crucial step of the construction is to get the Hartshorne-Rao module* $M = \mathrm{H}^1_* \mathcal{J}_C$ *of the curve.*

```
i1 : randomGenus14Curve = (R) -> (
        kappa:=koszul(3,vars R);
        kappakappa:=kappa++kappa;
```

```
        correctCodimAndDegreeAndSmooth:=false;
        count0:=0;count1:=0;count2:=0;
        while not correctCodimAndDegreeAndSmooth do (
             test:=false;
             while test==false do (
                  alpha=random(R^{5:-2},R^{12:-2});
                  beta=random(R^{5:-2},R^{3:-3});
                  M:=coker (alpha*kappakappa|beta);
                  fM:=res (M,DegreeLimit =>3);
             if (tally degrees fM_2)_{5}==3 then (
                       --further checks to pick up the right module
                       test=(tally degrees fM_2)_{4}==2 and
                       codim M==4 and degree M==23;);
                  count0=count0+1;);
        count1=count1+1;
             Gt:=transpose (res M).dd_3;
             I:=ideal syz (Gt_{5..17});
             correctCodimAndDegree=(codim I==2 and degree I==14);
        if correctCodimAndDegree then
              (singI:=I+minors(2,jacobian I);count2=count2+1;
                   s:=timing correctCodimAndDegreeAndSmooth= (codim singI==4));
             );
        <<"     -- "<<count0 <<" modules tested" <<endl;
        <<"     -- good modules tested:"<<count1 <<endl;
        <<"     -- "<<s#0<<" seconds used to check smoothness once" <<endl;
        <<"     -- curves tested:"<<count2 <<endl;
        I)

o1 = randomGenus14Curve
o1 : Function
i2 : R=ZZ/3[x_0..x_3]
o2 = R
o2 : PolynomialRing
i3 : time I=randomGenus14Curve(R);
     -- 44 modules tested
     -- good modules tested:2
     -- 4.49 seconds used to check smoothness once
     -- curves tested:2
     -- used 16.34 seconds
o3 : Ideal of R
i4 : hilbertPolynomial(R^1/I)
o4 = - 27*P  + 14*P
           0       1
o4 : ProjectiveHilbertPolynomial
i5 : betti(fI=res I)
```

```
o5 = total: 1 13 17 5
         0: 1 .  .  .
         1: . .  .  .
         2: . .  .  .
         3: . .  .  .
         4: . .  .  .
         5: . 13 17 4
         6: . .  .  1
o5 : Net
```

By reversing our construction we can rediscover the Hartshorne-Rao module M from the syzygies of the ideal of the curve. In fact, we compute the dual of the minimal free resolution of M:

```
i6 : betti res coker transpose fI.dd_3
o6 = total: 5 17 18 11 5
        -9: 1 .  .  .  .
        -8: 4 17 13 .  .
        -7: . .  3  .  .
        -6: . .  2  11 5
o6 : Net
```

The script `randomGenus14Curve` *is based on the following considerations. Assuming that the curve is a maximal rank curve, the Hilbert function of* $M = \bigoplus_{i\geq 0} M_i$ *takes the values* $(0,0,5,9,8,1,0,0,\ldots)$. *A general choice of the concatenated presentation matrix*

$$(\alpha\cdot(\kappa\oplus\kappa))|\beta\colon R^{11}(-3)\to R^5(-2)$$

in the script yields a module with Hilbert function $(0,0,5,8,9,0,0,0,\ldots)$, *having a resolution of type*

```
total: 5 11 21 23 8
   -2: 5 11  2  .  .
   -3: .  .  2  .  .
   -4: .  . 17 23  8
```

So within the space of presentation matrices, the subset of those giving a module with the desired Hilbert function has codimension 3, and, over the ground field with q elements, we can expect to find 1 good module in q^3 *trials. Thus the running time of this probabilistic algorithm depends heavily on the finite ground field (see* Schreyer and Tonoli, 2001*). In Section 4.1 we computed an example over the ground field* $\mathbb{Z}/5\mathbb{Z}$. □

For smaller genus $g \leq 13$ the situation is better, and random curves can be computed in a systematic way over any given finite ground field, thus verifying the uni-rationality of the moduli spaces for $g \leq 13$ by computer (Schreyer and Tonoli, 2001; for theoretical proofs see Arbarello and Sernesi, 1979; Sernesi, 1981; Mori and Mukai, 1983; Chang and Ran, 1984). It is a challenge to write similar programs for higher genus curves, say maybe even in the range where the moduli space of curves $\mathfrak{M}_g$ is of general type.

7. Enumerative geometry

The MAPLE package SCHUBERT "is designed to handle computations in intersection theory".

```
> with(SF): # Symmetric functions package; obtained from Schubert's web-page.
> with(schubert);
 [\&*, \&-!, \&-*, \&/, \&^*, End, Grass, Hom, POINT, Proj, Symm, adams,
     additivebasis, betti, blowup, blowuppoints, bundle, bundlesection, chern,
     chi, codimension, compose, curve, determinant, dimension, division, down,
     dual, grass, grobnerbasis, insertedge, integral, integral2, koszul,
     lowershriek, lowerstar, monomials, monomialvalues, morphism,
     multiplepoint, normalbundle, normalform, o, porteous, porteous2,
     productvariety, proj, rank, schur, schurfunctor, schurfunctor2, segre,
     setvariety, sheaf, strip, symm, tangentbundle, tensor, todd, toddclass,
     toricvariety, totalspace, twist, up, upperstar, variety, verifyduality,
     wedge, where, whichcone, wproj]
```

SCHUBERT deals with several data types, namely "sheaf objects", "variety objects", and "morphism objects", and it consists of the Maple procedures printed above, which "create these data types, modify these data types, and compute with these data types". A sheaf object is a polynomial in the reserved variable t, the Chern character polynomial of the sheaf. A variety object is a dynamic MAPLE table whose entries may include generators and relations of the intersection ring of the variety, or the Chern character polynomial of the tangent bundle. For instance:

```
> proj(5,h,tang);                   # tangent bundle needed for chi
                     currentvariety_ is Ph, DIM is 5
> Ph[variables_];
                                    [h]
> Ph[relations_];
                                      6
                                    [h ]
> FH:=sheaf(3,[0*h,3*h^2,0*h^3]);   # defines the Horrocks bundle
                                  2  2       4  4
                    FH := 3 - 3 h   t   + 3/4 h   t
> chi(FH&*o(m*h));                  # computes chi of its twists
                     5         3    103          4        2
              1/40 m   + 13/8 m   - --- m + 3/8 m   + 9/8 m   - 6
                                    20
```

```
> (subs(m=2,"));
                              8
```

A morphism object is also a dynamic MAPLE table with various possible entries (including source and target of the morphism).

EXAMPLE 7.1. *We compute the double point formula for smooth surfaces in $\mathbb{P}^4$.*

```
> DIM:=2:
> variety(X,tangentbundle=bundle(2,c)):
> integral(X,H^2):=d:
> proj(4,H,all):
> morphism(f,X,PH):
> multiplepoint(f,2):
> (subs(H^2=d,c1=-K,c2=12*X-K^2,"));

                 2     2
                d  - 2 K  - 5 K H - 10 d + 12 X
```

□

For more details we refer to the SCHUBERT manual which contains examples such as the following one:

EXAMPLE 7.2. *We compute the number of conics on a general quintic hypersurface in $\mathbb{P}^4$.*

```
# Conics on a quintic threefold. This is the top Chern class of the
# quotient of the 5th symmetric power of the universal quotient on the
# Grassmannian of 2 planes in P^5 by the subbundle of quintics containing
# the tautological conic over the moduli space of conics.

> grass(3,5,c):            # 2-planes in P^4.
> B:=Symm(2,Qc):           # The bundle of conics in the 2-plane.
> Proj(X,dual(B),z):       # X is the projective bundle of all conics.
> A:=Symm(5,Qc)-Symm(3,Qc)&*o(-z):  # The rank 11 bundle of quintics
>                                   # restricted to the universal conic.
> c11:=chern(rank(A),A):# its top Chern class.
> lowerstar(X,c11):        # push down to G(3,5).
> integral(Gc,");          # and integrate there.

                          609250
```

□

8. Testing and creating conjectures

Computer algebra has been used to construct counterexamples to conjectures, but also to find mathematical evidence by experiments. The celebrated conjectures of Birch and Swinnerton-Dyer (1963, 1965) in number theory, for example, are based on extensive computer calculations. In this section we discuss several conjectures in algebraic geometry which have presented a challenge to computer algebra.

8.1. THE MINIMAL RESOLUTION CONJECTURE

Let Γ be a general set of γ points in $\mathbb{P}^n$ over an infinite field K. Then the homogeneous coordinate ring S/I_Γ of Γ has a minimal free resolution of type

$$0 \leftarrow S/I_\Gamma \leftarrow S \leftarrow F_1 \leftarrow \cdots \leftarrow F_n \leftarrow 0,$$

with

$$F_i = S(-d-i)^{\beta_{i,d+i}} \oplus S(-d-1-i)^{\beta_{i,d+1+i}},$$

where d is the smallest integer such that $\gamma < \binom{d+n}{n}$. The differences $\beta_{i,d+i+1} - \beta_{i+1,d+i+1}$ only depend on the Hilbert function of Γ, that is, on i, n, d and γ. The minimal resolution conjecture states that the shape of the resolution is the minimal possible one with respect to the given Hilbert function:

CONJECTURE 8.1 (LORENZINI, 1987, 1993). *Let $\Gamma \subset \mathbb{P}^n$ be a general set of points as above. For each j at most one of the betti numbers $\beta_{i,j}$ of S/I_Γ is non-zero.* □

Thus this is a conjecture in the spirit of the maximal rank conjecture for space curves extended from the postulation to the syzygies.

If we print the betti numbers in a diagram as in the output of MACAULAY2 or SINGULAR, then the minimal resolution conjecture means that this diagram is of type

degree										
0	1	–	–		···	···		–	···	–
1	–	.	.					.		.
.	.	.	.					.		.
.	.	.	.					.		.
.	.	–	–							
$d-1$	–	$\beta_{1,d}$	$\beta_{2,d+1}$	···	*	*	?	–	···	–
d	–	–	–	···	–	?	*	*	···	*

with not both ? non-zero.

The analogous conjecture for the syzygies of a linearly normal non-special curve whose degree is large compared to the genus is false because pencils of low degree give a contribution to the syzygies beyond the range predicted by a maximal rank assumption on a suitable map (see Green and Lazarsfeld, 1984).

By the work of several authors the minimal resolution conjecture is known to be true for $n = 2, 3, 4$ and in other special cases (see Hirschowitz and Simpson, 1996 or Eisenbud and Popescu, 1999, for further information). Most striking, Hirschowitz and Simpson (1996) showed that the conjecture is true whenever γ is large compared to n.

When proving their result Hirschowitz and Simpson were aware of computer experiments done by Schreyer, who found that randomly chosen 11 points in $\mathbb{P}^6$, 12 points in $\mathbb{P}^7$, and 13 points in $\mathbb{P}^8$ did not have betti numbers as claimed by the conjecture. For instance, for 12 points in $\mathbb{P}^7$ the betti diagram of the resolution was of type

degree								
0	1	–	–	–	–	–	–	–
1	–	24	84	126	88	[4]	–	–
2	–	–	–	–	[4]	36	21	4

in all experiments.

For quite a while it was unclear whether these examples were indeed counter-examples to the minimal resolution conjecture (one could argue that the random points were not chosen general enough). Eisenbud and Popescu (1999) gave a geometric explanation. They found a novel structure inside the resolution of a set of points which accounts for the observed failures and provides a counterexample to the minimal resolution conjecture in $\mathbb{P}^n$ for every $n \geq 6$, $n \neq 9$. In fact, the geometry behind their construction occurs not in $\mathbb{P}^n$ but in a different projective space, in which there is a related set of points known as the "Gale transform" in coding theory. If M is the $(n+1) \times \gamma$ matrix whose columns represent the points in $\mathbb{P}^n$, then the kernel of M is generated by the columns of a $\gamma \times (\gamma - n - 1)$ matrix N. The set Γ' of the γ points represented by the rows of N in $\mathbb{P}^{\gamma-n-2}$ is the Gale transform of Γ. Γ' is determined by Γ up to a projectivity and vice versa. See also Eisenbud, Popescu, Schreyer, and Walter (2001).

8.2. GREEN'S CONJECTURE

One of the most outstanding conjectures about syzygies is Green's conjecture. Let $C \subset \mathbb{P}^{g-1}$ be a canonical curve. In the minimal free resolution of its coordinate ring each syzygy module has generators in at most two different degrees. The betti table looks like

degree												
0	1	–	–	–	–	–	–	–	–	–	–	–
1	–	$\beta_{1,2}$	$\beta_{2,3}$	$\cdots$	$\beta_{p,p+1}$	$\cdots$	$\beta_{g-2-p,g-1-p}$	–		–		–
2	–	–	–	–	$\beta_{p,p+2}$	$\cdots$	$\beta_{g-2-p,g-p}$		$\cdots$	$\beta_{g-3,g-1}$		–
3	–	–	–	–	–	–	–	–		–		1

The resolution is self-dual since C is arithmetically Cohen-Macaulay and $\omega_C = \mathcal{O}_C(1)$, in particular $\beta_{i,j} = \beta_{g-2-i,g+1-j}$. Green's conjecture gives a geometric interpretation for the range of the non-zero betti numbers:

CONJECTURE 8.2 (GREEN, 1984). *Let C be a canonical curve over $\mathbb{C}$, and let p be an integer. Then*

$$\beta_{p,p+2} \neq 0 \Leftrightarrow \operatorname{Cliff}(C) \geq p . \qquad \square$$

Here the Clifford index Cliff(C) of C is defined as

$$\text{Cliff}(C) = \min\{\deg L - 2(h^0 L - 1) \mid L \in \text{Pic}(C) \text{ with } h^0 L, h^1 L \geq 2\}.$$

Some evidence for the conjecture comes from certain positive results, namely the easy implication from geometry to syzygies, proved by Green and Lazarsfeld (1984), the cases of low genus $g \leq 8$ (Schreyer, 1986) and g=9 (Mukai, 1995), the cases of low p, namely $p = 1$ (Petri, 1923) and $p = 2$ (Voisin, 1988; Schreyer, 1991), and a result of Hirschowitz and Ramanan (1998) which says that if the conjecture holds for the generic curve of a given odd genus, then the set of curves with extra syzygies coincides with the expected divisor. Further evidence comes from partial results on generalizations by Green and Lazarsfeld (1986, 1988).

The conjecture does not hold over fields of finite characteristic, for instance, $\operatorname{char} K = 2$ fails for genus 7 (see Schreyer, 1986), and $\operatorname{char} K = 3$ for genus 9 (this follows from the work of Mukai, 1995).

Verfying the conjecture for the generic curve of small genus was a challenge to computer algebra in the early days of the pioneering system Macaulay. Machine computations allowed one prove the conjecture for the generic curve up to genus 15 (see Eisenbud, 1992).

Quite recently, Voisin (2001) anounced a proof for the generic curve of even genus.

8.3. ZARISKI'S CONJECTURE

We follow the presentations in Greuel and Pfister (1996) and Greuel (2000). Zariski (1971), in his retiring presidential address to the AMS, posed the following question, which is also known as Zariski's conjecture. If

$$f = \sum c_\alpha x^\alpha \in \mathbb{C}\{x_1, \ldots, x_n\},\ f(0) = 0,$$

is a hypersurface singularity, let

$$\text{mult}(f) := \min\{|\alpha| \mid c_\alpha \neq 0\}$$

be its multiplicity. Call two such hypersurface singularities f and g topologically equivalent, $f \overset{\text{top}}{\sim} g$, if there is a homeomorphism

$$(B, f^{-1}(0), 0) \overset{\cong}{\to} (B, g^{-1}(0), 0).$$

CONJECTURE 8.3 (ZARISKI, 1971). $f \overset{top}{\sim} g \Rightarrow \text{mult}(f) = \text{mult}(g)$.

The answer is known to be true for plane curve singularities (see the references in Greuel, 1986), and for weighted homogeneous singularities (see Greuel, 1986; O'Shea, 1987).

Greuel and Pfister tried to construct a counterexample to the general case along the following lines. Consider a deformation of $f = f_0$:

$$f_t(x) = f(x) + tg(x,t), \quad |t| \text{ small} .$$

Let

$$\mu(f_t) = \dim_{\mathbb{C}} \mathbb{C}\{x\}/(\partial f_t/\partial x_1, \ldots, \partial f_t/\partial x_n)$$

be the Milnor number of f_t. Suppose that $\mu(f_0)$ is finite (then also $\mu(f_t)$ is finite for small $|t|$). It is known, see for example (Teissier, 1972), that

$$f \overset{\text{top}}{\sim} g \Rightarrow \mu(f_0) = \mu(f_t) .$$

The converse is known to be true for $n \neq 3$ by work of Lê and Ramanujam (1976). The idea of Greuel and Pfister was to construct a deformation f_t of f_0 with constant Milnor number and non-constant multiplicity. They had examples in mind where the latter condition is fulfilled, and used SINGULAR to check the Milnor numbers (which cannot be computed by hand). They did not succeed, however, in finding a counterexample (see Example 3.4 for one case). Instead, a careful analysis of their examples let to a certain positive result (see Greuel and Pfister, 1996).

9. Visualization

General purpose systems such as MAPLE, MATHEMATICA, or MUPAD provide a variety of tools for visualization.

EXAMPLE 9.1. *We present a visualization of the Hopf fibration $S^3 \to S^2$, plotted with* MAPLE*:*

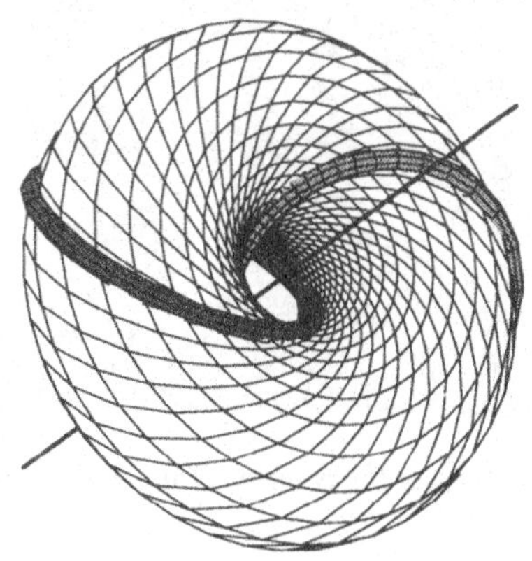

□

Packages such as CASA enable us to plot pictures of implicit curves with careful treatment of the singularities (see Section 5.4), implicit surfaces or intersections of surfaces.

EXAMPLE 9.2. CASA *plots Clebsch's diagonal cubic as follows:*

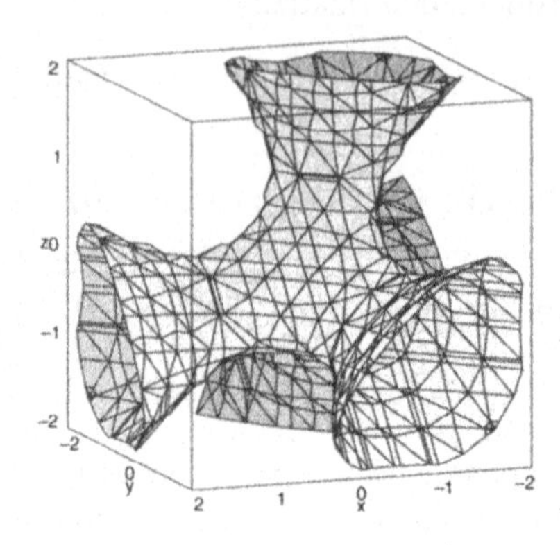

□

High quality pictures can be obtained from packages specialized on visualization. SURF (see `http://surf.sourceforge.net/index.shtml`), for example, allows one to visualize plane algebraic curves, algebraic surfaces in 3-space, and hyperplane sections of the surfaces.

EXAMPLE 9.3. *Here is how* Surf *visualizes the Barth sextic, a surface with 65 nodes (see* Barth, 1996*):*

For a further Surf *example see the Steiner Roman surface in Section 2.* □

POLYRAY (see `http://wims.unice.fr/~wims`) is another package for plotting algebraic surfaces given by their equation. It is based on POV-RAY (see `http://www.povray.org`). For further software and picture galleries we refer to the internet.

10. Complexity and benchmarks

The theoretical complexity of algebraic geometry problems is very bad (see Mayr, 1997 for a survey). The basic task ideal membership has a huge theoretical complexity:

THEOREM 10.1 (MAYR AND MEYER, 1982). *The ideal membership problem is exponentially space hard.* □

Note that the ideals constructed by Mayr and Meyer are binomial. Thus all primary components (including embedded ones) are again binomial and define rational varieties (see Eisenbud and Sturmfels, 1996). In practical terms, it seams reasonable to guess that Buchberger's algorithm is much worse behaved for general ideals than for binomial ideals.

The first to give an upper bound for ideal membership was Grete Hermann. Her bound is doubly exponential:

THEOREM 10.2 (GRETE HERMANN, 1926). *If $f_1,\dots,f_k \in K[x_1,\dots,x_n]$ are polynomials of degree $\leq d$ then $f \in (f_1,\dots,f_s)$ has an expression*

$$f = \sum_{j=1}^{k} g_j f_j$$

with $\deg g_j \leq \deg f + 2(kd)^{2^{n-1}}$. □

The examples of Mayr and Meyer show that the double exponential form of Hermann's bound cannot be improved. Similar statements hold for the computation of Gröbner bases and syzygies. A very remarkable result of Kollár says that the radical membership problem, which is a more geometric question, is much better behaved.

THEOREM 10.3 (KOLLÁR, 1988, 1999). *Let $f_1,\dots,f_k \in K[x_1,\dots,x_n]$ be polynomials of degrees $d_1 \geq \dots \geq d_k \geq 3$, and let $I = (f_1,\dots,f_k)$. Every $f \in \sqrt{I}$ has an expression*

$$f^s = \sum_{j=1}^{k} g_j f_j$$

with $s \leq N$ and each summand degree bounded by

$$\deg(g_j f_j) \leq N(1 + \deg f),$$

where $N = d_1 \cdot d_2 \cdot \ldots d_k$ if $k \leq n$ and $N = d_1 \cdot d_2 \cdot \ldots d_{n-1} \cdot d_k$ if $k > n > 1$. □

This bound is essentially simply exponential instead of doubly exponential, and it is sharp.

Because of this exponential behavior we should not expect too much from Gröbner bases. The range of problems which cannot be done on a piece of paper, but can be done by Gröbner bases is fairly narrow. For new users it can be wise to ask an experienced user for an opinion on whether a specific problem lies within reach or not.

To give an idea about the range in the number of variables and degrees in which computations can be completed, we present some benchmarks for somewhat typical tasks. The computations are done on a 450MHz Pentium II processor (with one of our favorite systems). All timings are in seconds.

Syzygies. The first example gives the time to compute the complete minimal free resolution of a random graded artinian gorenstein algebra in n variables with socle in degree d over a finite ground field. Since these algebras are in one-to-one correspondence with homogenous hypersurfaces of degree d in $\mathbb{P}^{n-1}$ (see for example Eisenbud, 1995, Theorem 21.6), it makes sense to speak of a random one.

$n \backslash d$	2	3	4	5	6	7	8	9	10	11
4	0.01	0.01	0.01	0.03	0.05	0.12	0.18	0.43	0.56	1.24
5	0.01	0.07	0.11	0.77	1.14	6.95	9.00	42.49	50.76	190.26
6	0.05	0.70	1.25	22.58	31.33	317.31				
7	0.21	11.05	16.54	587.44						
8	0.94	142.38								
9	5.06	2897.12								

In this and the following tables we compute up to the first degree for which more than 2 minutes of computing time is needed. □

Dimension. We take n random forms of degree d on $\mathbb{P}^{2n}$ and give the time for computing the dimension of the ideal they generate. Of course all these ideals will be complete intersections, so we always get codimension n. But the computer algebra system does not know this. For many applications it would be very valuable to have a faster way of detecting complete intersections than to bound the dimension from above.

$2n \backslash d$	2	3	4	5	6	7	8	9	10	11
4	0.01	0.01	0.02	0.06	0.15	0.38	0.88	2.02	4.21	9.06
6	0.03	0.88	12.47	170.21						
8	0.74	244.92								

□

Elimination. We consider a map $\mathbb{P}^n \to \mathbb{P}^{2n+1}$ defined by a collection of random forms of degree d and take the time to compute the ideal of the image. We do not use the Hilbert driven version of Buchberger's algorithm here (see Remark 2.17).

$n \backslash d$	2	3	4	5	6	7	8	9	10	11
1	0.01	0.01	0.05	0.09	0.39	0.30	2.44	12.83	5.75	223.20
2	0.09	17.47	489.74							

These are the running times for the computations over a finite ground field. Over $\mathbb{Q}$ much more time is needed:

$n \backslash d$	2	3	4	5	6
1	0.01	0.16	1.90	23.31	5422.89

□

REMARK 10.4. *It seems a bit odd that the elimination over a finite field in case* $n = 1, d = 10$ *is faster that in case* $n = 1, d = 9$. *We observed this in many experiments and believe that it is a true phenomenon.* □

Gröbner basis computations over $\mathbb{Q}$ are usually much more expensive than over finite fields, because the nominators and denominators of the coefficients can grow rapidly. Thus if we are not interested in the precise solutions of an actual system, but, say, only in the qualitative behaviour, a finite coefficient field is preferable. For some applications, flatness arguments can show that the results are typical also in characteristic 0.

In all these examples we worked with non-sparse systems. Examples which come up in actual applications are very often sparse. Already the fact that we are able to type the system of equations, is a hint for sparseness. For sparse systems computations are usually much faster.

References

Anick, D. (1980) Construction d'espaces de lacets et d'anneaux locaux à séries de Poincaré-Betti non rationnelles, *C. R. Acad. Sci., Paris, Ser. A* **290**, A729–A732.

Arbarello, E., and Sernesi, E. (1979) The equation of a plane curve, *Duke Math. J.* **46**, 469–485.

Avramov, L. (1998) Infinite free resolutions, in J. Elias et al. (eds.), *Six lectures on commutative algebra, Bellaterra, 1996*, Birkhäuser, Basel, pp. 1–118.

Avramov, L., and Grayson, D. (2001) Resolutions and cohomology over complete intersections, to appear in D. Eisenbud et al. (eds.), *Mathematical computations with Macaulay2*, Springer, New York.

Barth, W. (1996) Two projective surfaces with many nodes, admitting the symmetries of the icosahedron, *J. Algebr. Geom.* **5**, 173–186.

Bayer, D., and Mumford, D. (1993) What can be computed in algebraic geometry?, in D. Eisenbud et al. (eds.), *Computational algebraic geometry and commutative algebra. Proceedings of a conference held at Cortona, Italy, June 17-21, 1991*, Cambridge University Press, Cambridge, pp. 1–48.

Bayer, D., and Stillman, M. (1987) A theorem on refining division orders by the reverse lexicographic orders, *Duke J. Math.* **55**, 321–328.

Bayer, D., and M. Stillman. (1988) On the complexity of computing syzygies, *J. Symb. Comput.* **6**, 135–147.

Bernšteĭn, I. N., Gel'fand, I. M., and Gel'fand, S. I. (1978) Algebraic vector bundles on $\mathbb{P}^n$ and problems of linear algebra, *Functional Anal. Appl.* **12**, 212–214.

Birch, B. J., and Swinnerton-Dyer, H. P. F. (1963) Notes on elliptic curves. I, *J. Reine Angew. Math.* **212**, 7–25.

Birch, B. J., and Swinnerton-Dyer, H. P. F. (1965) Notes on elliptic curves. II, *J. Reine Angew. Math.* **218**, 79–108.

Buchberger, B. (1965) *Ein Algorithmus zum Auffinden der Basiselemente des Restklassenrings nach einem nulldimensionalen Polynomideal,* PhD thesis, Lepold-Franzens-Universität, Innsbruck, Austria.

Buchberger, B. (1970) Ein algorithmisches Kriterium für die Lösbarkeit eines algebraischen Gleichungssystems, *Aequationes mathematicae* **4**, 374–383.

Campillo, A. (1980) *Algebroid curves in positive characteristic*, Springer, Lecture Notes in Mathematics 813, Berlin.

Chang, M.-C., and Ran, Z. (1984) Unirationality of the moduli spaces of curves of genus 11, 13 (and 12), *Invent. Math.* **76**, 41–54.

Decker, W., Ein, L., and Schreyer, F.-O. (1993) Construction of surfaces in $\mathbb{P}^4$, *J. Algebraic Geom.* **2**, 85–237.

Decker, W., and Eisenbud, D. (2001) Sheaf algorithms using the exterior algebra, to appear in D. Eisenbud et al. (eds.), *Mathematical computations with Macaulay2*, Springer, New York.

Decker, W., Greuel, G.-M., and Pfister, G. (1999) Primary decomposition: algorithms and comparisons, in B.H. Matzat et al. (eds.), *Algorithmic algebra and number theory, Heidelberg 1997*, Springer, Berlin, pp. 187–220.

Decker, W., Heydtmann, A. E., and Schreyer, F.-O. (1998) Generating a noetherian normalization of the invariant ring of a finite group, *J. Symb. Comput.* **25**, 727–731.

Decker, W., and de Jong, T. (1998) Gröbner bases and invariant theory, in B. Buchberger and F. Winkler (eds.), *Gröbner bases and applications, Linz 1999*, pp. 61–89. Cambridge University Press, Cambridge, 1999

Decker, W., de Jong, T., Greuel, G.-M., and Pfister, G. (1999) The normalization: a new algorithm, implementation and comparisons, in P. Draexler et al. (eds.), *Computational methods for representations of groups and algebras. Proceedings of the Euroconference in Essen, Germany, April 1-5, 1997*, Birkhäuser, Basel, pp. 267–285.

Decker, W., Manolache, N., and Schreyer, F.-O. (1992) Geometry of the Horrocks bundle on $\mathbb{P}^5$, in G. Ellingsrud and C. Peskine (eds.), *Complex projective geometry, Sel. Pap. Conf. Proj. Var., Trieste/Italy 1989, and Vector Bundles and Special Proj. Embeddings, Bergen/Norway 1989*, Lond. Math. Soc. Lect. Note Ser. 179, London, pp. 128–148.

Decker, W., and Popescu, S. (1995) On surfaces in $\mathbb{P}^4$ and 3–folds in $\mathbb{P}^5$, in N. J. Hitchin et al. (eds.), *Vector bundles in algebraic geometry, Durham 1993*, Cambridge University Press, Cambridge, pp. 69–100.

Decker, W., and Schreyer, F.-O. (2000) Non-general type surfaces in $\mathbb{P}^4$: Some remarks on bounds and constructions, *J. Symb. Comput.* **29**, 545–582.

Derksen, H. (1999) Computation of invariants for reductive groups, *Adv. Math.* **141**, 366–384.

Eisenbud, D. (1992) Green's conjecture: An orientation for algebraists, in D. Eisenbud and C. Huneke (eds.), *Free resolutions in commutative algebra and algebraic geometry, Proc. Conf., Sundance/UT (USA) 1990*, Jones and Bartlett Publishers, Boston.

Eisenbud, D. (1993) Open problems in computational algebraic geometry, in D. Eisenbud (ed.), *Proceedings of a conference held at Cortona, Italy, June 17-21, 1991*, Cambridge University Press, Cambridge, pp. 49–70.

Eisenbud, D. (1995) *Commutative algebra. With a view toward algebraic geometry*, Springer, Berlin.

Eisenbud, D. (1998) Computing cohomology, a chapter in W. Vasconcelos, *Computational methods in commutative algebra and algebraic geometry*, Springer, Berlin.

Eisenbud, D., Fløystad, G., and Schreyer, F.-O. (2001) Sheaf cohomology and free resolutions over the exterior algebra, preprint, `math.AG/0104203`.

Eisenbud, D., Huneke, C., and Vasconcelos, W. (1992) Direct methods for primary decomposition, *Invent. Math.* **110**, 207–235.

Eisenbud, D., and Popescu, S. (1999) Gale duality and free resolutions of ideals of points, *Invent. Math.* **136**, 419–449.

Eisenbud, D., Popescu, S., Schreyer, F.-O., and Walter, C. (2000) Exterior algebra methods for the minimal resolution conjecture, preprint, `math.AG/0011236`.

Eisenbud, D., and Schreyer, F.-O. (2001) Chow forms via exterior syzygies, to appear.

Eisenbud, D., and Sturmfels, B. (1996) Binomial ideals, *Duke Math. J.* **84**, 1–45.

Ellingsrud, G., and Peskine, Ch. (1989) Sur les surfaces lisse de $\mathbb{P}^4$, *Invent. Math.* **95**, 1–12.

Gianni, P., and Trager, B. (1997) Integral closure of Noetherian rings, in W. W. Kuechlin (ed.), *Proceedings of the 1997 international symposium on symbolic and algebraic computation, ISSAC '97, Maui, HI, USA, July 21–23, 1997*, ACM Press, New York.

Gianni, P., Trager, B., and Zacharias, G. (1988) Gröbner bases and primary decomposition of polynomial ideals, *J. Symb. Comput.* **6**, 149–167.

Gordan, P. (1899) Neuer Beweis des Hilbertschen Satzes über homogene Funktionen, *Nachrichten König. Ges. der Wiss. zu Gött.*, 240–242.

Grauert, H. (1972) Über die Deformation isolierter Singularitäten analytischer Mengen, *Invent. Math.* **15**, 171–198.

Grauert, H., and Remmert, R. (1971) *Analytische Stellenalgebren*, Springer, Berlin.

Green, M. (1984) Koszul cohomology and the geometry of projective varieties, *J. Differ. Geom.* **19**, 125–167.

Green, M., and Lazarsfeld, R. (1984) The nonvanishing of certain Koszul cohomology groups, appendix to (Green, 1984).

Green, M., and Lazarsfeld, R. (1986) On the projective normality of complete linear series on an algebraic curve, *Invent. Math.* **83**, 73–90.

Green, M., and Lazarsfeld, R. (1988) Some results on the syzygies of finite sets and algebraic curves, *Compositio Math.* **67**, 301–314, 1988

Greuel, G.-M. (1985) Constant Milnor number implies constant multiplicity for quasihomogeneous singularities, *Manuscr. Math.* **56**, 159–166.

Greuel, G.-M. (2000) Computer algebra and algebraic geometry–achievements and perspectives, *J. Symb. Comput.* **30**, 253–289.

Greuel, G.-M., and Pfister, G. (1996) Advances and improvements in the theory of standard bases and syzygies, *Arch. Math.* **66**, 163–176.

Hermann, G. (1926) Die Frage der endlich vielen Schritte in der Theorie der Polynomideale, *Math. Ann.* **95**, 736–788.

Hilbert, D. (1890) Über die vollen Invariantensysteme, *Math. Ann.* **36**, 473–534.

Hilbert, D. (1893) Über die Theorie der algebraischen Formen, *Math. Ann.* **42**, 313–373.

Hironaka, H. (1964) Resolution of singularities of an algebraic variety over a field of characteristic zero. I, II, *Ann. Math., II. Ser.* **79**, 109–203, 205–326.

Hirschowitz, A., and Ramanan, S. (1998) New evidence for Green's conjecture on syzygies of canonical curves, *Ann. Sci. Ec. Norm. Super., IV. Ser.* **31**, 145–152.

Hirschowitz, A., and Simpson, C. (1996) La resolution minimale de l'ideal d'un arrangement general d'un grand nombre de points dans $\mathbb{P}^n$, *Invent. Math.* **126**, 467–503.

Hochster, M., and Roberts, J. (1974) Rings of Invariants of Reductive Groups Acting on Regular Rings are Cohen-Macaulay, *Adv. in Math.* **13**, 115–175.

de Jong, T. (1998) An algorithm for computing the integral closure, *J. Symb. Comput.* **26**, 273–277.

Kaltofen, E. (1982) Polynomial factorization, in B. Buchberger et al. (eds.), *Computer algebra*, Springer, Wien, pp. 95–113.

Kaltofen, E. (1990) Polynomial factorization 1982-1986, in I. Simon (ed.), *Computers in mathematics*, Marcel Dekker, New York, pp. 285–309.

Kaltofen, E. (1992) Polynomial factorization 1987-1991, in D.V. Chudnovsky and R.D. Jenks (eds.), *Proceedings of LATIN'92, Sao Paulo*, Springer, New York, pp. 294–313.

Kemper, G. (1996) Calculating invariant rings of finite groups over arbitrary fields, *J. Symb. Comput.* **21**, 351–366.

Kemper, G. (1999) An algorithm to calculate optimal homogeneous systems of parameters, *J. Symb. Comput.* **27**, 171–184.

Kemper, G., and Steel, A. (1999) Some algorithms in invariant theory of finite groups, in P. Draexler et al. (eds.), *Computational methods for representations of groups and algebras. Proceedings of the Euroconference in Essen, Germany, April 1-5, 1997*, Birkhäuser, Basel, pp. 267–285.

Kollár, J. (1988) Sharp effective Nullstellensatz, *J. Am. Math. Soc.* **1**, 963–975.

Kollár, J. (1999) Effective Nullstellensatz for arbitrary ideals, *J. Eur. Math. Soc.* **1**, 313–337.

Lê, D. T., and Ramanujam, C. P. (1976) The invariance of Milnor's number implies the invariance of the topological typ, *Am. J. Math.* **98**, 67–78.

Lorenzini, A. (1987) *On the betti numbers of points in projective space*, PhD thesis, Queens University, Kingston, Canada.

Lorenzini, A. (1993) The minimal resolution conjecture, *J. Algebra* **156**, 5-35.

Macaulay, F. (1927) Some properties of enumeration in the theory of modular systems, *Proc. London Math. Soc.* **26**, 531–555.

Madlener, K., and Reinert, B. (1998) String rewriting and Groebner bases – a general approach to monoid and group rings, in M. Bronstein et al. (eds.), *Symbolic rewriting techniques. Papers from the workshop held in Ascona, Switzerland, April 30 - May 4, 1995*, Birkhäuser, Basel, pp. 127–180.

Martin, B. (1999) Computing versal deformations with `SINGULAR`, in B.H. Matzat et al. (eds.), *Algorithmic algebra and number theory, Heidelberg 1997*, Springer, Berlin, pp. 283–293.

Mayr, E. W. (1997) Some complexity results for polynomial ideals, *J. of Complexity* **13**, 305–325.

Mayr, E. W., and Meyer, A. R. (1982) The complexity of the word problems for commutative semigroups and polynomial ideals, *Adv. Math.* **46**, 305–329.

Molien, T. (1897) Über die Invarianten der linearen Substitutionsgruppen, *Sitzungsber. König. Preuss. Akad. Wiss.*, 1152–1156.

Mora, T. (1982) An algorithm to compute the equations of tangent cones, in J. Calmet (ed.), *Computer algebra, EUROCAM '82, Conf. Marseille/France 1982*, Lecture Notes in Computer Science 144, Springer, Berlin, pp. 158–165.

Mori, S., and Mukai, S. (1983) The uniruledness of the moduli space of curves of genus 11, in M. Raynaud and T. Shioda (eds.), *Algebraic geometry, Proc. Jap.-Fr. Conf., Tokyo and Kyoto 1982*, Lecture Notes in Mathematics 1016, Springer, Berlin, pp. 334–353.

Mukai, S. (1995) Curves and symmetric spaces. I, *Am. J. Math.* **117**, 1627–1644.

Noether, E. (1916) Der Endlichkeitssatz der Invarianten endlicher Gruppen, *Math. Ann.* **77**, 89–92.

Noether, E. (1926) Der Endlichkeitssatz der Invarianten endlicher linearer Gruppen der Charakteristik p, *Nachr. v. d. Ges. d. Wiss. zu Göttingen*, 28–35.

Okonek, C., Schneider, M., and Spindler, H. (1980) *Vector bundles on complex projective spaces*, Birkhäuser, Boston.

Petri, K. (1923) Über die invariante Darstellung algebraischer Funktionen einer Veränderlichen, *Math. Ann.* **88**, 242–289.

Richman, D. R. (1990) On vector invariants of finite fields, *Adv. in Math.* **81**, 30–65.

Riemenschneider, O. (1974) Deformationen von Quotientensingularitaeten (nach zyklischen Gruppen), *Math. Ann.* **209**, 211–248.

Roos, J.-E. (1993) Commutative non-Koszul algebras having a linear resolution of arbitrarily high order. Applications to torsion in loop space homology. *C. R. Acad. Sci., Paris, Ser. I* **316**, 1123–1128.

Saito, M., Sturmfels, B., and Takayama, N. (2000) *Gröbner deformations of hypergeometric differential equations*, Springer, Berlin.

Schreyer, F.-O. (1980) Die Berechnung von Syzygien mit dem verallgemeinerten Weierstraßchen Divisionssatz und eine Anwendung auf analytische Cohen–Macaulay Stellenalgebren minimaler Multiplizität, Diplomarbeit, University of Hamburg, Germany.

Schreyer, F.-O. (1986) Syzygies of canonical curves and special linear series, *Math. Ann.* **275**, 105–137.

Schreyer, F.-O. (1991) A standard basis approach to syzygies of canonical curves, *J. Reine Angew. Math.* **421**, 83–123.

Schreyer, F.-O. (1996) Small fields in constructive algebraic geometry, in M. Maruyama (ed.), *Moduli of vector bundles*, Marcel Dekker, New York, pp. 221–228.

Schreyer, F.-O., and Tonoli, F. (2001) Needles in a haystack: Special varieties via small fields, to appear in D. Eisenbud et al. (eds.), *Mathematical computations with Macaulay2*, Springer, New York.

Seidenberg, A. (1975) Construction of the integral closure of a finite integral domain. II, *Proc. Am. Math. Soc.* **52**, 368–372.

Sernesi, E. (1981) L'unirazionalità della varieta dei moduli delle curve di genere dodici, *Ann. Sc. Norm. Super. Pisa, Cl. Sci., IV* **8**, 405–439.

O'Shea, D. B. (1987) Topologically trivial deformations of isolated quasihomogeneous hypersurface singularities are equimultiple, *Proc. Am. Math. Soc.* **101**, 260–262.

Shimoyama, T., and Yokoyama, K. (1996) Localization and primary decomposition of polynomial ideals, *J. Symb. Comput.* **22**, 247–277.

Smith, G. G. (2000) Computing global extension modules, *J. Symb. Comput.* **29**, 729–746.

Stanley, R. B. (1979) Invariants of Finite Groups and their Applications to Combinatorics, *Bull. Am. Math. Soc.* **1**, 475–511.

Stolzenberg, G. (1968) Constructive normalization of an algebraic variety, *Bull. Am. Math. Soc.* **74**, 595–599.

Sturmfels, B. (1993) *Algorithms in Invariant Theory*, Springer, Wien.

Teissier, B. (1974) Deformations a type topologique constant. II, *Asterisque* **16**, 228–249.

Traverso, C. (1996) Hilbert functions and the Buchberger algorithm, *J. Symb. Comput.* **22**, 355–376.

Vasconcelos, W. V. (1991) Computing the integral closure of an affine domain, *Proc. Am. Math. Soc.* **113**, 633–638.

Voisin, C. (1988) Courbes tétragonales et cohomologie de Koszul, *J. Reine Angew. Math.* **387**, 111–121.

Voisin, C. (2001) Green's generic syzygy conjecture for curves of even genus lying on a K3 surface, preprint, Paris, France.

Wahl, J. M. (1979) Simultaneous resolution of rational singularities, *Compos. Math.* **38**, 43–54.

Wang, D. (1989) Characteristic sets and zero structures of polynomial sets, preprint RISC-LINZ, Linz, Austria.

Walter, U. (2000) Cohomology with rational coefficients of complex varieties, preprint.

Walter, U. (2001) D-modules and the cohomology of varieties, to appear in D. Eisenbud et al. (eds.), *Mathematical computations with Macaulay2*, Springer, New York.

Wester, M. (ed., 1999) *Computer algebra systems. A practical guide*, Wiley, Chichester.

Winkler, F. (1996) *Polynomial algorithms in computer algebra*, Springer, Wien.

Zariski, O. (1971) Some open questions in the theory of singularities, *Bull. Am. Math. Soc.* **77**, 481–491.

SOME APPLICATIONS OF ALGEBRAIC CURVES TO COMPUTATIONAL VISION *

M. FRYERS
University of Hannover, Germany

J.Y. KAMINSKI and M. TEICHER
Bar-Ilan University, Ramat Gan, Israel

Abstract. We introduce a new formalism and a number of new results in the context of geometric computational vision. The classical scope of the research in geometric computer vision is essentially limited to static configurations of points and lines in $\mathbb{P}^3$. By using some well known material from algebraic geometry, we open new branches to computational vision.

We introduce algebraic curves embedded in $\mathbb{P}^3$ as the building blocks from which the tensor of a couple of cameras (projections) can be computed. In the process we address dimensional issues and as a result establish the minimal number of algebraic curves required for the tensor variety to be discrete as a function of their degree and genus.

We then establish new results on the reconstruction of an algebraic curves in $\mathbb{P}^3$ from multiple projections on projective planes embedded in $\mathbb{P}^3$. We address three different presentations of the curve: (i) definition by a set of equations, for which we show that for a generic configuration, two projections of a curve of degree d defines a curve in $\mathbb{P}^3$ with two irreducible components, one of degree d and the other of degree $d(d-1)$, (ii) the dual presentation in the dual space $\mathbb{P}^{3*}$, for which we derive a lower bound for the number of projections necessary for linear reconstruction as a function of the degree and the genus, and (iii) the presentation as an hypersurface of $\mathbb{P}^5$, defined by the set of lines in $\mathbb{P}^3$ meeting the curve, for which we also derive lower bounds for the number of projections necessary for linear reconstruction as a function of the degree (of the curve).

Moreover we show that the latter representation yields a new and efficient algorithm for dealing with mixed configurations of static and moving points in $\mathbb{P}^3$.

Key words: computational vision, algebraic curves.

Mathematics Subject Classification (2000): 14Q05.

* This research is partially supported by the Emmy Noether Research Institute for Mathematics and the Minerva Foundation of Germany and the excellency center (Group Theoretic Method In The Study Of Algebraic Varieties) of the Israeli Academy of Science.

C. Ciliberto et al. (eds.),
Applications of Algebraic Geometry to Coding Theory, Physics and Computation, **121–137.**

1. Introduction

Computational vision is the art of inferring three-dimensional properties of an object given several views of it. A view is concretely an image taken by a camera. From a mathematical point of view, a camera is a device that performs a projection from $\mathbb{P}^3$ to $\mathbb{P}^2$, through a center of projection. What are objects that we can deal with in this context? Until recently, for simplicity reasons, the objects that were considered in computer vision were essentially composed by static points and lines in $\mathbb{P}^3$. In the last two decades, intensive research has been carried out in this context, analyzing the 3D structure of the scene when given multiple views. A summary of the past decade of work in this area can be found in (Hartley and Zisserman; Faugeras and Luong) and references to earlier work in (Faugeras).

The theory of multiple-view geometry and 3D reconstruction is somewhat fragmented when it comes to curve features, especially to non-planar algebraic curves of any degree and genus. In (Quan; Ma and Chen; Ma and Li) it is shown how to recover the 3D position of a conic section from two and three views, and (Schmid and Zisserman) shows how to recover the homography induced by a conic between two views of it, and (Cross and Zisserman; Shashua and Toelg) show how to recover a quadric surface from projections of its occluding conics.

In all the above, the projection matrices are given but hardly any thing was done when this data is not available. (Kahl and Heyden; Kaminski and Shashua) show how to recover the fundamental matrix from matching conics with the result that 4 matching conics are minimally sufficient for a finite numbers of solutions. (Kaminski and Shashua) generalizes this result to higher order curves, but consider only planar curves.

In this paper we present a general theory for dealing with algebraic curves in computational vision, and more precisely in the multiple-view geometry and 3D reconstruction problem. Thus we investigate three of the fundamental questions of computer vision, using algebraic curves: (i) recovering camera geometry (fundamental matrix) from two views of the same algebraic curve, (ii) reconstruction of the curve from its projection across two or more views, and (iii) application of algebraic curves to the structure recovery of dynamic scenes from streams of projections (video sequences).

We start by giving a short presentation of the computer vision background related to our work. Then we show how one can recover the so-called projective stratum of the camera geometry (epipolar geometry) from two views of a (non-planar) algebraic curve. For that purpose we define and derive the *generalized Kruppa's equations* which are responsible for describing the epipolar constraint of two projections of the same algebraic curve. As part of the derivation of those constraints we address the issue of dimension analysis and as a result establish the minimal number of algebraic curves required for the epipolar geometry to be defined up to a finite-fold ambiguity as a function of their degree and genus.

On the reconstruction front, the curve admits three different representations: (i) as the solution of a set of equations in $\mathbb{P}^3$, for which we show that in a generic configuration, the reconstruction from two views of a curve of degree d has two irreducible components, one of degree d and the other of degree $d(d-1)$, (ii) as an hypersurface in the dual space $\mathbb{P}^{3*}$, for which we derive a lower bound of the minimal number of projections necessary for linear reconstruction as a function of the curve degree and genus, and (iii) as an hypersurface in $\mathbb{P}^5$, defined by the set of lines in $\mathbb{P}^3$ meeting the curve, for which we also derive lower bounds for the number of projections necessary for linear reconstruction as a function of curve degree alone. Moreover we show that the latter representation yields a new and efficient algorithm for dealing with mixed configurations of static and moving points in $\mathbb{P}^3$.

2. Foundation of linear computer vision

Projective algebraic geometry provides a natural framework to geometric computer vision. However one has to keep in mind that the geometric entities to be considered are in fact embedded into the physical three-dimension euclidian space. For this content, the euclidian space is provided with three structures defined by three groups of transformations: the orthogonal group $\mathbb{P}O_3$ (which defines the euclidian structure and which is included into the affine group), $\mathbb{P}A_3$ (defining the affine structure and itself included into the projective group), $\mathbb{P}G_3$ (defining the projective structure). We fix $[X,Y,Z,T]^T$, as homogeneous coordinates, and $T=0$ as the plane at infinity.

2.1. A SINGLE CAMERA SYSTEM

Computational vision starts with images captured by cameras. The camera components are the following:

- a plane $\mathcal{R}$, called the *retinal plane*, or *image plane*,
- a point $\mathbf{O}$, called either the *optical center* or the *camera center*, and which does not lie on the plane $\mathcal{R}$.

The plane $\mathcal{R}$ is regarded as a two dimension projective space embedded into $\mathbb{P}^3$. Hence it is also denoted by $i(\mathbb{P}^2)$. The camera is a projection machine: $\pi : \mathbb{P}^3 \setminus \{\mathbf{O}\} \to i(\mathbb{P}^2), \mathbf{P} \mapsto \overline{\mathbf{OP}} \cap i(\mathbb{P}^2)$. The projection π is determined (up to a scalar) by a 3×4 matrix $\mathbf{M}$ (the image of $\mathbf{P}$ being $\lambda\mathbf{MP}$).

The physical properties of a camera imply that $\mathbf{M}$ can be decomposed as follows:

$$\mathbf{M} = \begin{bmatrix} f & s & u_0 \\ 0 & \alpha f & v_0 \\ 0 & 0 & 1 \end{bmatrix} [\mathbf{R};\mathbf{t}],$$

where (f, α, s, u_0, v_0) are the so-called internal parameters of the camera, whereas the rotation $\mathbf{R}$ and the translation $\mathbf{t}$ are the external parameters.

It is easy to see that:

- The camera center $\mathbf{O}$ is given by $\mathbf{MO} = \mathbf{0}$.
- The matrix $\mathbf{M}^T$ maps a line in $i(\mathbb{P}^2)$ to the only plane containing both the line and $\mathbf{O}$.
- There exists a matrix $\widehat{\mathbf{M}} \in \mathcal{M}_{6\times 3}(\mathbb{R})$, which is a polynomial function of $\mathbf{M}$, that maps a point $\mathbf{p} \in i(\mathbb{P}^2)$ to the line $\overline{\mathbf{Op}}$ (optical ray), represented by its Plücker coordinates in $\mathbb{P}^5$. If the camera matrix is decomposed as follows:

$$\mathbf{M} = \begin{bmatrix} \Gamma^T \\ \Lambda^T \\ \Theta^T \end{bmatrix},$$

  then for $\mathbf{p} = [x, y, z]^T$, the optical ray $\mathbf{L_p} = \widehat{\mathbf{M}}\mathbf{p}$ is given by the extensor: $\mathbf{L_p} = x\Lambda \wedge \Theta + y\Theta \wedge \Gamma + z\Gamma \wedge \Lambda$, where $\wedge$ denotes the meet operator in the Grassman-Cayley algebra (Barnabei, Brini and Rota).
- The matrix $\widetilde{\mathbf{M}} = \widehat{\mathbf{M}}^T$ maps lines in $\mathbb{P}^3$ to lines in $i(\mathbb{P}^2)$.

Moreover we will need in the sequel to consider the projection of the **absolute conic** onto the image plane. The absolute conic is simply defined by the following equations:

$$\begin{cases} X^2 + Y^2 + Z^2 = 0 \\ T = 0 \end{cases}$$

By definition, the absolute conic is invariant by euclidian transformations. Therefore its projection onto the image plane, defined by the matrix ω, is a function of the internal parameters only. By Cholesky decomposition $\omega = \mathbf{LU}$, where $\mathbf{L}$ (respectively $\mathbf{U}$) is lower (respectively upper) triangular matrix. Hence it is easy to see that $\mathbf{U} = \overline{\mathbf{M}}^{-1}$, where $\overline{\mathbf{M}}$ is the 3×3 matrix of the internal parameters of $\mathbf{M}$.

2.2. A SYSTEM OF TWO CAMERAS

Given two cameras, let $(\mathbf{O}_j, i_j(\mathbb{P}^2))_{j=1,2}$ be their components where $i_1(\mathbb{P}^2)$ and $i_2(\mathbb{P}^2)$ are two generic projective planes embedded into $\mathbb{P}^3$, and $\mathbf{O}_1$ and $\mathbf{O}_2$ are two generic points in $\mathbb{P}^3$ not lying on the above planes. As in 2.1, let $\pi_j : \mathbb{P}^3 \setminus \{\mathbf{O}_j\} \to i_j(\mathbb{P}^2), \mathbf{P} \mapsto \overline{\mathbf{O}_j\mathbf{P}} \cap i_j(\mathbb{P}^2)$ be the respective projections. The camera matrices are $\mathbf{M}_i, i = 1, 2$.

2.2.1. *Homography between two images of the same plane*
Consider the case where the two cameras are looking at the same plane in space, denoted by Δ. Let

$$\mathbf{M}_i = \begin{bmatrix} \Gamma_i^T \\ \Lambda_i^T \\ \Theta_i^T \end{bmatrix}$$

be the camera matrices, decomposed as above. Let $\mathbf{P}$ be a point lying on Δ. We shall denote the projections of $\mathbf{P}$ by $\mathbf{p}_i = [x_i, y_i, z_i]^T \cong \mathbf{M}_i\mathbf{P}$, where $\cong$ means equality modulo multiplication by a non-zero scalar.

The optical ray generated by $\mathbf{p}_1$ is given by $\mathbf{L}_{\mathbf{p}_1} = x_1\Lambda_1 \wedge \Theta_1 + y_1\Theta_1 \wedge \Gamma_1 + z_1\Gamma_1 \wedge \Lambda_1$. Hence $\mathbf{P} = \mathbf{L}_{\mathbf{p}_1} \wedge \Delta = x_1\Lambda_1 \wedge \Theta_1 \wedge \Delta + y_1\Theta_1 \wedge \Gamma_1 \wedge \Delta + z_1\Gamma_1 \wedge \Lambda_1 \wedge \Delta$. Hence $\mathbf{p}_2 \cong \mathbf{M}_2\mathbf{P}$ is given by the following expression: $\mathbf{p}_2 \cong \mathbf{H}_\Delta\mathbf{p}_1$ where:

$$\mathbf{H}_\Delta = \begin{bmatrix} \Gamma_2^T(\Lambda_1 \wedge \Theta_1 \wedge \Delta) & \Gamma_2^T(\Theta_1 \wedge \Gamma_1 \wedge \Delta) & \Gamma_2^T(\Gamma_1 \wedge \Lambda_1 \wedge \Delta) \\ \Lambda_2^T(\Lambda_1 \wedge \Theta_1 \wedge \Delta) & \Lambda_2^T(\Theta_1 \wedge \Gamma_1 \wedge \Delta) & \Lambda_2^T(\Gamma_1 \wedge \Lambda_1 \wedge \Delta) \\ \Theta_2^T(\Lambda_1 \wedge \Theta_1 \wedge \Delta) & \Theta_2^T(\Theta_1 \wedge \Gamma_1 \wedge \Delta) & \Theta_2^T(\Gamma_1 \wedge \Lambda_1 \wedge \Delta) \end{bmatrix}.$$

This yield the expression of the collineation $\mathbf{H}_\Delta$ between two images of the same plane.

DEFINITION 1. *The previous collineation is called the homography between the two images, through the plane* Δ.

2.2.2. *Epipolar geometry*
DEFINITION 2. *Let* $(\mathbf{O}_j, i_j(\mathbb{P}^2), \mathbf{M}_j)_{j=1,2}$ *being defined as before. Given a pair* $(\mathbf{p}_1, \mathbf{p}_2) \in i_1(\mathbb{P}^1) \times i_2(\mathbb{P}^2)$, *we say that it is a pair of corresponding or matching points if there exists* $\mathbf{P} \in \mathbb{P}^3$ *such that* $\mathbf{p}_j = \pi_j(\mathbf{P})$ *for* $j = 1, 2$.

Consider a point $\mathbf{p} \in i_1(\mathbb{P}^2)$. Then $\mathbf{p}$ can be the image of any point lying on the fiber $\pi_1^{-1}(\mathbf{p})$. Then the matching point in the second image must lie on $\pi_2(\pi_1^{-1}(\mathbf{p}))$, which is, for a generic point $\mathbf{p}$, a line on the second image. Since π_1 and π_2 are both linear, there exists a matrix $\mathbf{F} \in \mathcal{M}_{3\times3}(\mathbb{R})$, such that: $\xi(\mathbf{p}) = \pi_2(\pi_1^{-1}(\mathbf{p}))$ is given by $\mathbf{Fp}$ for all but one point in the first image.

DEFINITION 3. *The matrix* $\mathbf{F}$ *is called the fundamental matrix, where as the line* $\mathbf{l}_\mathbf{p} = \mathbf{Fp}$ *is called the epipolar line of* $\mathbf{p}$.

Let $\mathbf{e}_1 = \overline{\mathbf{O}_1\mathbf{O}_2} \cap i_1(\mathbb{P}^2)$ and $\mathbf{e}_2 = \overline{\mathbf{O}_1\mathbf{O}_2} \cap i_2(\mathbb{P}^2)$. Those two points are respectively called the *first* and the *second epipole*. It is easy to see that $\mathbf{Fe}_1 = \mathbf{0}$, since $\pi_1^{-1}(\mathbf{e}_1) = \overline{\mathbf{O}_1\mathbf{O}_2}$ and $\pi_2(\overline{\mathbf{O}_1\mathbf{O}_2}) = \mathbf{e}_2$. Observe that by symmetry $\mathbf{F}^T$ is the fundamental matrix of the reverse couple of images. Hence $\mathbf{F}^T\mathbf{e}_2 = \mathbf{0}$. Since the only point in the first image that is mapped to zero by $\mathbf{F}$ is the first epipole, $\mathbf{F}$ has rank 2.

Now we want to deduce an expression of $\mathbf{F}$ as a function of the camera matrices. By the previous analysis, it is clear that: $\mathbf{F} = \widetilde{\mathbf{M}}_2\widehat{\mathbf{M}}_1$. Moreover we have the followings properties:

PROPOSITION 1. *For any plane Δ, not passing through the camera centers, the following equalities hold:*

1.

$$\mathbf{F} \cong [\mathbf{e}_2]_\times \mathbf{H}_\Delta,$$

where $[\mathbf{e}_2]_\times$ is the matrix associated with the cross-product as follows: for any vector $\mathbf{p}$, $\mathbf{e}_2 \times \mathbf{p} = [\mathbf{e}_2]_\times \mathbf{p}$. Hence we have:

$$[\mathbf{e}_2]_\times = \begin{bmatrix} 0 & -\mathbf{e}_{23} & \mathbf{e}_{22} \\ \mathbf{e}_{23} & 0 & -\mathbf{e}_{21} \\ -\mathbf{e}_{22} & \mathbf{e}_{21} & 0 \end{bmatrix}.$$

In particular, we have: $\mathbf{F} = [\mathbf{e}_2]_\times \mathbf{H}_\infty$, where $\mathbf{H}_\infty$ is the homography between the two images through the plane at infinity.

2.

$$\mathbf{H}_\Delta^T \mathbf{F} + \mathbf{F}^T \mathbf{H}_\Delta = \mathbf{0}. \tag{1}$$

PROOF. The first equality is clear according to its geometric meaning. Given a point $\mathbf{p}$ in the first image, $\mathbf{Fp}$ is its epipolar line in the second image. The optical ray $\mathbf{L_p}$ passing trough $\mathbf{p}$ meets the plane Δ in a point $\mathbf{Q}$, which projection in the second image is $\mathbf{H}_\Delta\mathbf{p}$. Hence the epipolar line must be $\mathbf{e}_2 \vee \mathbf{H}_\Delta\mathbf{p}$. This gives the required equality. The second equality is simply deduced from the first one by a short calculation. □

PROPOSITION 2. *For a generic plane Δ, the following equality holds:*

$$\mathbf{H}_\Delta \mathbf{e}_1 \cong \mathbf{e}_2.$$

PROOF. The image of the $\mathbf{e}_1$ by the homography must be the projection on the second image of the point defined as being the intersection of the optical ray generated by $\mathbf{e}_1$ and the plane Δ. Hence: $\mathbf{H}_\Delta\mathbf{e}_1 = \mathbf{M}_2(\mathbf{L}_{\mathbf{e}_1} \wedge \Delta)$. But $\mathbf{L}_{\mathbf{e}_1} = \overline{\mathbf{O}_1\mathbf{O}_2}$. Thus the result must be $\mathbf{M}_2\mathbf{O}_1$ (except when the plane is passing through $\mathbf{O}_2$), that is the second epipole $\mathbf{e}_2$. □

2.2.3. *Canonical stratification of the reconstruction*

Three-dimension reconstruction can be achieved from a system of two cameras, once the camera matrices are known. However a typical situation is that the camera matrices are unknown. Then we face a double problem: recovering the camera matrices and the actual object. There exists an inherent ambiguity. Consider a pair of camera matrices $(\mathbf{M}_1, \mathbf{M}_2)$. If you change the world coordinate system by a transformation $\mathbf{V} \in \mathbb{P}G_3$, the camera matrices are mapped to $(\mathbf{M}_1\mathbf{V}^{-1}, \mathbf{M}_2\mathbf{V}^{-1})$. Therefore we define the following equivalence relation:

DEFINITION 4. *Given a group of transformation G, two pairs of camera matrices, say $(\mathbf{M}_1, \mathbf{M}_2)$ and $(\mathbf{N}_1, \mathbf{N}_2)$ are said to be equivalent modulo multiplication by G if there exists $\mathbf{V} \in G$, such that: $\mathbf{M}_1 = \mathbf{N}_1\mathbf{V}$ and $\mathbf{M}_2 = \mathbf{N}_2\mathbf{V}$.*

Any reconstruction algorithm will always yields a reconstruction modulo multiplication by a certain group of transformations. More precisely there exist three levels of reconstruction according to the information that can be extracted from the two images and from a-priori knowledge of the world.

Projective Stratum

When the only available information is the fundamental matrix, then the reconstruction can be performed modulo multiplication by $\mathbb{P}G_3$. Indeed from $\mathbf{F}$, the so-called intrinsic homography $\mathbf{S} = -\frac{\mathbf{e}_2}{\|\mathbf{e}_2\|}\mathbf{F}$ is computed and the camera matrices are equivalent to: $([\mathbf{I}; \mathbf{0}], [\mathbf{S}; \mathbf{e}_2])$.

Affine Stratum

When, in addition of the epipolar geometry, the homography between the two images through the plane at infinity, denoted by $\mathbf{H}_\infty$, can be computed, the reconstruction can be done modulo multiplication by the group of affine transformations. Then the two camera matrices are equivalent to: $([\mathbf{I}; \mathbf{0}], [\mathbf{H}_\infty; \mathbf{e}_2])$.

Euclidian Stratum

The euclidian stratum is obtained by the data of the projection of the absolute conic Ω onto the image planes, which allows recovering the internal parameters of the cameras. Once the internal parameters of the cameras are known the relative motion between the cameras expressed by a rotation $\mathbf{R}$ and a translation $\mathbf{t}$ can be extracted from the fundamental matrix. However only the direction of $\mathbf{t}$ but not the norm can be recovered. Then the cameras matrices are equivalent, modulo multiplication by the group of similarity transformations, to: $(\overline{\mathbf{M}}_1[\mathbf{I}; \mathbf{0}], \overline{\mathbf{M}}_2[\mathbf{R}; \mathbf{t}])$, where $\overline{\mathbf{M}}_1$ and $\overline{\mathbf{M}}_2$ are the matrices of the internal parameters.

Note that the projection of the absolute conic on the image can be computed using some a-priori knowledge of the world. Moreover there exist famous equations linking ω_1 and ω_2 when the epipolar geometry is given. These are the so-called *Kruppa's equations*, defined in the following proposition.

PROPOSITION 3. *The projections of the absolute conic onto two images are related as follows. There exists a scalar λ such that:*

$$[\mathbf{e}_1]_\times^T \omega_1^* [\mathbf{e}_1]_\times = \lambda \mathbf{F}^T \omega_2^* \mathbf{F},$$

where $[\mathbf{e}_1]_\times$ is the matrix representing the cross-product by $\mathbf{e}_1$ and ω_i^* is the adjoint matrix of ω_i.

Let ε_i be the tangents to $\pi_i(\Omega)$ through $\mathbf{e}_i$. Kruppa's equations simply state that ε_1 and ε_2 are projectively isomorphic.

3. Recovering the epipolar geometry from curve correspondences

Recovering epipolar geometry from curve correspondences requires the establishment of an algebraic relation between the two image curves, involving the fundamental matrix. Hence such an algebraic relation may be regarded as an extension of Kruppa's equations. In their original form (see proposition 3), these equations have been introduced to compute the camera-intrinsic parameters from the projection of the absolute conic onto the two image planes (Maybank and Faugeras). However it is obvious that they still hold if one replaces the absolute conic by any conic that lies on a plane that does not meet any of the camera centers. In this form they can be used to recover the epipolar geometry from conic correspondences (Kahl and Heyden; Kaminski and Shashua). Furthermore it is possible to extend them to any planar algebraic curve (Kaminski and Shashua). Moreover a generalization for arbitrary algebraic spatial curves is possible and is a step toward the recovery of epipolar geometry from matching curves.

Therefore we shall prove and generalize Kruppa's equations to an arbitrary smooth irreducible curve that cannot be embedded in a plane and whose degree $d \geq 2$ (the case of line is excluded since one cannot deduce constraints on the epipolar geometry from a pair of matching lines and the case of planar curve has already been treated in (Kaminski and Shashua).

At this stage we recall a number of facts, that will be useful in the sequel, concerning the projection of a space curve onto a plane.

We shall mention that all our theoretical results are true when the ground field is the field of complex numbers. Finally we shall consider only the real solutions.

3.1. SINGLE VIEW OF A SPACE CURVE

Let $\mathbf{M}$ be the camera matrix, $\mathbf{O}$ the camera center, $\mathcal{R}$ the retinal plane (as in 2.1). Let X be a smooth irreducible embedded in $\mathbb{P}^3$ and which degree $d \geq 2$. Let Y be the projection of X by the camera. It is well known that:

1. The curve Y will always contain singularities. Furthermore for a generic position of the camera center, the only singularities of Y will be nodes.
2. The *class* of a planar curve is defined to be the degree of its dual curve. Let m be the class of Y. Then m is constant for a generic position of the camera center.
3. If d and g are the degree and the genus of X, they are respectively, for a generic position of $\mathbf{O}$, the degree and the genus of Y and the Plücker formula yields:

$$\begin{aligned} m &= d(d-1) - 2(\sharp\text{nodes}), \\ g &= \tfrac{(d-1)(d-2)}{2} - (\sharp\text{nodes}), \end{aligned}$$

where ♯nodes denotes the number of nodes of Y. Hence the genus, the degree and the class are related by:

$$m = 2d + 2g - 2.$$

3.2. GENERALIZED KRUPPA'S EQUATIONS

We are ready now to investigate the recovery of the epipolar geometry from matching curves. Let $\mathbf{M}_i$, $i = 1,2$, be the camera matrices. Let $\mathbf{F}$ and $\mathbf{e}_1$ be the fundamental matrix and the first epipole, $\mathbf{F}\mathbf{e}_1 = 0$. We will need to consider the two following mappings: $\mathbf{p} \overset{\gamma}{\mapsto} \mathbf{e}_1 \vee \mathbf{p}$ and $\mathbf{p} \overset{\xi}{\mapsto} \mathbf{F}\mathbf{p}$. Both are defined on the first image plane; γ associates a point to its epipolar line in the first image, while ξ sends it to its epipolar line in the second image.

Let Y_1 and Y_2 be the image curves (projections of X onto the image planes). Assume that they are defined by the polynomials f_1 and f_2. Let $Y_1^\star$ and $Y_2^\star$ denote the dual image curves, whose polynomials are respectively ϕ_1 and ϕ_2. Roughly speaking, the generalized Kruppa's equations state that the sets of epipolar lines tangent to the curve in each image are projectively equivalent.

THEOREM 1 (GENERALIZED KRUPPA'S EQUATIONS). *For a generic position of the camera centers with respect to the curve in space, there exists a non-zero scalar λ, such that for all points $\mathbf{p}$ in the first image, the following equality holds:*

$$\phi_2(\xi(\mathbf{p})) = \lambda \phi_1(\gamma(\mathbf{p})). \tag{2}$$

Remark: Observe that if X is a conic and $\mathbf{C}_1$ and $\mathbf{C}_2$ the matrices that respectively represent Y_1 and Y_2, the generalized Kruppa's equations reduce to the classical Kruppa's equations, that is: $[\mathbf{e}_1]_\times^T \mathbf{C}_1^\star [\mathbf{e}_1]_\times \cong \mathbf{F}^T \mathbf{C}_2^\star \mathbf{F}$, where $\mathbf{C}_1^\star$ and $\mathbf{C}_2^\star$ are the adjoint matrices of $\mathbf{C}_1$ and $\mathbf{C}_2$.

PROOF. Let ε_i be the set of epipolar lines tangent to the curve in image i. We start by proving the following lemma.

LEMMA 1. *The two sets ε_1 and ε_2 are projectively equivalent. Furthermore for each corresponding pair of epipolar lines, $(\mathbf{l},\mathbf{l}') \in \varepsilon_1 \times \varepsilon_2$, the multiplicity of $\mathbf{l}$ and $\mathbf{l}'$ as points of the dual curves $Y_1^\star$ and $Y_2^\star$ are the same.*

PROOF. Consider the following three pencils:

1. $\sigma(\mathbf{L}) \cong \mathbb{P}^1$, the pencil of planes containing the baseline, generated by the camera centers,
2. $\sigma(\mathbf{e}_1) \cong \mathbb{P}^1$, the pencil of epipolar lines through the first epipole,
3. $\sigma(\mathbf{e}_2) \cong \mathbb{P}^1$, the pencil of epipolar lines through the second epipole.

Thus we have $\varepsilon_i \subset \sigma(\mathbf{e}_i)$. Moreover if E is the set of plane in $\sigma(\mathbf{L})$ tangent to the curve in space, there exist a one-to-one mapping from E to each ε_i. This mapping also leaves the multiplicities unchanged. This completes the lemma. □

Lemma 1 implies that both sides of the equation (2) define the same algebraic set, that the union of eipolar lines through $\mathbf{e}_1$ tangent to Y_1. Since ϕ_1 and ϕ_2, in the generic case, have same degree (as stated in 3.1), each side of equation (2) can be factorized into linear factors, satisfying the following:

$$\begin{aligned}\phi_1(\gamma(x,y,z)) &= \prod_i(\alpha_{1i}x+\alpha_{2i}y+\alpha_{3i}z)^{a_i}\\ \phi_2(\xi(x,y,z)) &= \prod_i\lambda_i(\alpha_{1i}x+\alpha_{2i}y+\alpha_{3i}z)^{b_i},\end{aligned}$$

where $\sum_i a_i = \sum_j b_j = m$. By Lemma 1, we must also have $a_i = b_i$ for i. □

By eliminating the scalar λ from the generalized Kruppa's equations (2) we obtain a set of bi-homogeneous equations in $\mathbf{F}$ and $\mathbf{e}_1$. Hence they define a variety in $\mathbb{P}^2 \times \mathbb{P}^8$. This gives rise to an important question. How many of those equations are algebraically independent, or in other words what is the dimension of the set of solutions? This is the issue of the next section.

3.3. DIMENSION OF THE SET OF SOLUTIONS

Let $\{E_i(\mathbf{F},\mathbf{e}_1)\}_i$ be the set of bi-homogeneous equations on $\mathbf{F}$ and $\mathbf{e}_1$, extracted from the generalized Kruppa's equations (2). Our first concern is to determine whether all solutions of equation (2) are admissible, that is whether they satisfy the usual constraint $\mathbf{F}\mathbf{e}_1 = 0$. Indeed we prove the following statement:

PROPOSITION 4. *As long as there are at least 2 distinct lines through* $\mathbf{e}_1$ *tangent to* Y_1, *equation (2) implies that* $\operatorname{rank}\mathbf{F} = 2$ *and* $\mathbf{F}\mathbf{e}_1 = \mathbf{0}$.

PROOF. The variety defined by $\phi_1(\gamma(\mathbf{p}))$ is then a union of at least 2 distinct lines through $\mathbf{e}_1$. If equation (2) holds, $\phi_2(\xi(\mathbf{p}))$ must define the same variety.

There are 2 cases to exclude: if $\operatorname{rank}\mathbf{F} = 3$, then the curve defined by $\phi_2(\xi(\mathbf{p}))$ is projectively equivalent to the curve defined by ϕ_2, which is $Y_1^\star$. In particular, it is irreducible.

If $\operatorname{rank}\mathbf{F} < 2$ or $\operatorname{rank}\mathbf{F} = 2$ and $\mathbf{F}\mathbf{e}_1 \neq \mathbf{0}$, then there is some $\mathbf{a}$, not a multiple of $\mathbf{e}_1$, such that $\mathbf{F}\mathbf{a} = \mathbf{0}$. Then the variety defined by $\phi_2(\xi(\mathbf{p}))$ is a union of lines through $\mathbf{a}$. In neither case this variety can contain two distinct lines through $\mathbf{e}_1$, so we must have $\operatorname{rank}\mathbf{F} = 2$ and $\mathbf{F}\mathbf{e}_1 = \mathbf{0}$. □

As a result, in a generic situation every solution of $\{E_i(\mathbf{F},\mathbf{e}_1)\}_i$ is admissible. Let V be the subvariety of $\mathbb{P}^2 \times \mathbb{P}^8 \times \mathbb{P}^2$ defined by the equations $\{E_i(\mathbf{F},\mathbf{e}_1)\}_i$ together with $\mathbf{F}\mathbf{e}_1 = \mathbf{0}$ and $\mathbf{e}_2{}^T\mathbf{F} = \mathbf{0}^T$, where $\mathbf{e}_2$ is the second epipole. We next compute the lower bound on the dimension of V, after which we would be ready for the calculation itself.

PROPOSITION 5. *If V is non-empty, the dimension of V is at least $7-m$.*

PROOF. Choose any line $\mathbf{l}$ in $\mathbb{P}^2$ and restrict $\mathbf{e}_1$ to the affine piece $\mathbb{P}^2 \setminus \mathbf{l}$. Let (x,y) be homogeneous coordinates on $\mathbf{l}$. If $\mathbf{F}\mathbf{e}_1 = \mathbf{0}$, the two sides of equation (2) are both unchanged by replacing $\mathbf{p}$ by $\mathbf{p}+\alpha\mathbf{e}_1$. So equation (2) will hold for all $\mathbf{p}$ if it holds for all $\mathbf{p} \in \mathbf{l}$. Therefore equation (2) is equivalent to the equality of 2 homogeneous polynomials of degree m in x and y, which in turn is equivalent to the equality of $(m+1)$ coefficients. After eliminating λ, we have m algebraic conditions on $(\mathbf{e}_1, \mathbf{F}, \mathbf{e}_2)$ in addition to $\mathbf{F}\mathbf{e}_1 = \mathbf{0}$, ${\mathbf{e}_2}^T\mathbf{F} = \mathbf{0}^T$.

The space of all epipolar geometries, that is, solutions to $\mathbf{F}\mathbf{e}_1 = \mathbf{0}$, ${\mathbf{e}_2}^T\mathbf{F} = \mathbf{0}^T$, is irreducible of dimension 7. Therefore, V is at least $(7-m)$-dimensional. □

For the calculation of the dimension of V we introduce some additional notations. Given a triplet $(\mathbf{e}_1, \mathbf{F}, \mathbf{e}_2) \in \mathbb{P}^2 \times \mathbb{P}^8 \times \mathbb{P}^2$, let $\{\mathbf{q}_{1\alpha}(\mathbf{e}_1)\}$ (resp. $\{\mathbf{q}_{2\alpha}(\mathbf{e}_2)\}$) be the tangency points of the epipolar lines through $\mathbf{e}_1$ (respectively $\mathbf{e}_2$) to the first (respectively second) image curve. Let $\mathbf{Q}_\alpha(\mathbf{e}_1, \mathbf{e}_2)$ be the 3d points projected onto $\{\mathbf{q}_{1\alpha}(\mathbf{e}_1)\}$ and $\{\mathbf{q}_{2\alpha}(\mathbf{e}_2)\}$. Let $\mathbf{L}$ be the baseline joining the two camera centers. We next provide a sufficient condition for V to be discrete.

PROPOSITION 6. *For a generic position of the camera centers, the variety V will be discrete if, for any point $(\mathbf{e}_1, \mathbf{F}, \mathbf{e}_2) \in V$, the union of $\mathbf{L}$ and the points $\mathbf{Q}_\alpha(\mathbf{e}_1, \mathbf{e}_2)$ is not contained in any quadric surface.*

PROOF. For generic camera positions, there will be m distinct points $\{\mathbf{q}_{1\alpha}(\mathbf{e}_1)\}$ and $\{\mathbf{q}_{2\alpha}(\mathbf{e}_2)\}$, and we can regard $\mathbf{q}_{1\alpha}$, $\mathbf{q}_{2\alpha}$ locally as smooth functions of $\mathbf{e}_1$, $\mathbf{e}_2$.

We let W be the affine variety in $\mathbb{C}^3 \times \mathbb{C}^9 \times \mathbb{C}^3$ defined by the same equations as V. Let $\Theta = (\mathbf{e}_1, \mathbf{F}, \mathbf{e}_2)$ be a point of W corresponding to a non-isolated point of V. Then there is a tangent vector $\vartheta = (\mathbf{v}, \mathbf{\Phi}, \mathbf{v}')$ to W at Θ with $\mathbf{\Phi}$ not a multiple of $\mathbf{F}$.

If χ is a function on W, $\nabla_{\Theta,\vartheta}(\chi)$ will denote the derivative of χ in the direction defined by ϑ at Θ. For

$$\chi_\alpha(\mathbf{e}_1, \mathbf{F}, \mathbf{e}_2) = \mathbf{q}_{2\alpha}(\mathbf{e}_2)^T \mathbf{F} \mathbf{q}_{1\alpha}(\mathbf{e}_1),$$

the generalized Kruppa's equations imply that χ_α vanishes identically on W, so its derivative must also vanish. This yields

$$\nabla_{\Theta,\vartheta}(\chi_\alpha) = (\nabla_{\Theta,\vartheta}(\mathbf{q}_{2\alpha}))^T \mathbf{F}\mathbf{q}_{1\alpha} + \mathbf{q}_{2\alpha}^T \mathbf{\Phi} \mathbf{q}_{1\alpha} + \mathbf{q}_{2\alpha}^T \mathbf{F}(\nabla_{\Theta,\vartheta}(\mathbf{q}_{1\alpha})) = 0. \quad (3)$$

We shall prove that $\nabla_{\Theta,\vartheta}(\mathbf{q}_{1\alpha})$ is in the linear span of $\mathbf{q}_{1\alpha}$ and $\mathbf{e}_1$. (This means that when the epipole moves slightly, $\mathbf{q}_{1\alpha}$ moves along the epipolar line.) Consider $\kappa(t) = f(\mathbf{q}_{1\alpha}(\mathbf{e}_1 + t\mathbf{v}))$, where f is the polynomial defining the image curve Y_1. Since $\mathbf{q}_{1\alpha}(\mathbf{e}_1 + t\mathbf{v}) \in Y_1$, $\kappa \equiv 0$, so the derivative $\kappa'(0) = 0$. On the other hand, $\kappa'(0) = \nabla_{\Theta,\vartheta}(f(\mathbf{q}_{1\alpha})) = \mathrm{grad}_{\mathbf{q}_{1\alpha}}(f)^T \nabla_{\Theta,\vartheta}(\mathbf{q}_{1\alpha})$.

Thus we have $\mathrm{grad}_{\mathbf{q}_{1\alpha}}(f)^T\nabla_{\Theta,\vartheta}(\mathbf{q}_{1\alpha}) = 0$. But also $\mathrm{grad}_{\mathbf{q}_{1\alpha}}(f)^T\mathbf{q}_{1\alpha} = 0$ and $\mathrm{grad}_{\mathbf{q}_{1\alpha}}(f)^T\mathbf{e}_1 = 0$. Since $\mathrm{grad}_{\mathbf{q}_{1\alpha}}(f) \neq \mathbf{O}$ ($\mathbf{q}_{1\alpha}$ is not a singular point of the curve), this shows that $\nabla_{\Theta,\vartheta}(\mathbf{q}_{1\alpha})$, $\mathbf{q}_{1\alpha}$, and $\mathbf{e}_1$ are linearly dependent. $\mathbf{q}_{1\alpha}$ and $\mathbf{e}_1$ are linearly independent, so $\nabla_{\Theta,\vartheta}(\mathbf{q}_{1\alpha})$ must be in their linear span.

We have that $\mathbf{q}_{2\alpha}^T\mathbf{F}\mathbf{e}_1 = \mathbf{q}_{2\alpha}^T\mathbf{F}\mathbf{q}_{1\alpha} = 0$, so $\mathbf{q}_{2\alpha}^T\mathbf{F}\nabla_{\Theta,\vartheta}(\mathbf{q}_{1\alpha}) = 0$: the third term of equation (3) vanishes. In a similar way, the first term of equation (3) vanishes, leaving

$$\mathbf{q}_{2\alpha}^T\Phi\mathbf{q}_{1\alpha} = 0.$$

The derivative of $\chi(\mathbf{e}_1, \mathbf{F}, \mathbf{e}_2) = \mathbf{F}\mathbf{e}_1$ must also vanish, which yields:

$$\mathbf{e}_2{}^T\Phi\mathbf{e}_1 = 0.$$

From the first equality, we deduce that for every $\mathbf{Q}_\alpha$, we have:

$$\mathbf{Q}_\alpha^T\mathbf{M}_2^T\Phi\mathbf{M}_1\mathbf{Q}_\alpha = 0.$$

From the second equality, we deduce that every point $\mathbf{P}$ lying on the baseline must satisfy:

$$\mathbf{P}^T\mathbf{M}_2^T\Phi\mathbf{M}_1\mathbf{P} = 0.$$

The fact that Φ is not a multiple of F implies that $\mathbf{M}_2^T\Phi\mathbf{M}_1 \neq 0$, so together these two last equations mean that the union $\mathbf{L}\cup\{\mathbf{Q}_\alpha\}$ lies on a quadric surface. Thus if there is no such quadric surface, every point in V must be isolated. □

Observe that this result is consistent with the previous proposition, since there always exist a quadric surface containing a given line and six given points. However in general there is no quadric containing a given line and seven given points. Therefore we can conclude with the following theorem.

THEOREM 2. *For a generic position of camera centers, the generalized Kruppa's equations define the epipolar geometry up to a finite-fold ambiguity if and only if* $m \geq 7$.

Since different curves in generic position give rise to independent equations, this result means that the sum of the classes of the image curves must be at least 7 for V to be a finite set.

4. 3D Reconstruction

We turn our attention to the problem of reconstructing an algebraic curve from two or more views, given known camera matrices (epipolar geometries are known). The basic idea is to intersect together the cones defined by the camera centers and the image curves. However this intersection can be computed in three different spaces, giving rise to different algorithms and applications.

We shall mention that in (Forsyth) a scheme is proposed to reconstruct an algebraic curve from a single view by blowing-up the projection. This approach results in a spatial curve defined up to an unknown projective transformation. In fact the only computation this reconstruction allows is the recovery of the properties of the curve that are invariant to projective transformation. Moreover this reconstruction is valid for irreducible curves only. However reconstructing from two projections not only gives the projective properties of the curve, but also the relative depth of it with respect to others objects in the scene and furthermore the relative position between irreducible components.

4.1. RECONSTRUCTION $\mathbb{P}^3$

Let the camera projection matrices be $\mathbf{M}_1$ and $\mathbf{M}_2$. Hence the two cones defined by the image curves and the camera centers are given by: $\Delta_1(\mathbf{P}) = f_1(\mathbf{M}_1\mathbf{P})$ and $\Delta_2(\mathbf{P}) = f_2(\mathbf{M}_2\mathbf{P})$. The reconstruction is defined as the curve whose equations are $\Delta_1 = 0$ and $\Delta_2 = 0$. This curves has two irreducible components as the following theorem states.

THEOREM 3. *For a generic position of the camera centers, that is when no epipolar plane is tangent twice to the curve X, the curve defined by $\{\Delta_1 = 0, \Delta_2 = 0\}$ has two irreducible components. One has degree d and is the actual solution of the reconstruction. The other one has degree $d(d-1)$.*

PROOF. For a line $\mathbf{l} \subset \mathbb{P}^3$, we write $\sigma(\mathbf{l})$ for the pencil of planes containing $\mathbf{l}$. For a point $\mathbf{p} \in \mathbb{P}^2$, we write $\sigma(\mathbf{p})$ for the pencil of lines through $\mathbf{p}$. There is a natural isomorphism between $\sigma(\mathbf{e}_i)$, the epipolar lines in image i, and $\sigma(\mathbf{L})$, the planes containing both camera centers. Consider the following covers of $\mathbb{P}^1$:

1. $X \xrightarrow{\eta} \sigma(\mathbf{L}) \cong \mathbb{P}^1$, taking a point $x \in X$ to the epipolar plane that it defines with the camera centers.
2. $Y_1 \xrightarrow{\eta_1} \sigma(\mathbf{e}_1) \cong \sigma(\mathbf{L}) \cong \mathbb{P}^1$, taking a point $y \in Y_1$ to its epipolar line in the first image.
3. $Y_2 \xrightarrow{\eta_2} \sigma(\mathbf{e}_2) \cong \sigma(\mathbf{L}) \cong \mathbb{P}^1$, taking a point $y \in Y_2$ to its epipolar line in the second image.

If ρ_i is the projection $X \to Y_i$, then $\eta = \eta_i \rho_i$. Let $\mathcal{B}$ the union set of branch points of η_1 and η_2. It is clear that the branch points of η are included in $\mathcal{B}$. Let $S = \mathbb{P}^1 \setminus \mathcal{B}$, pick $t \in S$, and write $X_S = \eta^{-1}(S), X_t = \eta^{-1}(t)$. Let μ_{X_S} be the monodromy: $\pi_1(S,t) \longrightarrow \mathrm{Perm}(X_t)$, where $\mathrm{Perm}(Z)$ is the group of permutation of a finite set Z. It is well known that the path-connected components of X are in one-to-one correspondence with the orbits of the action of $\mathrm{im}(\mu_{X_S})$ on X_t. Since X is assumed to be irreducible, it has only one component and $\mathrm{im}(\mu_{X_S})$ acts transitively on X_t. Then if $\mathrm{im}(\mu_{X_S})$ is generated by transpositions, this will imply that $\mathrm{im}(\mu_{X_S}) = \mathrm{Perm}(X_t)$. In order to show that $\mathrm{im}(\mu_{X_S})$ is actually generated by transpositions, consider a loop in $\mathbb{P}^1$ centered at t, say l_t. If l_t does not go round any branch point, then l_t is homotopic to the constant path in S and then $\mu_{X_S}([l_t]) = 1$. Now in $\mathcal{B}$, there are three types of branch points:

1. branch points that come from nodes of Y_1: these are not branch points of η,
2. branch points that come from nodes of Y_2: these are not branch points of η,
3. branch points that come from epipolar lines tangent either to Y_1 or to Y_2: these are genuine branch points of η.

If the loop l_t goes round a point of the first two types, then it is still true that $\mu_{X_S}([l_t]) = 1$. Now suppose that l_t goes round a genuine branch point of η, say b (and goes round no other points in $\mathcal{B}$). By genericity, b is a simple two-fold branch point, hence $\mu_{X_S}([l_t])$ is a transposition. This shows that $\mathrm{im}(\mu_{X_S})$ is actually generated by transpositions and so $\mathrm{im}(\mu_{X_S}) = \mathrm{Perm}(X_t)$.

Now consider $\tilde{X}$, the curve defined by $\{\Delta_1 = 0, \Delta_2 = 0\}$. By Bezout's Theorem $\tilde{X}$ has degree d^2. Let $\tilde{x} \in \tilde{X}$. It is projected onto a point y_i in Y_i, such that $\eta_1(y_1) = \eta_2(y_2)$. Hence $\tilde{X} \cong Y_1 \times_{\mathbb{P}^1} Y_2$; restricting to the inverse image of the set S, we have $\tilde{X}_S \cong X_S \times_S X_S$. We can therefore identify $\tilde{X}_t$ with $X_t \times X_t$. The monodromy $\mu_{\tilde{X}_S}$ can then be given by $\mu_{\tilde{X}_S}(x,y) = (\mu_{X_S}(x), \mu_{X_S}(y))$. Since $\mathrm{im}(\mu_{X_S}) = \mathrm{Perm}(X_t)$, the action of $\mathrm{im}(\mu_{\tilde{X}_S})$ on $X_t \times X_t$ has two orbits, namely $\{(x,x)\} \cong X_t$ and $\{(x,y) | x \neq y\}$. Hence $\tilde{X}$ has two irreducible components. One has degree d and is X, the other has degree $d^2 - d = d(d-1)$. □

This result provides an algorithm to find the right solution for the reconstruction in a generic configuration, except in the case of conics, where the two components of the reconstruction are both admissible.

4.2. RECONSTRUCTION IN THE DUAL SPACE

Let $X^\star$ be the dual variety of X. Since X is supposed not to be a line, the dual variety $X^\star$ must be a hypersurface of the dual space (Harris). Hence let Υ be a minimal degree polynomial that represents $X^\star$. Our first concern is to determine the degree of Υ.

PROPOSITION 7. *The degree of Υ is m, that is, the common degree of the dual image curves.*

PROOF. Since $X^\star$ is a hypersurface of $\mathbb{P}^{3\star}$, its degree is the number of points where a generic line in $\mathbb{P}^{3\star}$ meets $X^\star$. By duality it is the number of planes in a generic pencil that are tangent to X. Hence it is the degree of the dual image curve. Another way to express the same fact is the observation that the dual image curve is the intersection of $X^\star$ with a generic plane in $\mathbb{P}^{3\star}$. Note that this provides a simple proof that the degree of the dual image curve is constant for a generic position of the camera center. □

For the reconstruction of X^* from multiple views, we will need to consider the mapping from a line $\mathbf{l}$ of the image plane to the plane that it defines with the camera center. Let $\mu : \mathbf{l} \mapsto \mathbf{M}^T\mathbf{l}$ denote this mapping. There exists a link involving Υ, μ and ϕ, the polynomial of the dual image curve: $\Upsilon(\mu(\mathbf{l})) = 0$ whenever $\phi(\mathbf{l}) = 0$. Since these two polynomials have the same degree (because μ is linear) and ϕ is irreducible, there exist a scalar λ such that

$$\Upsilon(\mu(\mathbf{l})) = \lambda\phi(\mathbf{l}),$$

for all lines $\mathbf{l} \in \mathbb{P}^{2*}$. Eliminating λ, we get $\binom{m+2}{m} - 1$ linear equations on Υ. Since the number of coefficients in Υ is $\binom{m+3}{m}$, we can state the following result:

PROPOSITION 8. *The reconstruction in the dual space can be done linearly using at least* $k \geq \frac{m^2+6m+11}{3(m+3)}$ *views.*

From a practical point of view, it is worth noting that the fitting of the dual image curve is not necessary. It is sufficient to extract tangents to the image curves at distinct points. Each tangent $\mathbf{l}$ contributes to one linear equation on Υ: $\Upsilon(\mu(\mathbf{l})) = 0$. However one cannot obtain more than $\binom{m+2}{m} - 1$ linearly independent equations per view.

4.3. RECONSTRUCTION IN $\mathcal{G}(1,3)$

Let $\mathcal{G}(1,3)$ be the Grassmanian of lines of $\mathbb{P}^3$. Consider the set of lines in $\mathbb{P}^3$ meeting the curve X of degree d. This defines an irreducible subvariety of $\mathcal{G}(1,3)$ which is the intersection of $\mathcal{G}(1,3)$ with an irreducible hypersurface of degree d in $\mathbb{P}^5$ (Harris), given by a homogeneous polynomial Γ, defined modulo the dth graded piece $I(\mathcal{G}(1,3))_d$ of the ideal of $\mathcal{G}(1,3)$ and modulo scalars. However picking one representative of this equivalence class is sufficient to reconstruct entirely without any ambiguity the curve X. Hence we need to compute the class of Γ in the homogeneous coordinate ring of $\mathcal{G}(1,3)$, or more precisely in its dth graded piece, $S(\mathcal{G}(1,3))_d$, which dimension is $\binom{d+5}{d} - \binom{d-2+5}{d-2}$.

Let f be the polynomial defining the image curve, Y. Consider the mapping that associates to an image point its optical ray: $\nu : \mathbf{p} \mapsto \widehat{\mathbf{M}}\mathbf{p}$. Hence the polynomial $\Gamma(\nu(\mathbf{p}))$ vanishes whenever $f(\mathbf{p})$ does. Since they have same degree and f is irreducible, there exists a scalar λ such as for every point $\mathbf{p} \in \mathbb{P}^2$, we have:

$$\Gamma(\nu(\mathbf{p})) = \lambda f(\mathbf{p}).$$

This yields $\binom{d+2}{d} - 1$ linear equations on Γ.

Hence a similar statement to that in Proposition 8 can be made:

PROPOSITION 9. *The reconstruction in* $\mathcal{G}(1,3)$ *can be done linearly using at least* $k \geq \frac{1}{6}\frac{d^3+8d^2+23d+28}{d+3}$ *views.*

As in the case of reconstruction in the dual space, it is not necessary to explicitly compute f. It is enough to pick points on the image curve. Each point yields a linear equation on Γ: $\Gamma(\nu(\mathbf{p})) = 0$. However for each view, one cannot extract more than $\frac{1}{2}d^2 + \frac{3}{2}d$ independant linear equations.

5. Applications to dynamic configurations of points

In this section we show that the reconstructions in $\mathcal{G}(1,3)$ can be applied to trajectory recovery. Consider a point moving along a smooth trajectory. The motion is assumed to be well approximated by a low degree irreducible algebraic curve. This requirement is in fact very natural and has a widely broad validity in practice. Now we proceed to show how trajectory recovery can be achieved.

A set of cameras $\mathbf{M}_i, i = 1, ..., m$ which are either static or moving is viewing at a set of points $\mathbf{P}_j, j = 1, ..., n$ either static or moving. The cameras are independant and in particular they are not supposed to be synchronized. Let $\mathbf{p}_{ijk}$ be the projection of the point $\mathbf{P}_j$ onto the camera i at time k.

For a given point $\mathbf{P}_j$, for all i and k, the optical rays, $\mathbf{L}_{ijk} = \widehat{\mathbf{M}}_i\mathbf{p}_{ijk}$ meet the trajectory of $\mathbf{P}_j$. Then according to the geometric entity those rays generate, the motion of $\mathbf{P}_j$ can be recovered. Here we provide a table that gives the correspondence between this entity and the motion of $\mathbf{P}_j$.

Motion of $\mathbf{P}_j$	Geometry entity generated by $\{\mathbf{L}_{ijk}\}$ in $\mathbb{P}^5$
Static point	Plane included in $\mathcal{G}(1,3)$
Point moving on a line	Intersection of $\mathcal{G}(1,3)$ with an hyperplane
Point moving on a conic	Intersection of $\mathcal{G}(1,3)$ with a quadric
...	...
Point moving on a curve of degree d	Intersection of $\mathcal{G}(1,3)$ with an hypersurface of degree d

References

Barnabei, M., Brini, A., and Rota, G.-C. (1985) On the Exterior Calculus of Invariant Theory, *Journal of Algebra* **96**, 120–160.

Faugeras, O.D. (1993) *Three-Dimensional Computer Vision, A geometric approach*, MIT Press.

Faugeras, O.D., and Luong, Q.T. (2001) *The Geometry of Multiple Images*, MIT Press.

Faugeras, O.D., and Papadopoulo, T. (1997) Grassman-Cayley algebra for modeling systems of cameras and the algebraic equations of the manifold of trifocal tensors, Technical Report - INRIA 3225.

Forsyth, F., Recognizing algebraic surfaces from their outlines.

Cross and Zisserman, A. (1998) Quadric Reconstruction from Dual-Space Geometry.

Harris, J. (1992) Algebraic Geometry, a first course, Springer-Verlag.

Hartley, R., and Zisserman, A. (2000) *Multiple View Geometry in computer vision*, Cambridge University Press.

Kahl, F., and Heyden, A. (1998) Using Conic Correspondence in Two Images to Estimate the Epipolar Geometry, inn *Proceedings of the International Conference on Computer Vision.*

Kaminski, J.Y., and Shashua, A. (2000) On Calibration and Reconstruction from Planar Curves, in *Proceedings European Conference on Computer Vision.*

Luong, Q.T., and Vieville, T. (1994) Canonic Representations for the Geometries of Multiple Projective Views, in *Proceedings European Conference on Computer Vision.*

Ma, S.D. and Chen, X. (1994) Quadric Reconstruction from its Occluding Contours, in *Proceedings International Conference of Pattern Recognition.*

Ma, S.D., and Li, L. (1996) Ellipsoid Reconstruction from Three Perspective Views, in *Proceedings International Conference of Pattern Recognition.*

Maybank, S.J., and Faugeras, O.D. (1992) A theory of self-calibration of a moving camera, *International Journal of Computer Vision* **8** (2), 123–151.

Quan, L. (1996) Conic Reconstruction and Correspondence from Two Views, *IEEE Transactions on Pattern Analysis and Machine Intelligence* **18** (2).

Schmid, C., and Zisserman, A. (1998) The Geometry and Matching of Curves in Multiple Views, in *Proceedings European Conference on Computer Vision.*

Shashua, A., and Toelg, S. (1997) The Quadric Reference Surface: Theory and Applications, *International Journal of Computer Vision* **23** (2), 185–198.

CODING THEORY AND ALGEBRAIC CURVES OVER FINITE FIELDS

A Survey and Questions

G. VAN DER GEER
University of Amsterdam, The Netherlands

Abstract. We give a survey on recent developments in the field and list a number of questions.

Key words: code, curve, finite field, number of rational points.

Mathematics Subject Classification (2000): 11G, 94B.

Introduction

In the last two decades two remarkable and unexpected applications of the theory of algebraic curves emerged. The first was the discovery by Goppa in 1981 that curves defined over finite fields can be used for constructing impressingly good error-correcting codes. The second was the idea that Riemann surfaces can be used to describe the life-lines of particles in physics (strings).

Both discoveries, though quite different in depth and significance, have had tremendous consequences inside pure mathematics. The discovery of Goppa stimulated renewed interest for the question how many points a curve of fixed genus over a finite field of given cardinality can have. Witten's work on strings led to an intensive study of the moduli spaces of Riemann surfaces (or algebraic curves) and gave rise to the formulation of the spectacular Witten conjecture on the intersection numbers of the tautological classes on the moduli space and the not less spectacular proof by Kontsevich of this conjecture.

Both these applications outside mathematics raised new questions, and created a new intuition about curves. It is interesting to note a parallell between the work of Goppa and that of Deligne-Lusztig on the representation theory of finite Lie groups which is just slightly older. Deligne and Lusztig showed that irreducible representations of finite Lie groups can be found in a natural way in the cohomology of certain degeneracy loci inside Grassmannians, namely the Deligne-Lusztig varieties, cf. (Deligne and Lusztig). Goppa showed that good codes can be found using the cohomology (linear systems) of algebraic curves.

C. Ciliberto et al. (eds.),
Applications of Algebraic Geometry to Coding Theory, Physics and Computation, 139–159.

With hindsight it seems remarkable that many mathematicians were surprised by the fact that curves–in some sense the simplest non-linear algebraic objects–could be applied to fields like information transmission and physics. And more recently, a new application has been added: the use of curves over finite fields in cryptography. Is it that mathematicians underestimate as a rule the applicability of their trade? For me it seems a safe bet that many more surprise applications of curves and their moduli are in store.

1. The Hasse-Weil Upper Bound

Although curves over finite fields (or at least the equations defining them) occur already in the work of Gauss, the systematic interest for curves over finite fields dates back to the beginning of the 20th century. Inspired by the analogy between function fields over finite fields and number fields E. Artin introduced in 1924 a zeta function $\zeta_F(s)$ for hyperelliptic function fields $F = \mathbb{F}_q(x,y)$ with q odd and y satisfying an equation $y^2 = f(x)$. He made the observation that on substituting $t = q^{-s}$ in this function he obtained a rational function $Z_F(t)$ of t and that $\zeta_F(s)$ satisfied a functional equation relating $\zeta_F(s)$ and $\zeta_F(1-s)$. He proposed as an analogue of the Riemann hypothesis the conjecture that the zeros of $Z_F(t)$ satisfy $|t| = q^{-1/2}$, cf. (Artin). A little later F.K. Schmidt reconsidered this from a more geometric point of view and wrote the zeta function for a smooth absolutely irreducible projective curve defined over $\mathbb{F}_q$ in the form

$$Z_X(t) = \exp(\sum_{r=1}^{\infty} \#X(\mathbb{F}_{q^r})\frac{t^r}{r}).$$

Moreover, he observed that the Riemann-Roch theorem implies that $Z_X(t)$ is of the form

$$Z_X(t) = \frac{P_X(t)}{(1-t)(1-qt)},$$

where $P_X(t)$ is a polynomial of degree $2g$ in t and that it satisfies a functional equation

$$Z_X(1/qt) = q^{1-g}t^{2-2g}Z_X(t).$$

Around 1932 Hasse became interested in the conjecture posed by Artin and noticed that it implied the upper bound

$$|\#X(\mathbb{F}_q) - (q+1)| \leq 2g\sqrt{q}$$

and he proved it for $g = 1$ by a now deceptively simple argument. He lifted the curve (an elliptic curve) together with the Frobenius endomorphism to characteristic zero. Since the degree of the Frobenius is q one finds a lifted endomorphism ϕ with $\phi\bar{\phi} = q$ from which the bound results, cf. (Hasse). The use of the relation between the curve in characteristic 0 and characteristic p was a bold step. Deuring noticed then that in order to extend this result to higher genera one needed a theory of correspondences in arbitrary characteristic. This theory was developed by Weil. He showed that $P_X(t)$ is a polynomial with integral coefficients of the form

$$P_X(t) = \prod_{i=1}^{2g}(1-\alpha_i t),$$

where the α_i are algebraic integers with $|\alpha_i| = \sqrt{q}$. This implies

$$\#X(\mathbb{F}_{q^r}) = q^r + 1 - \sum_{i=1}^{2g} \alpha_i^r$$

and it also implies the famous Hasse-Weil bound

$$\#X(\mathbb{F}_q) \leq q + 1 + 2g\sqrt{q}.$$

After this breakthrough interest for the number of rational points on curves over finite fields seemed to disappear as quickly as it had appeared two decades earlier, only flaring up when Stepanov came forward with an elementary proof of the Hasse-Weil bound using only Riemann-Roch, cf. (Bombieri).

2. Coding Theory

One can view coding theory as a very early spin-off of the development of the computer. In fact, it was born out of the frustration over the frequent stops by which early computers were plagued. We refer to (Thompson) for the history.

Error correcting codes are now used ubiquitously in digital communication. In this section we just give the definitions and we refer to (MacWilliams and N. Sloane) or to (van Lint) for a more extensive treatment.

Let $\mathbb{F}_q$ be a finite field of cardinality q and fix a positive integer n which will be called the *word length*. We consider the $\mathbb{F}_q$-vector space $\mathbb{F}_q^n$. The set $\mathbb{F}_q$ is called the alphabet and the elements of $\mathbb{F}_q^n$ are called *words*.

On $\mathbb{F}_q^n$ we have a distance function, the Hamming distance, named after Hamming, one of the early pioneers of coding theory

$$d(x,y) = \#\{i : 1 \leq i \leq n, x_i \neq y_i\}.$$

The *weight* of a word $x \in \mathbb{F}_q^n$ is defined as

$$w(x) = d(x,0).$$

A *code* is by definition a subset of $\mathbb{F}_q^n$ and its elements are called code words. We shall restrict ourselves to *linear* codes, which are by definition linear subspaces of $\mathbb{F}_q^n$. To a linear code $C \subset \mathbb{F}_q^n$ we can associate a triple (n,k,d) which gives the word length n, the dimension $k = \dim_{\mathbb{F}_q}(C)$ and the *minimum distance* d, that is, the minimum weight of a non-zero word in C.

The quotients $R = k/n$ and $\delta = d/n$ are called the *transmission rate* and the *relative distance* of the code C and are a rough indication of the quality of the code.

Depending on the application one has in mind one tries to find an optimum for these invariants and coding theory can be seen as a theory how to optimize these invariants under boundary conditions.

For a linear code C one defines the dual code $C^\perp$ by

$$C^\perp = \{x \in \mathbb{F}_q^n : \langle x,y \rangle = 0 \quad \text{for all } y \in C\},$$

where $\langle x,y \rangle = \sum x_i y_i$ is the usual pairing on $\mathbb{F}_q^n$.

To a code we can associate a polynomial

$$\sum_{c \in C} X^{w(c)} \quad \in \mathbb{Z}[X],$$

called the *weight distribution* of C. It tells us which weights occur in C and which are the frequencies. The weight distribution of C and the dual $C^\perp$ are determined by each other via the MacWilliams identities, a discrete analogue of the Poisson summation formula.

It is a central and difficult problem in coding theory to determine the weight distributions of many classical codes.

We can also associate a weight to every subcode (=subspace) of a code. If $D \subset C$ is an r-dimensional linear subspace we define the weight of D by

$$w(D) = \frac{1}{q^r - q^{r-1}} \sum_{c \in D} w(c),$$

or equivalently, $w(D)$ is the number of coordinate places for which D contains a word with a non-vanishing coordinate at that place.

The minimum distance d of C then also admits generalizations $d_r(C)$ defined by

$$d_r(C) = \min\{w(D) : D \subset C, \dim(D) = r\}.$$

We have $d_1 = d$. The set $\{d_r(C) : 1 \leq r \leq n\}$ is called the weight hierarchy and these numbers serve in practice as a measure for the reliability of a transmission channel where hostile eavesdroppers might capture part of the transmitted data, cf. (Wei).

Already long before Goppa's work it was clear that coding theory is linked to algebraic geometry. To illustrate this we give the example of the *Reed-Muller codes*. If P_s is the $\mathbb{F}_q$-vectorspace

$$P_s = \{f \in \mathbb{F}_q[X_1, \ldots, X_r] : \deg(f) \leq s\}$$

of polynomials of total degree $\leq s$ we can evaluate the elements of P_s at the points of affine r-space over $\mathbb{F}_q$ via

$$\beta : P_s \longrightarrow \mathbb{F}_q^n, \qquad f \mapsto (f(P)_{P \in \mathbb{F}_q^n})$$

with $n = q^r$.

By definition the q-ary Reed-Muller code $R_q(s,r)$ is the image of this map. If $\beta(f)$ is a codeword, then the weight $w(\beta(f))$ is equal to $n - \#H_f(\mathbb{F}_q)$, where H_f is the hypersurface in affine r-space over $\mathbb{F}_q$ defined by $f = 0$. We see that determining the weight distribution of this code comes down to determining the distribution of the number of $\mathbb{F}_q$-rational points in the family of hypersurfaces of degree $\leq s$ in affine r-space over $\mathbb{F}_q$. Elementary questions in coding theory thus translate into difficult questions in algebraic geometry over finite fields. Strangely enough, it seems that a stimulus from outside was needed to bring such questions into focus.

3. Goppa Codes

The invariants of codes satisfy certain inequalities. Early coding theorists were interested in constructing asymptotically good codes, i.e. sequences of codes C_i such that the relative distance d_i/n_i and transmission rate k_i/n_i converge to a δ and R with $\delta R > 0$.

Gilbert and Varshamov proved by an averaging argument that for $0 < \delta \leq (q-1)/q$ there exist asymptotically good sequences of codes with $R \geq 1 - H_q(\delta)$, where we view R as a function of δ and $H_q(\delta)$ is the entropy function defined by

$$\begin{cases} 0 & \delta = 0, \\ \delta \log_q(q-1) - \delta \log_q(\delta) - (1-\delta)\log_q(1-\delta) & 0 < \delta \leq (q-1)/q. \end{cases}$$

The bound $R \geq 1 - H_q(\delta)$ is called the Gilbert-Varshamov bound. For a long time coding theorists were not able to construct (explicit) sequences of codes with limit point on or above the Gilbert-Varshamov bound.

Goppa succeeded in 1973 in constructing sequences of codes (usually called classical Goppa codes) attaining the Gilbert-Varshamov bound by taking codes defined by parity check matrices whose entries were the values of rational functions (on the affine line). He arrived at his fundamental discovery of the geometric Goppa codes in 1981 when he tried to generalize the classical Goppa codes by taking the values of rational functions on algebraic curves, cf. (Goppa, 1977 and 1981).

A *(geometric) Goppa code* is defined as follows. Let X be a smooth absolutely irreducible projective curve of genus g defined over $\mathbb{F}_q$ and let $P = \{P_1, \ldots, P_n\} \subset X(\mathbb{F}_q)$ be a set of n distinct $\mathbb{F}_q$-rational points. If $L \subset \mathbb{F}_q(X)$ denotes a finite-dimensional $\mathbb{F}_q$-vector space of functions, no element of which has a pole in one of the points P_i of P, then one can evaluate each $f \in L$ at the P_i and one obtains a map

$$\alpha : L \longrightarrow \mathbb{F}_q^n, \qquad f \mapsto (f(P_i)_{i=1}^n).$$

The image of L under α is called a Goppa code. If $L = L(D) = H^0(X, O(D))$ for a $\mathbb{F}_q$-rational divisor D on X then we write $C(D,P)$ for $\alpha(L(D))$.

Algebraic geometry immediately yields bounds for these codes. For example, one has:

PROPOSITION 1. *Let D be a divisor on X defined over $\mathbb{F}_q$ with $g \leq \deg(D) \leq n$ and* $\mathrm{supp}(D) \cap P = \emptyset$. *Then the dimension k of $C(D,P)$ satisfies*

$$k \geq \deg(D) + 1 - g,$$

with equality if $\deg(D) \geq 2g-1$. *Moreover, the minimum distance d of $C(D,P)$ satisfies $d \geq n - \deg(D)$.*

In particular, for divisors as in the proposition we find

$$k + d \geq n + 1 - g,$$

while Singleton's bound, an elementary result in coding theory, gives $k + d \leq n + 1$. The proposition shows that if we fix the relative distance d/n then the transmission rate improves if n/g increases. So one should use curves with as many points as possible.

If X_ℓ is now a sequence of curves defined over $\mathbb{F}_q$ such that the genus $g_\ell = g(X_\ell)$ tends to ∞ and such that

$$\lim_{\ell \to \infty} \#X_\ell(\mathbb{F}_q)/g_\ell = \gamma > 0,$$

then we can construct a sequence of Goppa codes $C(D_\ell, X_\ell(\mathbb{F}_q))$, with D_ℓ a suitably chosen divisor of degree $\deg(D_\ell) = [\#X_\ell(\mathbb{F}_q))(1-\delta)]$ on X_ℓ, such that

$$R_\ell + \delta_\ell \geq 1 + (1 - g_\ell)/\#X_\ell(\mathbb{F}_q)$$

and this tends to $1 - 1/\gamma$. By using modular curves (classical or Shimura curves) and the rational points on them provided by the 'supersingular' points one obtains very good codes as Tsfasman, Vladuts and Zink showed with the following result for the case that q is a square (cf. also (Ihara, 1985)).

THEOREM 1. (Tsfasman, Vladuts and Zink) *Let q be a square. Then there exists a sequence of Goppa codes over $\mathbb{F}_q$ with limit transmission rate R and limit relative distance δ with $R + \delta \geq 1 - 1/(\sqrt{q} - 1)$.*

For $q \geq 49$ the line $R + \delta = 1 - 1/(\sqrt{q} - 1)$ rises above the Gilbert-Varshamov bound. This theorem showed in a spectacular way that Goppa codes can be a powerful tool in coding theory. A side effect of this was renewed interest in the topic of curves over finite fields.

4. Families of Curves and Codes

Many codes constructed in the early days of coding theory are trace codes of the form $\mathrm{Tr}_{q^s/q}(C')$, where C' is a code obtained by evaluating polynomials or rational functions on the points of the affine line over $\mathbb{F}_{q^s}$. Here $\mathrm{Tr}_{q^s/q}$ denotes the usual trace map from $\mathbb{F}_{q^s}$ to $\mathbb{F}_q$ and it is applied to every coordinate of a code word of C'.

An easy example are the classical dual Melas codes $M(q)^{\perp}$ of length $q-1$ over a prime field $\mathbb{F}_p$. The words in $M(q)^{\perp}$ are of the form

$$c_{a,b} = \mathrm{Tr}_{q/p}((ax+b/x)_{x\in\mathbb{F}_q^*}),$$

where a, b run through $\mathbb{F}_q$. The weight of the word $c_{a,b}$ can be expressed as

$$w(c_{a,b}) = q-1-\frac{1}{p}\#(X_{a,b}(\mathbb{F}_q)-2), \tag{1}$$

where $X_{a,b}$ is the smooth projective curve defined over $\mathbb{F}_q$ by

$$y^p - y = ax + b/x.$$

Indeed, we can solve an equation $v^p - v = u$ in $\mathbb{F}_q$ if and only if $\mathrm{Tr}_{q/p}(u) = 0$. So knowing the weight distribution in $M(q)^{\perp}$ is equivalent to knowing the distribution of the number of points in the family of curves $X_{a,b}$, $a, b \in \mathbb{F}_q$.

This can be generalized. Consider a Goppa code $C' = C(D,P)$ defined over $\mathbb{F}_{q^s}$ associated to a triple (X,D,P) consisting of a curve X, a divisor D and a set of rational points P as in the preceding section. For P we usually take all or almost all of $X(\mathbb{F}_q)$. The words of the trace code

$$C = \mathrm{Tr}_{q^s/q}(C')$$

are given as

$$c_f = \mathrm{Tr}_{q^s/q}(f(P_i)_{i=1}^n) \qquad \text{for } f \in L(D).$$

Then we also have a simple relation between the weight $w(c_f)$ and the number of $\mathbb{F}_{q^s}$-rational points on the curve X_f defined by

$$y^q - y = f$$

similar to (1). Therefore, the trace codes of Goppa codes lead immediately to families of curves and the weight distribution of the trace code is intimately related to the distribution of the number of points in the family X_f with $f \in L(D)$.

In this way coding theory leads to the problem of determining the behavior of the number of points, or more generally of the behavior of the zeta function in families of curves. It also points to certain families of curves with interesting properties.

5. Reed-Muller Codes and Families of Curves

The relations between codes and curves are not limited to those sketched above. We give here another one. It uses the phenomenon that a hypersurface in affine r-space over $\mathbb{F}_q$ can be translated into a curve defined over the extension $\mathbb{F}_{q^r}$.

We illustrate this with the Reed-Muller codes that we encountered before. In order to explain the connection we view the field $\mathbb{F}_{q=p^m}$ as an m-dimensional vector space over $\mathbb{F}_p$ with coordinates $X_1,\ldots,X_m$. We may replace these coordinates by other linearly independent linear forms on this $\mathbb{F}_p$-vector space.

Using the trace map $\mathrm{Tr}_{q/p}:\mathbb{F}_q\to\mathbb{F}_p$ given by

$$x\mapsto x+x^p+\ldots+x^{p^{m-1}}$$

we can make such linear forms: choose m elements $a_1,\ldots,a_m\in\mathbb{F}_q$ such that the linear forms

$$\mathrm{Tr}_{q/p}(a_ix):\mathbb{F}_q\longrightarrow\mathbb{F}_p,\qquad x\mapsto\mathrm{Tr}_{q/p}(a_ix)$$

are linearly independent over $\mathbb{F}_p$.

Now we can substitute the expression $\mathrm{Tr}_{q/p}(a_ix)$ for X_i in any polynomial $f\in\mathbb{F}_q[X_1,\ldots,X_m]$. Repeated application of the identity

$$\mathrm{Tr}_{q/p}(ax)\mathrm{Tr}_{q/p}(bx)=\mathrm{Tr}_{q/p}(\mathrm{Tr}_{q/p}(ax)bx)=\mathrm{Tr}_{q/p}(\sum_{j=0}^{m-1}a^{p^j}bx^{p^j+1})$$

transforms $f\in\mathbb{F}_q[X_1,\ldots,X_m]$ via $f(\mathrm{Tr}_{q/p}(ax),\ldots,\mathrm{Tr}_{q/p}(ax))$ into an expression $\mathrm{Tr}_{q/p}F_f$ with F_f a polynomial in one variable in $\mathbb{F}_q[x]$.

If $f\in\mathbb{F}_q[X_1,\ldots,X_m]$ has degree $\leq s$ then F_f is a polynomial in which the terms have an exponent whose q-adic expansion has weight $\leq s$, i.e., of the form $q^{i_1}+q^{i_2}+\ldots+q^{i_r}$ with $i_1\geq i_2>\ldots\geq i_r$ and $r\leq s$. We may even assume that $i_r=1$.

If we now carry this out for the Reed-Muller code $R_q(s,m)$ we associate to each code word $\beta(f)$ an expression

$$\mathrm{Tr}_{q/p}(F_f)$$

and we have the following relation for the weight

$$w(\beta(f))=q^m-\#H_f(\mathbb{F}_q)=q^m-\frac{1}{p}\#X_f(\mathbb{F}_q).$$

Here H_f is the hypersurface defined by $f = 0$ and X_f is the curve defined over $\mathbb{F}_q$ by the equation

$$y^p - y = F_f.$$

For example, if we carry this procedure out for the second order Reed-Muller codes $R_q(2,m)$ we are led to the family of curves

$$y^p - y = xR(x),$$

where R runs through the linearized polynomials of the form

$$R = \sum_{i=0}^{h} a_i x^{p^i+1} \qquad a_i \in \mathbb{F}_{q=p^m}$$

and we can take $h \leq m$. It turns out that these curves are all supersingular, again illustrating the fact that coding theory points to interesting families of curves.

6. The Function $A(q)$

After Goppa's discovery of the use of algebraic curves for coding theory had renewed the interest in curves over finite fields, Ihara presented a simple but elegant argument that improves the Hasse-Weil bound if the genus satisfies $g > \sqrt{q}(\sqrt{q}-1)/2$. Starting from the formula

$$\#X(\mathbb{F}_{q^r}) = q^r + 1 - \sum_{i=1}^{2g} \alpha_i^r$$

for the number of $\mathbb{F}_{q^r}$-rational points on a curve of genus g over $\mathbb{F}_q$ he compares $\#X(\mathbb{F}_q)$ and $\#X(\mathbb{F}_{q^2})$ and by using the Cauchy-Schwarz inequality he gets

$$\#X(\mathbb{F}_q) \leq q + 1 + [(\sqrt{(8q+1)g^2 + 4(q^2-q)g} - g)/2],$$

cf. (Ihara, 1982), and this is better than the Hasse-Weil bound for $g > \sqrt{q}(\sqrt{q} - 1)/2$. This argument was exploited systematically by Drinfeld and Vladuts, who not only used the quadratic extension $\mathbb{F}_{q^2}$ but all extension fields $\mathbb{F}_{q^r}$ and they obtained the following asymptotic result, cf. (Vladuts and Drinfeld).

We put

$$N_q(g) := \text{maximum value of } \#X(\mathbb{F}_q),$$

where X runs over all curves X of genus g defined over $\mathbb{F}_q$. Moreover, we set

$$A(q) := \limsup_{g \to \infty} N_q(g)/g.$$

The result of Drinfeld and Vladuts is:

$$A(q) \leq \sqrt{q} - 1.$$

(Compare this to what the Hasse-Weil bound implies: $A(q) \leq 2\sqrt{q}$.) If q is a square then this bound is best possible as follows from a result of Ihara (which was obtained independently by Tsfasman, Vladuts and Zink). They used modular curves to construct a sequence of curves for which $\#X(\mathbb{F}_q)/g(X)$ converges to $\sqrt{q}-1$. We refer to (Ihara, 1985) for the story.

THEOREM 2. *For q a square we have $A(q) = \sqrt{q}-1$.*

For q not a square the situation is yet unclear, though there are partial results. In (Zink) curves on certain degenerate Shimura surfaces has been used to get a bound for $q = p^3$ with p prime:

$$A(p^3) \geq \frac{2(p^2-1)}{p+2}.$$

Serre proved by employing towers of Hilbert class fields that

$$A(q) \geq c \log q,$$

where $c > 0$ is an absolute effective constant. In (Niederreiter and Xing, 1998c) it was proved for q odd and $m \geq 3$ that

$$A(q^m) \geq \frac{2q}{[2\sqrt{2q+1}]+1}.$$

which improves the Serre lower bound for many q, and they have also a result for even q. Temkine obtains the lower bound

$$A(q^n) \geq cn^2 \log q \frac{\log q}{\log n + \log q}$$

for an effective absolute constant $c > 0$, cf. (Temkine). This reproduces Serre's lower bound for $n = 1$. For specific values of q some lower bounds have been given: $A(2) \geq 81/317$, cf. (Niederreiter and Xing, 1998c), $A(3) \geq 12/25$, cf. (Hajir and Maire) and $A(4) \geq 8/11$ cf. (Temkine). We refer to the paper of (Hajir and Maire) for recent asymptotic results using class fields.

In (Garcia and Stichtenoth, 1996) it was showed that there exists a completely explicit tower of curves X_ℓ defined over $\mathbb{F}_{q^2}$ for which $g(X_\ell) \to \infty$ and the ratio $\#X_\ell(\mathbb{F}_{q^2})/g(X_\ell)$ tends to $q-1$. The tower is defined as follows. One starts with X_1 the projective line with (affine) coordinate x_1 and one defines X_ℓ recursively via the Artin-Schreier extension

$$y_{\ell+1}^q + y_{\ell+1} = x_\ell^{q+1},$$

with $x_\ell = y_\ell / x_{\ell-1}$. Later, in (Elkies, 1998 and 2001), it was showed that this is a tower consisting of modular curves.

An example of an explicit tower over $\mathbb{F}_8$ which reaches the same limit $3/2 = 2(p^2-1)/(p+2)$ as the sequence of Zink is given in (van der Geer and van der Vlugt, 2001).

For a recent survey on the work on explicit asymptotic towers we refer to the paper (Stichtenoth).

7. The Function $N_q(g)$

Nature provides us with a constant $N_q(g)$ for every pair (q,g), namely the maximum value of the number of $\mathbb{F}_q$-rational points on a curve of genus g over $\mathbb{F}_q$. The Hasse-Weil bound gives a first upper bound for $N_q(g)$:

$$N_q(g) \leq q+1+[2g\sqrt{q}].$$

Serre improved this upper bound in (Serre, 1983) slightly to

$$N_q(g) \leq q+1+g[2\sqrt{q}]$$

by applying some arithmetic to the Frobenius roots α_i. As we saw, the bound by Ihara is in general better for g large compared to q. Serre showed that the method of 'formules explicites' from number theory can be applied with success to the case of curves over finite fields. The result is as follows. Take a trigonometric polynomial

$$f(\theta) = 1 + 2\sum_{n\geq 1} u_n \cos n\theta$$

which is even, has real coefficients $u_n \in \mathbb{R}_{\geq 0}$ and satisfies $f(\theta) \geq 0$ for all $\theta \in \mathbb{R}$. Set $\psi = \sum_{n\geq 1} u_n t^n$. Then the number of $\mathbb{F}_q$-rational points on a curve of genus g over $\mathbb{F}_q$ is bounded by

$$\#X(\mathbb{F}_q) \leq a_f g + b_f$$

with $a_f = 1/\psi(1/\sqrt{q})$ and $b_f = 1+\psi(\sqrt{q})/\psi(1/\sqrt{q})$.

One then faces the problem to find the optimal choice for the function f. This problem was solved by Oesterlé and the resulting bound is called the Oesterlé bound. We refer to (Schoof) for an exposition of this. The Oesterlé bound is better than the Hasse-Weil bound for $g > \sqrt{q}(\sqrt{q}-1)$.

With an eye to practical applications in coding theory or cryptography or just to satisfy our mathematical curiosity we may ask for the actual value of $N_q(g)$, especially for low values of q and g.

The Hasse-Weil bound and its improvements (Ihara, Serre, Oesterlé) imply an upper bound for $N_q(g)$; however, we have no idea in general how good this bound is. What we have in fact is an interval $[a,b]$ in which $N_q(g)$ has to lie. Here b is the best upper bound we know and a is the largest number of points we know to occur for a curve of genus g over $\mathbb{F}_q$. The interval maybe quite large. See (van der Geer and van der Vlugt, 1998b) for tables of these intervals.

Serre started the study of the actual value of $N_q(g)$ for a number of small values of g and q. For example, he determined $N_q(2)$ for all q and $N_q(g)$ for certain pairs (q,g) with $q=2$. Another result of Serre says that for a curve of genus ≥ 3 with

$$\#X(\mathbb{F}_q) < q+1+g[2\sqrt{q}]$$

one has

$$\#X(\mathbb{F}_q) \leq q-1+g[2\sqrt{q}].$$

For some small genera certain values of $N_q(g)$ can be eliminated by analyzing the possibilities of the zeta function and eliminate certain zeta functions because these imply a decomposition of the Jacobian as a product of principally polarized abelian varieties, which contradicts the irreducibility of the theta divisor of the curve, cf. (Serre, 1985; Lauter, 1999 and 2000).

Sometimes one can rule out that $N_q(g)$ equals the Serre bound by a specific argument, like Galois descent. This works e.g. for $(q=27, g=3)$ and $(q=8, g=4)$, cf. (Serre, 1985; Lauter, 1999). For example, one knows that $N_{27}(3)=56$.

The tables for $N_q(g)$ (see (van der Geer and van der Vlugt, 1998c)) suggests the following question.

QUESTION 1. *Fix q. Is the function $N_q(g)$ a non decreasing function of g?*

The results on $A(q)$ make use of towers with genera that can be rather sparse. One can ask what happens if one uses all sufficiently large genera.

In (Kresch, Wetherell and Zieve) it is shown that $N_q(g)$ goes to ∞ with g:

THEOREM 3. (Kresch, Wetherell and Zieve) *For fixed q we have* $\lim_{g\to\infty} N_q(g) = \infty$.

The following question suggests itself:

QUESTION 2. *What is* $\liminf_{g\to\infty} N_q(g)/g$?

In (Elkies, Kresch, Poonen, Wetherell and Zieve) it was proved that if q is a square, then $\liminf_{g\to\infty} N_q(g)/g \geq (\sqrt{q}-1)/3$.

The following results gives restrictions for the number $N_q(g)$ for all g:

THEOREM 4. (Kresh, Wetherell and Zieve) *For fixed q there are constants e_q and f_q depending on q with $0 < e_q < f_q$ such that for every $g > 0$ one has $e_q g < N_q(g) < f_q g$.*

Instead of studying the maximum value of $\#X(\mathbb{F}_q)$ for all curves of genus g one could restrict to curves which belong to certain strata in the moduli space. For example, hyperelliptic curves have at most $\#X(\mathbb{F}_q) \leq 2(q+1)$ rational points. One could try to generalize this to bounds for the maximum number of points on a curve of genus g with given gonality. Recall that the gonality vector $\gamma(X) = (\gamma_1, \gamma_2, \ldots)$ of a curve X over an algebraically closed field k is given by

$$\gamma_r(X) = \min\{d : 1 \leq d \leq g-1, \text{ there exists a } g_d^r \text{ on } X\},$$

where g_d^r stands for a linear system of degree d and dimension r. It is well-known that a curve of genus g admits a map of degree $\leq [(g+3)/2]$ to the projective line, so the (geometric) gonality $\gamma_1(X)$ is bounded by $\leq [(g+3)/2]$, but we have to consider the gonality over the ground field $\mathbb{F}_q$. For example, a non-hyperelliptic curve of genus 4 over $\mathbb{F}_q$ has a map of degree 3 to $\mathbb{P}^1$ over $\mathbb{F}_{q^2}$, but not necessarily over $\mathbb{F}_q$, depending on whether the quadric containing the canonical curve is split over over $\mathbb{F}_q$ or not.

QUESTION 3. *What is the maximum number of rational points on a curve of genus g and gonality γ defined over $\mathbb{F}_q$?*

The important paper (Stöhr and Voloch) presents an approach to this question, namely an upper bound for the maximum number of points on a curve over $\mathbb{F}_q$ which does not only depend on the genus g, but also on a given linear system defined over $\mathbb{F}_q$. The approach of Stöhr and Voloch is a sort of infinitesimal approach which counts points on a curve, embedded in projective space with such a linear system, such that the Frobenius image of a point lies in the osculating hyperplane of the curve at that point. As a special case this provides a new proof of the Hasse-Weil bound. This approach deserves further exploitation.

Quite a lot of people have tried to construct curves with many points in order to probe the upper bounds on $N_q(g)$. A variety of methods have been used for this, like methods from class field theory (Serre, Schoof, Lauter, Niederreiter, Xing and Auer), methods from Drinfeld modules (Niederreiter and Xing), fibre products of Artin-Schreier curves (van der Geer and van der Vlugt, Shabat), Kummer curves (van der Geer and van der Vlugt, Garcia and Quoos) and various other methods. We refer to (van der Geer and van der Vlugt, 1997 and 1998b) for a summary of the methods and results.

8. Maximal Curves

Since a curve X of genus g defined over a finite field $\mathbb{F}_q$ satisfies the Hasse-Weil bound

$$\#X(\mathbb{F}_q) \leq q+1+2g\sqrt{q}$$

for its number of $\mathbb{F}_q$-rational points, it is natural to ask for which curves this bound is attained. Curves X over $\mathbb{F}_q$ with $\#X(\mathbb{F}_q) = q+1+2g\sqrt{q}$ are called *maximal curves*. Of course, for such a curve the cardinality q of $\mathbb{F}_q$ has to be a square.

An example of a curve defined over $\mathbb{F}_{q^2}$ which is maximal is the so-called Hermitian curve defined by the equation

$$x^{q+1} + y^{q+1} + z^{q+1} = 0 \tag{2}$$

in the projective plane. This is a curve of genus $g = q(q-1)/2$ with q^3+1 points rational over $\mathbb{F}_{q^2}$. Note that we can write (2) as $x\bar{x}+y\bar{y}+z\bar{z} = 0$, where $\bar{x} = F(x) = x^q$, with F the Frobenius morphism, which explains the name.

If a curve C is maximal then any curve D dominated by it is so too, because its Jacobian is an abelian subvariety of $\mathrm{Jac}(C)$, so the eigenvalues of Frobenius are also $-\sqrt{q}$.

It follows from the fact that Ihara's bound is better than the Hasse-Weil bound for a curve over $\mathbb{F}_{q^2}$ if $g > q(q-1)/2$ that for a maximal curve over $\mathbb{F}_{q^2}$ we must have

$$g \leq q(q-1)/2.$$

In (Rück and Stichtenoth) it was proved that a maximal curve of genus $q(q-1)/2$ over $\mathbb{F}_{q^2}$ is isomorphic to the Hermitian curve (2).

This characterization of maximal curves was extended to the interval

$$[(q-1)^2/4] \leq g \leq q(q-1)/2$$

by Fuhrmann and Torres and again further to the interval

$$[(q^2-q+4)/6] \leq g \leq [(q-1)^2/4]$$

by Korchmaros-Torres. Collecting their results we obtain the following theorem.

THEOREM 5. (Rück and Stichtenoth; Fuhrmann and Torres; Korchmaros and Torres) *If X is a maximal curve defined over $\mathbb{F}_{q^2}$ then either*

i) $g = q(q-1)/2$ and X is $\mathbb{F}_{q^2}$-isomorphic to the hermitian curve (2), or

ii) $g = [(q-1)^2/4]$ and if q is odd X is $\mathbb{F}_{q^2}$-isomorphic to the curve defined by

$$y^q + y = x^{(q+1)/2}, \quad \text{or}$$

iii) $g \leq [(q^2-q+4)/6]$.

For $g = [(q^2-q+4)/6]$ one has examples of maximal curves, so that $g = [(q^2-q+4)/6]$ is the third largest genus of a maximal curve over $\mathbb{F}_{q^2}$. For example, for $q \equiv 2 \pmod 3$ one has the maximal curve given by

$$x^{(q+1)/3} + x^{2(q+1)/3} + y^{q+1} = 0.$$

It was conjectured by Stichtenoth that every maximal curve is dominated by the Hermitian curve (2). This raises the following problem.

PROBLEM 1. *Determine all curves dominated by the hermitian curve given by equation (2).*

The Hermitian curve and also the curves defined by $y^q + y = x^{(q+1)/2}$ are extremal in more than one sense. To explain this we need some terminology.

We introduce two numbers, the *Castelnuovo number* $c_0(d,r)$ and the *Halphen number* $c_1(d,r)$ both depending on two integral parameters d and r. The Castelnuovo number $c_0(d,r)$ is defined by

$$c_0(d,r) = \frac{d-1-\varepsilon_0}{2(r-1)}(d-r+\varepsilon_0),$$

where the integer ε_0 is uniquely determined by

$$0 \leq \varepsilon_0 \leq r-2 \quad \text{and} \quad \varepsilon_0 \equiv d-1 \,(\operatorname{mod}(r-1)).$$

Castelnuovo proved that the genus g of any curve of degree d in projective space $\mathbb{P}^r$ satisfies

$$g \leq c_0(d,r).$$

The Halphen number $c_1(d,r)$ is defined as

$$c_1(d,r) = \frac{d-1-\varepsilon_1}{2r}(d-r+\varepsilon_1+1) + \begin{cases} 0 & \text{if } \varepsilon_1 \leq r-1, \\ 1 & \text{if } \varepsilon_1 = r-1, \end{cases}$$

where ε_1 is the unique integer with $0 \leq \varepsilon_1 \leq r-1$ and $\varepsilon_1 \equiv d-1 \,(\bmod\, r)$. Halphen proved for $r=3$ and Eisenbud and Harris for general r that if X is a curve in $\mathbb{P}^r$ of degree d and genus g and $d \geq 2^{r+1}$ for $r \geq 8$ (resp. $d \geq 36r$ if $r \leq 6$ and $d \geq 288$ if $r=7$) then X lies on a surface of degree $\leq r-1$ provided $g > c_1(d,r)$. We refer to Thm (3.15) in (Harris).

As one may expect, curves in projective space $\mathbb{P}^r$ which attain the Castelnuovo bound possess special features and these make them candidates for having many rational points. Korchmaros and Torres propose in their paper the following conjecture.

CONJECTURE 8.1. (Korchmaros and Torres) *There is no maximal curve over* $\mathbb{F}_{q^2}$ *whose genus satisfies*

$$c_1(q+1,r) < g < c_0(q+1,r),$$

where $c_0(q+1,r)$ *and* $c_1(q+1,r)$ *are the Castelnuovo and Halphen number defined above.*

How does one connect the property that a curve is maximal with the geometry of the curve? If X is a maximal curve defined over $\mathbb{F}_{q^2}$, then the eigenvalues of Frobenius acting on $H^1_{\text{et}}(X,\mathbb{Q}_\ell)$ or the Tate module $T_\ell(\text{Jac}(X))$ of the Jacobian are equal to $-q$. This implies that if F is the (relative) Frobenius map on the Jacobian of X (raising the coordinates of a point to the q-th power) we have

$$q+F \qquad \text{annihilates the class group } \text{Jac}(X).$$

That is, if $\overline{\mathbb{F}}$ denotes an algebraic closure of $\mathbb{F}_{q^2}$ then for any $\overline{\mathbb{F}}$-rational point P of X the divisor $(q+F)(P)$ is linearly equivalent to $(q+1)Q$, where Q is a fixed $\mathbb{F}_{q^2}$-rational point of X. So the curve X carries a base-point free linear series

$$|(q+1)Q|$$

which does not depend on the choice of our point $Q \in X(\mathbb{F}_{q^2})$. Korchmaros and Torres prove that this linear series embeds X into projective space as a curve of degree $q+1$. Applying Castelnuovo's theorem yields $g \leq q(q-1)/2$ which we already knew from Ihara's result. But it also shows that if $\dim|(q+1)Q| \geq 3$ then $g \leq (q-1)^2/4$. They show that every maximal curve over $\mathbb{F}_{q^2}$ of genus $g > [(q^2-q+4)/6]$ can be embedded in $\mathbb{P}^3$ such that it has degree $q+1$ and lies on a $\mathbb{F}_{q^2}$-rational quadratic cone with vertex on the curve. Together with the earlier work of Rück and Stichtenoth and Fuhrmann and Torres this yields the theorem of Korchmaros and Torres.

9. Stratifications on Moduli Spaces

Some of the questions in algebraic geometry inspired by coding theory are directly or indirectly related to stratifications on the moduli spaces of principally polarized abelian varieties and on the moduli spaces of curves.

For example, the bounds on the number of rational points on a curve defined over a finite field can sometimes be used to show that a certain principally polarized abelian variety (A,Θ) is not a jacobian variety. Indeed, if the eigenvalues α_i with $i=1,\ldots,2g$ of the action of Frobenius on $H^1_{et}(A,\mathbb{Q}_\ell)$ are such that

$$q+1-\sum_{i=1}^{2g}\alpha_i$$

violates the Oesterlé bound then (A,Θ) cannot be a jacobian. For example, if all eigenvalues are $-\sqrt{q}$ and $g > \sqrt{q}(\sqrt{q}-1)/2$ then (A,Θ) is not a jacobian. Or to give an even simpler example, a hyperelliptic curve can have at most $2q+2$ $\mathbb{F}_q$-rational points, so on the hyperelliptic locus the Hasse-Weil or Oesterlé bound can in general be improved.

The jacobian locus, i.e., the (closure of the) image of the Torelli map

$$t : \mathcal{M}_g \longrightarrow \mathcal{A}_g$$

which associates to a curve its jacobian, is just one of the strata of a stratification on $\mathcal{A}_g$. If (B,Σ) is a principally polarized abelian variety then we say that it has exponent e if there exists a curve of B which generates B and such that B is an abelian subvariety of $\mathrm{Jac}(C)$ via say $i : B \subset \mathrm{Jac}(C)$ and $i^*(\Theta) = e\Sigma$ and such that e is minimal, cf. (Birkenhake and Lange). The jacobian locus is the smallest stratum corresponding to $e=1$, and the locus of Prym varieties is the second stratum. In general the stratum to which an abelian variety belongs indicates how far the abelian variety is from being a jacobian.

One now may ask whether one can find bounds on the trace of Frobenius (acting on $H^1(A,\mathbb{Q}_\ell)$) which generalize the bounds that we found on the jacobian locus. For example, does the trace satisfy a bound for Prym varieties which is stronger than the Hasse-Weil bound? Or are there other properties of the eigenvalues of Frobenius which are somehow restricted by this parameter e?

On the moduli space $\mathcal{A}_g \otimes \mathbb{F}_p$ of principally polarized abelian varieties in characteristic $p > 0$ there are two very natural stratifications.

The first stratification is given by the Newton polygon of the formal group (or of the characteristic polynomial of Frobenius acting on $H^1(A, \mathbb{Q}_\ell)$). Grothendieck proved that the Newton polygon goes up under specialization. In this way one gets a stratification in which the largest stratum is the whole moduli space and the smallest is the supersingular locus (corresponding to slope $1/2$ Newton polygon). The dimensions of these strata are $g(g+1)/2$ and $[g^2/4]$ respectively, cf. (Li and Oort). Intermediate strata are for example the strata where the p-rank is $\leq f$. These strata have codimension $g-f$.

Under pull-back by the Torelli-map this induces a stratification on the moduli space $\mathcal{M}_g \otimes \mathbb{F}_p$. Very little is know about this stratification. Faber and I proved that the stratum of curves with p-rank f has codimension $g-f$ in the moduli space $\mathcal{M}_g \otimes \mathbb{F}_p$, cf. (Faber and van der Geer). It is unknown whether for every genus g and every prime p there exists a supersingular curve of genus g in characteristic p. It is known that in characteristic 2 for every given genus g one can explicitly construct a supersingular curve of genus g over $\mathbb{F}_2$, see the paper (van der Geer and van der Vlugt, 1998c). In (Scholten and Zhu) it was proved that in characteristic 2 there are no hyperelliptic supersingular curves of genus $2^h - 1$ for $h \geq 2$.

In view of this we pose explicitly the following questions:

QUESTION 4. *Which Newton polygons do occur for Jacobians?*

In characteristic zero one might also ask the related question: which endomorphism algebras (rings) occur as the endomorphism algebra (ring) of a jacobian? The answer seems unknown.

QUESTION 5. *For which genera does there exists a supersingular curve in positive characteristic?*

Since the Newton polygon is an isogeny invariant we can ask:

QUESTION 6. *Which isogeny classes of abelian varieties do contain a jacobian variety?*

In (Re) bounds on the genus of a curve in characteristic $p > 0$ were given under the condition that the Cartier operator on the differentials has low rank or is nilpotent.

Another more subtle stratification on $\mathcal{A}_g \otimes \mathbb{F}_p$ was introduced by Ekedahl and Oort. Its strata correspond to the type of the kernel of multplication by p for an abelian variety (as a group scheme). Here we have strata in all dimensions between $g(g+1)/2$ and 0. The smallest stratum corresponds to so-called superspecial abelian varieties. We refer to (Oort) and (van der Geer) for an exposition.

Again this induces a stratification on the moduli of curves and we may ask what it means for the zeta function of a curve that it belongs to a certain stratum.

Above we also met the concept of gonality. It defines also a stratification on $\mathcal{M}_g$. Here hyperelliptic curves are the second stratum and there are obvious bounds on the numbers of points for curves in a stratum. One may therefore ask more generally for the intersection of the gonality stratification and the Newton polygon stratification. The results of (Re) and of (Scholten and Zhu) can be viewed as results in this direction.

10. Distributions

Coding theory asks explicitly for the distribution of the weights of words and more generally of subcodes of a given codes. Using the connections between coding theory and curves over finite fields this question translates into the question how the number of rational points varies in families of curves. The most basic family is the universal family over (a cover of) the moduli space of curves $\mathcal{M}_g$. But the question is maybe more amenable for certain specific families of curves.

PROBLEM 2. *Determine for a given pair (g,q) which values the number of rational points on a smooth irreducible projective curve of genus g over $\mathbb{F}_q$ can assume.*

PROBLEM 3. *Determine the frequencies of the number of points in the universal family and other interesting families of curves.*

Of course, this is a very difficult question. For example, the answer for a fixed genus g enables one to determine the number of $\mathbb{F}_q$-rational points on the moduli space $\mathcal{M}_{g,n}$ of n-pointed genus g curves. In joint work with Carel Faber and Sebastian del Baño we have determined the frequencies for the genus 2 moduli spaces $\mathcal{M}_2 \otimes \mathbb{F}_p$ for all primes $p \leq 233$. We try to use this to obtain information on the cohomology of $\mathcal{M}_{2,n}$ counting points on curves over finite fields. We have a heuristic answer for the motivic Serre polynomial of $\mathcal{M}_{2,n}$ for $n \leq 16$.

References

Auer, R. (1998) Ray class fields of global function fields with many rational places, Report, University of Oldenburg.

Artin, E. (1924) Quadratische Körper im Gebiet der höheren Kongruenzen I, II, *Math. Zeitschrift* **19**, 153–246.

Birkenhake, Ch., and Lange, H. (1991) The exponent of an abelian subvariety, *Math. Ann.* **290**, 801–814.

Bombieri, E. (1974) Counting points on curves over finite fields (d'après S. A. Stepanov), in *Séminaire Bourbaki*, 25ème année, Exp. No. 430, Lecture Notes in Math., Vol. 383, Springer, Berlin, pp. 234–241.

Deligne, P., and Lusztig, G. (1976) Representations of reductive groups over finite fields, *Ann. of Math.* **103**, 103–161.

Elkies, N. (1998) Explicit Modular Towers, in T. Basar and A. Vardy (eds.), *Proceedings of the Thirty-Fifth Annual Allerton Conference on Communication, Control and Computing (1997)*, Univ. of Illinois at Urbana-Champaign, pp. 23–32.

Elkies, N. (2001) Explicit towers of Drinfeld modular curves, in *Proceedings of the Third European Math. Congress*, Birkhäuser.

Elkies, N., Kresch, A., Poonen, B., Wetherell, J., and Zieve, M. (2001) Curves of every genus with many points II: asymptotically good families, in preparation.

Fuhrmann, R., and Torres, F. (1996) The genus of curves over finite fields with many rational points, *Manuscripta Math.* **89**, 103–106.

Fuhrmann, R., Garcia, A., and Torres, F. (1997) On maximal curves, *J. Number Theory* **67**, 29–51.

Faber, C., and van der Geer, G. (2001) *Complete subvarieties of the moduli space of curves*, Manuscript in preparation.

Garcia, A., and Quoos, L. (2000) A construction of curves over finite fields, Preprint 2000.

Garcia, F., and Stichtenoth, H. (1995) A tower of Artin-Schreier extensions of function fields attaining the Drinfeld-Vladut bound, *Invent. Math.* **121**, 211–222.

van der Geer, G. (1999) Cycles on the moduli space of abelian varieties, in C. Faber and E. Looijenga (eds.), *Moduli of curves and abelian varieties*, Aspects Math., E33, Vieweg, Braunschweig, pp. 65–89.

van der Geer, G., and van der Vlugt, M. (1997) How to construct curves over finite fields with many points, in F. Catanese (ed.), *Arithmetic Geometry (Cortona 1994)*, Cambridge Univ. Press, Cambridge, pp. 169–189.

van der Geer, G., and van der Vlugt, M. (1998a) On generalized Reed-Muller codes and curves with many points, *J. of Number Theory* **72**, 257–268.

van der Geer, G., and van der Vlugt, M. (1998b) Tables for the function $N_q(g)$, *Math. of Computation.* Regularly updated tables can be found at the address http://www.science.uva.nl/˜geer.

van der Geer, G., and van der Vlugt, M. (1998c) On the existence of supersingular curves of given genus, *Journal für die Reine und angewandte Math.* **458**, 53–61.

van der Geer, G., and van der Vlugt, M. (2000) Kummer covers with many points, *Finite and their Appl.* **6**, 327–341.

van der Geer, G., and van der Vlugt, M. (2001) An asymptotically good tower of curves over the field with eight elements, preprint, math.AG/0102158.

Goppa, V.D. (1977) Codes associated with divisors, (Russian), *Problemy Peredavci Informacii* **13**, 33–39.

Goppa, V.D. (1981) Codes on algebraic curves, (Russian), *Dokl. Akad. Nauk SSSR* **259**, 1289–1290.

Hajir and Maire (2001) Asymptotically good towers of global fields, *Proceedings of the Third European Congress*, Birkhäuser.

Harris, J. (1982) *Curves in projective space*, Université de Montreal.

Hasse, H. (1934) Abstrakte Begründung der komplexen Multiplikation und Riemannsche Vermutung in Funktionenkörper, *Abh. Math. Sem. Hamburg* **10**, 325–348.

Ihara, Y. (1979) Congruence relations and Shimura curves. II, *J. Fac. Sci. Univ. Tokyo* **25**, 301–361.

Ihara, Y. (1982) Some remarks on the number of points of algebraic curves over finite fields, *J. Fac. Sci. Tokyo* **28**, 721–724.

Ihara, Y. (1985) Review of (Tsfasman, Vladuts and Zink, 1982). Math. Reviews # 85i:11108.

Korchmaros, G., and Torres, F. (2001) On the genus of a maximal curve, Preprint.

Kresch, A., Wetherell, J., and Zieve, M. (2000) Curves of every genus with many points, I: Abelian and toric families, Preprint.

Lauter, K. (1996) Ray class field constructions of curves over finite fields with many rational points, in H. Cohen (ed.), *Algorithmic Number Theory (Talence, 1996)*, Lecture Notes in Computer Science 1122, Springer, Berlin, pp. 187–195.

Lauter, K. (1999) Improved upper bounds for the number of rational points on algebraic curves over finite fields, *Comptes Rend. Acad. Sci. Paris Sér. I Math.* **328**, 1181–1185.

Lauter, K. (2000) Non-existence of a curve over $\mathbb{F}_3$ of genus 5 with 14 rational points, *Proc. Amer. Math. Soc.* **128**, 369–374.

Lauter, K. (2001) Geometric methods for improving the upper bounds on the number of rational points on algebraic curves over finite fields, With an appendix in French by J.-P. Serre, *J. Algebraic Geom.* **10**, 19–36.

Li, K.-Z., and Oort, F. (1998) *Moduli of supersingular abelian varieties*, Lecture Notes in Mathematics 1680, Springer Verlag, Berlin.

van Lint, J. (1998) *Introduction to Coding Theory*, Graduate Texts in Mathematics, Springer Verlag.

MacWilliams, F., and Sloane, N. (1977) *The theory of error-correcting codes*, North-Holland Publishing Company.

Manin, Yu.I. (1981) What is the maximum number of points on a curve over $\mathbb{F}_2$?, *J. Fac. Sci. Tokyo* **28**, 715–720.

Niederreiter, H., and Xing, C.P. (1997) Drinfeld modules of rank 1 and algebraic curves with many rational points II, *Acta Arithm.* **81**, 81–100.

Niederreiter, H., and Xing, C.P. (1997) Algebraic curves with many rational points over finite fields of characteristic 2, to appear in *Proc. Number Theory Conference (Zakopane 1997)*, de Gruyter, Berlin.

Niederreiter, H., and Xing, C.P. (1998a) Global function fields fields with many rational points over the ternary field, *Acta Arithm.* **83**, 65–86.

Niederreiter, H., and Xing, C.P. (1998b) A general method of constructing global function fields with many rational places, to appear in *Algorithmic Number Theory (Portland 1998)*, Lecture Notes in Comp. Science, Springer, Berlin.

Niederreiter, H., and Xing, C.P. (1998c) Towers of global function fields with asymptotically many rational places and an improvement on the Gilbert-Varshamov bound, *Math. Nachr.* **195** (1998), 171–186.

Oort, F. (199) A stratification of a moduli space of polarized abelian varieties in positive characteristic, in C. Faber and E. Looijenga (eds.), *Moduli of curves and abelian varieties*, Aspects Math., E33, Vieweg, Braunschweig, pp. 47–64.

Re, R. (1999) The rank of the Cartier operator and linear systems on curves, Preprint University of Amsterdam, To appear in *Journal of Algebra*.

Rück, H.-G., and Stichtenoth, H. (1994) A characterization of Hermitian function fields over finite fields, *J. Reine Angew. Math.* **457**, 185–188.

Scholten, J., and Zhu, H.J. (2000) Hyperelliptic supersingular curves over fields of characteristic 2, math.AG/0012178.

Schoof, R. (1990) *Algebraic curves and coding theory*, UTM 336, University of Trento.

Serre, J.-P. (1983) Sur le nombre des points rationnels d'une courbe algébrique sur un corps fini, *Comptes Rendus Acad. Sci. Paris* **296**, 397–402.

Serre, J.-P. (1982/3) Nombre de points des courbes algébriques sur $\mathbb{F}_q$, Sém. de Théorie des Nombres de Bordeaux, 1982/83, exp. no. 22. (= Oeuvres III, No. 129, pp. 664–668).

Serre, J.-P. (1983/4) Quel est le nombre maximum de points rationnels que peut avoir une courbe algébrique de genre g sur un corps fini $\mathbb{F}_q$?, Résumé des Cours de 1983-1984. (= Oeuvres III, No. 132, pp. 701–705).

Serre, J.-P. (1985) Rational points on curves over finite fields, Notes of lectures at Harvard University.

Shabat, V. (2001) Thesis, University of Amsterdam.

Stichtenoth, H. (1993) *Algebraic function fields and codes*, Springer Verlag, Berlin.

Stichtenoth, H. (2001) Explicit constructions of towers of function fields with many rational places, in *Proceedings of the Third European Congress of Math.*, Birkhäuser.
Stöhr, K.O., and Voloch, J.F. (1986) Weierstrass points and curves over finite fields, *Proc. London Math. Soc.* **52**, 1–19.
Temkine, A. (2000) Hilbert class field towers of function fields over finite fields and lower bounds for $A(q)$, Preprint.
Thompson, T. (1983) *From error-correcting codes through sphere packings to simple groups*, Carus Math. Monographs 21, MAA.
Tsfasman, M.A., Vladuts, S.G., and Zink, Th. (1982) On Goppa codes which are better than the Varshamov-Gilbert bound, *Math. Nachrichten* **109**, 21–28.
Vladuts, S.G., and Drinfeld, V.G. (1983) Number of points of an algebraic curve, *Funct. Anal.* **17**, 68–69.
Wei, V.K. (1991) Generalized Hamming weights for linear codes, *IEEE Trans. Inform. Theory* **37**, 1412–1418.
Weil, A. (1948) *Variétés abéliennes et courbes algébriques*, Hermann, Paris.
Zink, Th. (1985) Degeneration of Shimura surfaces and a problem in coding theory, in *Fundamentals of computation theory (Cottbus, 1985)*, Lecture Notes in Comput. Sci. 199, Springer, Berlin, pp. 503–511.

THREE ALGORITHMS IN ALGEBRAIC GEOMETRY, CODING THEORY AND SINGULARITY THEORY

G.-M. GREUEL, C. LOSSEN and M. SCHULZE
University of Kaiserslautern, Germany

Abstract. We describe three algorithms in algebraic geometry, coding theory and singularity theory, which are new, resp. have new ingredients. Moreover, we put special emphasis on their implementation in the computer algebra system SINGULAR. The first algorithm computes the normalization of an affine reduced ring, an ideal defining the non-normal locus and, as an application, the integral closure of an ideal. The second is devoted to the computation of the places of a projective plane curve, of bases of adjoint forms, and of the linear system of a given rational divisor on the normalization of the curve. Finally, the third algorithm provides a method to compute the V-filtration, the monodromy and the singularity spectrum of an arbitrary isolated hypersurface singularity.

Key words: Normalization, integral closure, Rees algebra, AG-codes, Hamburger-Noether expressions, Brill-Noether theorem, monodromy, singularity spectrum, V-filtration, Gauß-Manin connection.

Mathematics Subject Classification (2000): 13, 14, 13P, 11T, 32S, 34M.

Introduction

Algorithmic and computational aspects have become a major, still growing issue in mathematical research and teaching. This has various deeper reasons in the cultural and technological development of today's society but also quite practical reasons. One of these is certainly the existence and maturity of software systems which, via implemented algorithms, provide easy access to hard and sometimes sophisticated computations, assisting and supporting mathematical research.

In this article, we describe three algorithms in algebraic geometry, coding theory, and singularity theory, which are new, resp. have new ingredients. The first one describes how to compute the normalization of an affine reduced ring, an ideal defining the non-normal locus and, as an application, the integral closure of an ideal. The second is devoted to the computation of the places of a projective plane curve defined over a finite field, and the computation of bases of adjoint forms and of the linear system of a given rational divisor on the normalization

C. Ciliberto et al. (eds.),
Applications of Algebraic Geometry to Coding Theory, Physics and Computation, 161–194.

of the curve. Finally, the third one provides a method to compute the V-filtration, the monodromy and the singularity spectrum of an arbitrary isolated hypersurface singularity.

All three algorithms require non-trivial up to deep mathematical knowledge and go beyond foundational algorithms in computer algebra. Indeed, one of the purposes of this note is to show that highly complex mathematical objects can nowadays be represented in a computer and, thus, can be used in mathematical research on a higher level than ever before. All the algorithms described in this paper are implemented in the computer algebra system SINGULAR (Greuel, Pfister and Schönemann) as a free service to the mathematical community.

The normalization algorithm is based on an old criterion of Grauert and Remmert and has already been published (De Jong; Decker, Greuel, de Jong and Pfister). From this it is not difficult to derive the principle for algorithms to compute the non-normal locus and the integral closure of an ideal, however, the concrete description and its realization appear to be new. The proposed algorithm for computing the places of plane curves is based on the Hamburger-Noether development and has been described on a theoretical level in (Campillo and Farrán), as well as the Brill-Noether algorithm for computing bases of linear systems. Since then, this algorithm has been implemented, together with a full coding and decoding algorithm, and we mainly concentrate on new algorithmic and computational aspects. It should be mentioned that the construction of AG codes, using quadratic transformations instead of Hamburger-Noether expansions, has been described and implemented before (Le Brigand and Risler; Haché and Le Brigand). Finally, the algorithm to compute the V-filtration and the singularity spectrum is very recent, a theoretical description of parts of it are to be published in (Schulze and Steenbrink). Here, we give a short description of the theory together with a description of some computational aspects. A paper with full details will be published by the third author.

Acknowledgements. We would like to thank very much the German Israeli Foundation for Research and Development (Grant No. G-616-15.6/1999) and the Hermann Minkowski – Minerva Center for Geometry at Tel Aviv University for support during the preparation of this paper.

1. Integral closure of rings and ideals

In this section we present an algorithm to compute, for a reduced affine ring A, the normalization, that is, the integral closure of A in its total ring of fractions. This algorithm can be used to describe an algorithm for computing the integral closure of an ideal I in A.

The normalization of a ring is an important construction in commutative algebra and algebraic geometry, as well as the integral closure of an ideal. Both have many applications in commutative algebra, algebraic geometry and singularity theory, e.g., in the theory of equisingularity and for resolution of singularities. Hence it is desirable to have a sufficiently good implementation of the algorithms which can compute interesting examples. Such an implementation is distributed with SINGULAR, version 2.0 (Greuel, Pfister and Schönemann).

The algorithm for the normalization is based on a criterion of Grauert and Remmert and was first described in (De Jong), see also (Decker, Greuel, de Jong and Pfister). The Grauert-Remmert criterion provides an algorithm to compute the non-normal locus of an affine ring, which we also describe. Both seem to be the only known general algorithms. To compute the integral closure of an ideal I, we use the normalization of the Rees algebra of I together with several extra procedures, which make this basically a new algorithm.

A good reference for computational aspects in connection with integral closure is found in the textbooks (Vasconcelos, 1994 and 1997) and in the articles (Vasconcelos, 1991), (Corso, Huneke and Vasconcelos). Other references are (Stolzenberg) and (Gianni and Trager) or can be found in (Vasconcelos, 1994 and 1997).

Let $Q(A)$ denote the total ring of fractions of A and $\overline{A}$ the integral closure of A in $Q(A)$. There have been many attempts to construct a bigger ring A', $A \subset A' \subset Q(A)$, which is finite over A and then continuing with A', in order to approximate $\overline{A}$. The problem of this approach is to know when to stop, that is, to have an effective criterion for a ring to be normal. It had escaped the computer algebra community that such a criterion has been known for more than thirty years, having been discovered by Grauert and Remmert (1971). It was rediscovered by De Jong (1998) and says that A is normal if and only if the natural embedding $A \subset \mathrm{Hom}_A(J,J)$, where J denotes the ideal of the singular locus, is an isomorphism. To be able to continue with $A' = \mathrm{Hom}_A(J,J)$ we must present A' as a polynomial ring modulo some ideal, together with the embedding $A \hookrightarrow A'$, which is not difficult. By the theorem of Grauert and Remmert, we know when to stop to reach the normalization of A (for affine rings the algorithm stops by a classical theorem of E. Noether).

If $I \subset A$ is an ideal, then it is well-known that the Rees algebra

$$\mathcal{R}(I) = \bigoplus_{k \geq 0} I^k t^k \subset A[t]$$

of I has, as normalization, $\overline{\mathcal{R}(I)} = \bigoplus_{k \geq 0} \overline{I^k} t^k$. That is, the component of t-degree 1 of $\overline{\mathcal{R}(I)}$ is the integral closure $\overline{I}$ of I in A. However, to obtain the component of t-degree k of $\overline{\mathcal{R}(I)}$ we need a careful analysis of the morphism $\mathcal{R}(I) \subset \overline{\mathcal{R}(I)} \subset A[t]$. The embedding $\overline{\mathcal{R}(I)} \subset A[t]$ is extra information which has to be added to the normalization algorithm.

Unfortunately, this algorithm computes too much, namely all $\overline{I^k}$ and, therefore, its application is restricted only to examples of moderate size. However, already a first implementation in SINGULAR, cf. (Hirsch, 2000a and 2000b), shows that we can compute interesting examples (and it is the only existing implementation). It is still an open problem to find a better algorithm, not using the Rees algebra.

1.1. RING NORMALIZATION

Let $A \subset B$ be a ring extension and $I \subset A$ an ideal. Recall that $b \in B$ is called (strongly) *integral over I* if b satisfies a relation

$$b^n + a_1 b^{n-1} + \ldots + a_n = 0 \text{ with } a_i \in I^i.$$

The set $C(I,B) = \{b \in B \mid b \text{ is integral over } I\}$ is called the *integral closure of I* in B, it is a $C(A,B)$-module, where $C(A,B)$, the integral closure of A in B, is a ring. We are interested in the two most interesting cases:

- A reduced, $B = Q(A)$ the total quotient ring of A, and $I = A$. In this case $I^i = I$ and $C\big(A, Q(A)\big) =: \overline{A}$ is the *normalization* of A.
- $A = B$ and $I \subset A$ arbitrary. Then $C(I,A) =: \overline{I}$ is the integral closure of I in A.

Let us consider the normalization first.

LEMMA 1.1 (KEY-LEMMA). *Let A be a reduced Noetherian ring and $J \subset A$ an ideal containing a non-zero divisor u of A. Then there are natural inclusions of rings*

$$A \subset \operatorname{Hom}_A(J,J) \cong \frac{1}{u}(uJ : J) \subset \overline{A},$$

where $uJ : J = \{h \in A \mid hJ \subset uJ\}$, *and inclusions* $\operatorname{Hom}_A(J,J) \subset \operatorname{Hom}_A(J,A) \cap \overline{A} \subset \operatorname{Hom}_A\big(J, \sqrt{J}\big)$ *of A-modules.*

PROOF. If $\varphi \in \operatorname{Hom}_A(J,A)$, then $\frac{\varphi(u)}{u}$ is independent of the non-zero-divisor u and, hence, $\varphi \mapsto \frac{\varphi(u)}{u}$ defines an embedding

$$\operatorname{Hom}_A(J,A) \xrightarrow{\cong} \{h \in Q(A) \mid hJ \subset A\} \hookrightarrow Q(A).$$

The inclusion $A \subset \operatorname{Hom}_A(J,J) \cong \frac{1}{u}(uJ : J) = \frac{1}{u}\{h \in Q(A) \mid hJ \subset uJ\}$ is given by the multiplication with elements of A. To see that the image is contained in $\overline{A}$, consider $\varphi \in \operatorname{Hom}_A(J,J)$. The characteristic polynomial of φ defines (by Cayley-Hamilton) an integral relation for φ. To see the last inclusion, consider $h \in \overline{A}$ such that $hJ \subset A$, and let $h^n + a_1 h^{n-1} + \cdots + a_n = 0$, $a_i \in A$, be an integral relation for h. For a given $g \in J$ multiply the relation for h with g^n. This shows that $(gh)^n \in J$, hence $gh \in \sqrt{J}$. □

Since A is normal if and only if the localization A_P is normal for all $P \in \operatorname{Spec} A$, we define the *non-normal locus* of A as

$$\mathrm{NN}(A) := \{P \in \operatorname{Spec} A \mid A_P \text{ is not normal}\}.$$

It is easy to see that $\mathrm{NN}(A) = V(C)$ where $C = \mathrm{Ann}_A(\overline{A}/A)$ is the *conductor* of A in $\overline{A}$, in particular, $\mathrm{NN}(A)$ is closed in $\mathrm{Spec}\,A$. However, since we cannot yet compute $\overline{A}$, we cannot compute C either.

The following proposition is the basis for the algorithm to compute the normalization as well as for an algorithm to compute an ideal with zero set $\mathrm{NN}(A)$. It is basically due to Grauert and Remmert (1971).

PROPOSITION 1.2 (CRITERION FOR NORMALITY). *Let A be a reduced Noetherian ring and $J \subset A$ an ideal satisfying*

1. *J contains a non-zerodivisor of A,*
2. *$J = \sqrt{J}$,*
3. *$\mathrm{NN}(A) \subset V(J)$.*

Then $A = \overline{A}$ if and only if $A = \mathrm{Hom}_A(J,J)$.

An ideal $J \subset A$ satisfying 1. – 3. is called a *test ideal* for the normalization.

PROOF. If $A = \overline{A}$ then $\mathrm{Hom}_A(J,J) = A$, by Lemma 1.1. For the converse, notice that 3. implies $J \subset \sqrt{C}$, hence there exists a minimal $d \geq 0$ such that $J^d \in C$, that is, $J^d\overline{A} \subset A$. If $d > 0$, choose $h \in \overline{A}$ and $a \in J^{d-1}$ such that $ha \notin A$. Since $ah \in \overline{A}$ and $ahJ \subset hJ^d \subset A$ we have $ah \in \mathrm{Hom}_A(J,A) \cap \overline{A}$ and hence, using 2., $ah \in \mathrm{Hom}_A(J,J)$, by Lemma 1.1. By assumption $\mathrm{Hom}_A(J,J) = A$ and, hence, $ah \in A$. This is a contradiction, and we conclude $d = 0$ and $A = \overline{A}$. □

REMARK 1.3. As the proof shows, condition 2. can be weakened to

2'. $\mathrm{Hom}_A(J,J) = \mathrm{Hom}_A(J,A) \cap \overline{A}$.

However, this cannot be used in practice, since we do not know $\overline{A}$, while, on the other hand, we can always pass from J to $\sqrt{J}$ without violating the conditions 1 and 3, and $\sqrt{J}$ is computable.

COROLLARY 1.4. *Let $J \subset A$ be an ideal as in Proposition 1.2 and define the ideal $I_{NN} := Ann_A\big(\mathrm{Hom}_A(J,J)\big)$. Then $\mathrm{NN}(A) = V(I_{NN})$.*

PROOF. This follows from 1.2, 1.1 and the fact that the operations which define the annihilator are compatible with localization. □

Suppose that we have a test ideal J and a non-zero divisor $u \in J$ of A. Then we can compute A-module generators of $\mathrm{Hom}_A(J,J) \cong uJ : J$ and A-module generators for $I_{NN} = \mathrm{Ann}_A\big(\mathrm{Hom}_A(J,J)\big) \cong \langle u \rangle : (uJ : J)$, since we can compute ideal quotients using Gröbner basis methods, cf. (Greuel and Pfister, 2001b).

Let us describe the *ring structure* of $\mathrm{Hom}_A(J,J)$. Let $u_0 = u,\ u_1, \dots, u_s$ be generators of $uJ : J$ as A-module, and let $(\alpha_0^i, \dots, \alpha_s^i)$ be generators of the module of syzygies of $u_0, \dots, u_s$. Since $\mathrm{Hom}_A(J,J)$ is a ring, we have $u_i \cdot u_j = \sum_{\ell=0}^{s} \beta_\ell^{ij} u_\ell$,

$1 \leq i, j \leq s$ for certain $\beta_\ell^{ij} \in A$, the quadratic relations between the u_i. Define $\mathrm{Ker} \subset A[t_1, \ldots, t_s]$ as the ideal generated by the linear and quadratic relations,

$$\alpha_0^i + \alpha_1^i t_1 + \cdots + \alpha_s^i t_s, \quad t_i t_j - (\beta_0^{ij} + \beta_1^{ij} t_1 + \cdots + \beta_s^{ij} t_s).$$

We get an isomorphism $\mathrm{Hom}_A(J,J) \cong uJ : J \cong A[t_1, \ldots, t_s]/\mathrm{Ker}$ of A-algebras, by sending t_i to u_i. This presentation is needed to continue the algorithm.

To compute I_{NN}, we only need the A-module structure of $uJ : J$.

1.2. TEST IDEALS

It remains to find a test ideal. For this we consider the singular locus

$$\mathrm{Sing}(A) = \{P \in \mathrm{Spec}\, A \mid A_P \text{ is not regular}\}.$$

Since every regular local ring is normal, $\mathrm{NN}(A) \subset \mathrm{Sing}(A)$. For general Noetherian rings, however, $\mathrm{Sing}(A)$ may not be closed in the Zariski topology. Therefore, we pass to more special rings.

Let $S = K[x_1, \ldots, x_n]$ and $A = S/I$ be an affine ring where K is a perfect field. If A is equidimensional of codimension c, that is, all minimal primes P of I have the same height c, then the *Jacobian ideal* of I defines $\mathrm{Sing}(A)$. That is, if $I = \langle f_1, \ldots, f_k \rangle$ and

$$J = \left\langle c\text{-minors of } \left(\frac{\partial f_i}{\partial x_j}\right), f_1, \ldots, f_k \right\rangle \subset S$$

is the Jacobian ideal of I, then $\mathrm{Sing}(A) = V(J)$. If, on the other hand, A is not equidimensional, then $V(J)$ may be strictly contained in $\mathrm{Sing}(A)$, if we define J as above with c the minimal height of minimal primes of I. In this case we need another ideal. There are several alternatives to compute an ideal I_{Sing} such that $V(I_{\mathrm{Sing}}) = \mathrm{Sing}(A)$. Either we compute an equidimensional decomposition $I = \bigcap_i I_i$ (Eisenbud, Huneke and Vasconcelos; Greuel and Pfister, 2001b) of I, compute the Jacobian ideal J_i for each equidimensional ideal I_i and compute the ideal describing the intersection of any two equidimensional parts. The same works for a primary decomposition (Gianni, Trager and Zacharias; Eisenbud, Huneke and Vasconcelos; Greuel and Pfister, 2001b) instead of an equidimensional decomposition.

We can avoid an equidimensional, resp. primary, decomposition if we compute the ideal of the non-free locus of the module of *Kähler differentials*,

$$\Omega^1_{A/K} = \Omega^1_{S/K} \Big/ \left(\sum_{i=1}^k f_i \Omega^1_{S/K} + \sum_{i=1}^k S df_i \right),$$

where $\Omega^1_{S/K} = \bigoplus_{i=1}^n S dx_i$. $\Omega^1_{A/K}$ is isomorphic to A^n modulo the submodule generated by the rows of the Jacobian matrix of $(f_1, \ldots, f_k)$, hence it is finitely presented by the transpose of the Jacobian matrix.

For any finitely presented A-module M we can compute the non-free locus

$$\mathrm{NF}(M) = \{P \in \operatorname{Spec} A \mid M_P \text{ is not } A_P\text{-free}\}$$

just by Gröbner basis and syzygy computations, cf. (Greuel and Pfister, 2001b).

Let I_{Sing} be an ideal defining the singular locus of $A = S/I$. Since A is reduced, I_{Sing} contains a non-zerodivisor of A. Indeed, a general linear combination u of the generators of I_{Sing} will be a non-zerodivisor. Hence, any radical ideal of S which contains I and u will be a test ideal for normality. Two extreme choices for test ideals are $\sqrt{I_{\mathrm{Sing}}}$ or $\sqrt{\langle I, u\rangle}$.

Since the radical of an ideal in an affine ring can be computed (Eisenbud, Huneke and Vasconcelos; Krick and Logar; Greuel and Pfister, 2001b), we have all ingredients to compute the normalization of A.

In the remaining part of this section, we describe algorithms to compute

- the *normalization* $\overline{A}$ of A, that is, we represent $\overline{A}$ as affine ring and describe the map $A \hookrightarrow \overline{A}$,
- generators for an ideal $I_{NN} \subset A$ describing the *non-normal locus*, that is, $V(I_{NN}) = \mathrm{NN}(A) = \{P \in \operatorname{Spec} A \mid A_P \text{ is not normal}\}$,
- for any ideal $I \subset A$, generators for the *integral closure* $\overline{I}$ of I in A.

1.3. NORMALIZATION ALGORITHM

Let us first describe the algorithm to compute the non-normal locus. K denotes a perfect field.

1.3.1. *Computing the non-normal locus*

INPUT: $f_1, \ldots, f_k \in S = K[x_1, \ldots, x_n]$, $I := \langle f_1, \ldots, f_k\rangle$.
We assume that I is a radical ideal.

OUTPUT: Generators for I_{NN} such that $V(I_{NN}) = \mathrm{NN}(S/I)$.

1. Compute an ideal I_{Sing} as described in Sect. 1.2.
2. Compute a non-zerodivisor $u \in I_{\mathrm{Sing}}$: choose an S-linear combination u of the generators of I_{Sing} and test if $(I : u) := \{g \in S \mid gu \in I\}$ is zero, by using that u is a non-zerodivisor iff $(I : u) = 0$.
   A sufficiently general linear combination gives a non-zerodivisor.
3. Compute a test ideal J, e.g., $J = \sqrt{\langle u, I\rangle}$ or $J = \sqrt{I_{\mathrm{Sing}}}$.
4. Compute generators $g_1, \ldots, g_\ell$ for $\langle u, I\rangle : \big((uJ + I) : J\big)$ as S-module.
5. Return $\{g_1, \ldots, g_\ell\}$.

1.3.2. *Computing the normalization*

The idea for computing the normalization of $A = S/I$ is as follows:

- We construct an increasing sequence of rings

$$A \subset A_1 \subset A_2 \subset \cdots \subset A_k \subset A_{k+1} \subset \cdots \subset Q(A)$$

with $J_0 \subset A_0 := A$ a test ideal for A, $A_i := \mathrm{Hom}_{A_{i-1}}(J_{i-1}, J_{i-1})$, and $J_i \subset A_i$ a test ideal for A_i, $i \geq 1$.
- If $A_k = A_{k+1}$ then $\overline{A} = A_k$.

For performance reasons we do not look for a non-zerodivisor in J_i but choose any non-zero element u. If u is a zerodivisor then it gives a splitting of the ring which makes the subsequent computations easier.

We obtain the following (highly recursive) algorithm for computing the normalization:

INPUT: $f_1, \ldots, f_k \in S = K[x_1, \ldots, x_n]$, $I = \langle f_1, \ldots, f_k \rangle$
We assume that I is a radical ideal.

OUTPUT: Polynomial rings $S_1, \ldots, S_\ell$, ideals $I_j \subset S_j$, and maps $\pi_j : S \to S_j$ such that $S/I \to S_1/I_1 \oplus \cdots \oplus S_\ell/I_\ell$, induced by $(\pi_1, \ldots, \pi_\ell)$ is the normalization map of S/I.

1. Compute an ideal I_{Sing} describing the singular locus.
2. Compute the radical $J = \sqrt{I_{\mathrm{Sing}}}$.
3. Choose $u \in J \setminus \{0\}$ and compute $I : u$. If $I : u = I$ go to 4 (then u is a non-zerodivisor of S/I). Otherwise, set $R = A/\langle u \rangle \oplus A/(I : u)$ and go to 6.
4. Compute $R := \mathrm{Hom}_A(J, J)$ and $\pi : A \to R$.
5. If $A = R$ then return (R, π), otherwise go to 1.
6. Suppose $R = R_1/I_1 \oplus \cdots \oplus R_\ell/I_\ell$. Then, for each $i = 1, \ldots, \ell$, set $A = R_i/I_i$ and go to 1.

1.4. INTEGRAL CLOSURE ALGORITHM

Let A be a ring, $I \subset A$ an ideal. We propose an algorithm to compute the *integral closure* $\overline{I} = \{b \in A \mid b \text{ is integral over } I\}$ of I. An ideal I is called *integrally closed* if and only if $I = \overline{I}$. It is called *normal* if $I^k = \overline{I^k}$ for all $k > 0$. Note that $I \subset \overline{I} \subset \sqrt{I}$. We are mainly interested in the case $A = K[x_1, \ldots, x_n]$.

In the following, we describe an algorithm to compute $\overline{I^k}$ for all k, simultaneously. Consider the *Rees algebra* $\mathcal{R}(I) = \bigoplus_{k \geq 0} I^k t^k \subset A[t]$, and let $\widetilde{\mathcal{R}(I)}$ denote the integral closure of $\mathcal{R}(I)$ in $A[t]$. Then

$$\widetilde{\mathcal{R}(I)} = \bigoplus_{k \geq 0} \overline{I^k} t^k \subset A[t].$$

If A is normal, then $A[t]$ is normal and hence, the normalization of $\mathcal{R}(I)$, that is, the integral closure of $\mathcal{R}(I)$ in $Q(\mathcal{R}(I))$, satisfies

$$\overline{\mathcal{R}(I)} = \widetilde{\mathcal{R}(I)} = \bigoplus_{k\geq 0} \overline{I^k} t^k .$$

Hence, computing the normalization of $\mathcal{R}(I)$ provides the integral closure of $\overline{I^k}$ for all k.

To be specific, let $A = K[\mathbf{x}] = K[x_1,\ldots,x_n]$, $I = \langle f_1,\ldots,f_k\rangle \subset A$ with K a perfect field. Then

$$\mathcal{R}(I) = K[\mathbf{x}, tf_1,\ldots,tf_k] \underset{\varphi}{\overset{\cong}{\longleftarrow}} K[\mathbf{x}, U_1,\ldots,U_k]/(\operatorname{Ker}\varphi)$$

where $\varphi : K[\mathbf{x},\mathbf{U}] \to K[\mathbf{x},t]$ maps $x_i \mapsto x_i$, $U_j \mapsto tf_j$. $\operatorname{Ker}\varphi$ can be computed by eliminating t from

$$J := \langle U_1 - tf_1,\ldots,U_k - tf_k\rangle \subset K[\mathbf{x},\mathbf{U},t],$$

that is, $\operatorname{Ker}\varphi = J \cap K[\mathbf{x},\mathbf{U}]$. For the integral closure of I we need to compute

$$\begin{array}{ccc}
\mathcal{R}(I) \cong K[\mathbf{x},\mathbf{U}]/(\operatorname{Ker}\varphi) & \xrightarrow{\ \varphi\ } & K[\mathbf{x},t] \\
\big\downarrow & \nearrow_{T_i \mapsto \frac{\varphi(a_i)}{\varphi(b_i)},\ T_i = \frac{a_i}{b_i} \in Q(\mathcal{R}(I))} & \\
Q(\mathcal{R}(I)) \supset \overline{\mathcal{R}(I)} = K[T_1,\ldots,T_s]/I' . & &
\end{array}$$

This means that we compute $\overline{\mathcal{R}(I)}$ as an affine ring $K[\mathbf{T}]/I'$ and, in each inductive step during the computation of $\overline{\mathcal{R}(I)}$, we also compute the map from the intermediate ring to $K[\mathbf{x},t]$.

The algorithm then reads as follows:

INPUT: $f_1,\ldots,f_\ell \in K[x_1,\ldots,x_n]$, $k \geq 1$ an integer, $I := \langle f_1,\ldots,f_\ell\rangle$.

OUTPUT: Generators for $\overline{I^k} \subset K[x_1,\ldots,x_n]$.

1. Compute the Rees algebra $\mathcal{R}(I) \subset K[\mathbf{x},t]$.
2. Compute the normalization $\overline{\mathcal{R}(I)}$, together with maps φ, ψ, so that

$$\begin{array}{ccc}
\mathcal{R}(I) & \overset{\varphi}{\hookrightarrow} & K[\mathbf{x},t] \\
\big\downarrow & & \big\uparrow \psi \\
Q(\mathcal{R}(I)) \supset \overline{\mathcal{R}(I)} & = & K[T_1,\ldots,T_s]/J
\end{array}$$

   commutes.

3. Determine $a_i, b_i \in \mathcal{R}(I)$, so that $T_i = \frac{a_i}{b_i} \in Q(\mathcal{R}(I))$ compute $\frac{\varphi(a_i)}{\varphi(b_i)} \in K[\mathbf{x},t]$ (indeed, we find a universal denominator $b = b_i$ for all i).
4. Determine generators $g_1,\ldots,g_s$ of the $K[\mathbf{x}]$-ideal which is mapped to the component of t-degree k of the subalgebra $\psi(\overline{\mathcal{R}(I)}) \subset K[\mathbf{x},t]$.
5. Return $g_1,\ldots,g_s$.

The algorithms described above are implemented in SINGULAR and contained in the libraries `normal.lib` (Greuel and Pfister, 2001a) and `reesclos.lib` (Hirsch, 2000b) contained in the distribution of SINGULAR 2.0 (Greuel, Pfister and Schönemann). Similar procedures can be used to compute the conductor ideal of A in $\overline{A}$. An implementation will be available soon.

1.5. EXAMPLES

1.5.1. *Ring normalization*

We first load the library `normal.lib`

```
LIB "normal.lib";
```

We define the ring and an ideal, describing two transversal cusps.

```
ring S  = 0,(x,y),dp;
ideal I = (x2-y3)*(x3-y2);
```

If the printlevel is sufficiently high, the algorithm will display intermediate results and pause during the computation until the user hits the return button in order to be able to follow what is going on. Here we will reproduce only part of the comments and slightly change the SINGULAR output.

```
printlevel = 5;
```

Now we start the computation using the procedure `normal` from `normal.lib`:

```
list nor = normal(I);
```

The first normalization loop starts with $\langle x^3y^3-x^5-y^5+x^2y^2\rangle$, and computes the radical of the singular locus as

$$J = \langle xy^2-y,\, x^2y-x,\, y^4-x,\, x^4-y\rangle,$$

chooses the non-zerodivisor $u = 3x^2y^3-5x^4+2xy^2$ in J, computes

$$uJ:J = \langle 3x^2y^3-5x^4+2xy^2,\, 2x^5-3y^5+x^2y^2,\, 3y^6-5x^4+2xy^2,\, xy^5-x^3y^2,\, x^4y^2-xy^4\rangle,$$

and checks that $A \neq \mathrm{Hom}_A(J,J)$. Therefore the ring structure of $\mathrm{Hom}_A(J,J)$ has to be computed. The result is the affine ring A_1 with variables $T_1,\ldots,T_4$ (after eliminating linear equations) modulo the ideal

```
==> T(3)*T(4)-T(4)
==> T(1)^4+4*T(1)^2*T(2)*T(4)+4*T(2)^2*T(4)^2
              -T(2)*T(3)+T(2)
==> T(1)^2*T(2)+2*T(2)^2*T(4)-T(1)*T(3)+T(1)
==> T(1)^3+2*T(1)*T(2)*T(4)-T(2)^2-2*T(4)
==> T(1)^5+4*T(1)^3*T(2)*T(4)+4*T(1)*T(2)^2*T(4)^2
              -T(3)^2+2*T(3)-1
==> T(1)^3*T(2)+2*T(1)*T(2)^2*T(4)-T(1)^2*T(3)
              -2*T(2)*T(3)*T(4)+T(1)^2+2*T(2)*T(4)
==> T(1)^2*T(4)+2*T(2)*T(4)^2
```

with map $A = S/I \to A_1, x \mapsto T_1, y \mapsto T_2$.

Now the second normalization loop has to be started with A_1. Again the criterion for stopping is not fullfilled, that is, A_1 is not equal to $\mathrm{Hom}_{A_1}(J_1,J_1)$, and the ring A_2 will be computed as affine ring in 4 variables modulo 8 equations. Again, the criterion is not fullfilled, and A_3 is a ring in 3 variables modulo 9 equations. Now in J_3 a zero-divisor of A_3 is found and the ring splits into two rings. Both rings are isomorphic to the polynomial ring in one variable and the algorithm stops with the message:

```
==>// 'normal' created a list of 2 ring(s).
==>// To see the rings, type (if the name of your list
==>// is nor):
==>      show(nor);
==>// To access the 1-st ring and map (similar for the
==>// others), type:
==>      def R = nor[1]; setring R;  norid; normap;
==>// R/norid is the 1-st ring of the normalization and
==>// normap the map from the original basering to
==>// R/norid
def R1 = nor[1]; setring R1;  norid; normap;
==> norid[1]=0
==> normap[1]=T(1)^2   normap[2]=T(1)^3
def R2 = nor[2]; setring R2;  norid; normap;
==> norid[1]=0
==> normap[1]=-T(1)^3  normap[2]=T(1)^2
```

Hence one finds out that $R_1 = \mathbb{Q}[T]$, $R_2 = \mathbb{Q}[T]$, and the normalization of A is given as $A \longrightarrow \mathbb{Q}[T] \oplus \mathbb{Q}[T]$, $x \mapsto (T^2, -T^3)$, $y \mapsto (T^3, T^2)$.

1.5.2. *Non-normal locus*

Let us start with a plane A_k-singularity given by $I = \langle x^2 + y^{k+1} \rangle$ in $\mathbb{Q}[x,y]$, which can easily be done by hand.

We get $I_{\text{Sing}} = \langle x, y^k \rangle$, as test ideal $J = \sqrt{I_{\text{Sing}}} = \langle x, y \rangle$ and $u = x$ as a non-zerodivisor of $S/I = \mathbb{Q}[x,y]/\langle x^2 + y^{k+1} \rangle$. Now

$$(uJ+I) : J = \langle x^2, xy, y^{k+1} \rangle : \langle x, y \rangle = \langle x, y^k \rangle,$$

hence $I_{NN} = \langle x, x^2 + y^{k+1} \rangle : \langle x, y^k \rangle = \langle x, y \rangle$, as expected. If we do the same with the A_k-surface singularity $I = \langle x^2 + z^2 + y^{k+1} \rangle \subset \mathbb{Q}[x,y,z]$, we get $J = \langle x,y,z \rangle$, $u = z$,

$$(uJ+I) : J = \langle z^2, yz, xz, y^{k+1} + x^2 \rangle : \langle x,y,z \rangle = \langle z, y^{k+1} + x^2 \rangle$$

and, finally, $I_{NN} = \langle u, I \rangle : ((uJ+I) : J) = \langle 1 \rangle$ which is true, since any isolated hypersurface singularity of dimension ≥ 2 is normal.

Let us compute the nonnormal locus of two transversal cusps in the plane, using the procedure `nnlocus` from `normal.lib` in SINGULAR.

```
ring S = 0,(x,y),dp;
ideal I = (x2-y3)*(x3-y2);
ideal NN = nnlocus(I);
```

The radical J of the singular locus is computed as

```
==> J[1]=xy2-y   J[2]=x2y-x   J[3]=y4-x   J[4]=x4-y
```

$u = xy^2 - y$ is chosen as a non-zerodivisor in J and $(uJ+I) : J$ is

```
==> _[1]=xy2-y   _[2]=y4-x2y   _[3]=x3y-y3   _[4]=x4-y
```

Typing `NN;` we get as result the following ideal defining the non-normal locus (equal to J, but with different generators):

```
==> NN[1]=y3-x2   NN[2]=xy2-y   NN[3]=x2y-x   NN[4]=x3-y2
```

1.5.3. *Integral closure of an ideal*

We want to compute the integral closure of the Jacobian ideal I of the singularity E_8, defined by $x^5 + y^3 + z^2 = 0$. Again, we set the printlevel sufficiently high to get intermediate results and comments, which are partly reproduced.

```
LIB "reesclos.lib";
ring A = 0,(x,y,z),dp;
ideal I = jacob(x5+y3+z2);   // the Jacobian ideal
```

Let us first compute the Rees algebra of I.

```
list rees = ReesAlgebra(I);
def Rees = rees[1];
setring Rees;
```

```
reesid;
==> reesid[1]=3*y^2*U(3)-2*z*U(2)
==> reesid[2]=5*x^4*U(3)-2*z*U(1)
==> reesid[3]=5*x^4*U(2)-3*y^2*U(1)
def At = rees[2]; setring At;
reesmap;
==> reesmap[1]=x      reesmap[2]=y      reesmap[3]=z
==> reesmap[4]=5x4t   reesmap[5]=3y2t   reesmap[6]=2zt
```

$\mathbb{Q}[x,y,z,U_1,U_2,U_3]/\text{reesid}$ is isomorphic to the Rees algebra $\mathcal{R}(I)$ as subalgebra of $\mathbb{Q}[x,y,z,t]$, under the map

$$(x,y,z,U_1,U_2,U_3) \longmapsto \big(\text{reesmap}(1),\ldots,\text{reesmap}(6)\big)\,.$$

Let us now compute the integral closure of I:

```
list norI = normalI(I);
```

After 3 iterations we reach the normalization of the Rees algebra as $Q[T_1,\ldots,T_7]$ modulo the ideal

```
==> 4*T(4)*T(5)-15*T(7)^2
==> 5*T(1)^2*T(7)-2*T(2)*T(4)
==> 3*T(2)^2*T(6)-2*T(3)*T(5)
==> 2*T(1)^2*T(5)-3*T(2)*T(7)
==> 4*T(1)*T(4)*T(5)*T(7)-15*T(1)*T(7)^3
==> T(1)^2*T(2)*T(6)-T(3)*T(7)
==> 5*T(1)^4*T(6)-2*T(3)*T(4)
```

Now we have to determine the map $Q[T_1,\ldots,T_7] \to A[t]$. This is computed by representing the ring variables $T_1,\ldots,T_7$ as fractions in the variables of the Rees algebra. We get

```
==> T(1)  : 25*x^9*z
==> T(2)  : 25*x^8*y*z
==> T(2)  : 25*x^8*z^2
==> T(2)  : 25*x^8*z*U(1)
==> T(2)  : 15*x^4*y^2*z*U(1)
==> T(2)  : 10*x^4*z^2*U(1)
==> T(2)  : 10*x^6*y*z*U(1)
```

with the "universal" denominator: $25x^8z$. Since $\mathcal{R}(I)$ is the image under the map $\mathbb{Q}[x,y,z,U_1,U_2,U_3] \longrightarrow \mathbb{Q}[x,y,z,t]$, where $U_1 \mapsto 5x^4t$, $U_2 \mapsto 3y^2t$, $U_3 \mapsto 2zt$, then $\mathcal{R}(I)$ is generated in $Q[x,y,z,t]$ by

```
==> generator 1 : x          generator 2 : y
==> generator 3 : z          generator 4 : 5x4t
==> generator 5 : 3y2t       generator 6 : 2zt
==> generator 7 : 2x2yt
```

That is, $\mathcal{R}(I)$ is generated in t-degree 1 and, hence, $\overline{I^k} = (\overline{I})^k$ for all $k > 1$. In particular, the integral closure of $I = \langle 5x^4, 3x^2, 2z \rangle$ is generated by 4 elements, the extra element being x^2y. This result is stored in the first entry of the list `norI`:

```
norI[1];
==> _[1]=5x4   _[2]=3y2   _[3]=2z   _[4]=2x2y
```

2. Effective Construction of Algebraic Geometry Codes

Goppa's construction of linear codes using algebraic geometry, the so-called *geometric Goppa* or *AG codes*, was a major breakthrough in the history of coding theory. In particular, it was the first (and only) construction leading to a family of codes with parameters above the Gilbert-Varshamov bound (Tsfasman, Vlăduţ and Zink).

There exist several (essentially equivalent) ways to construct AG codes starting from a smooth projective curve $\widetilde{C}$ defined over a finite field $\mathbb{F}$. Mainly, we should like to mention the L-, resp. the *Ω-construction.*

Given rational points $Q_1, \ldots, Q_n \in \widetilde{C}$ and a rational divisor G on $\widetilde{C}$ having disjoint support with the divisor $D = Q_1 + \ldots + Q_n$, the AG code $C_L(G, D, \widetilde{C})$, resp. $C_\Omega(G, D, \widetilde{C})$, is the image of the $\mathbb{F}$-linear map

$$ev_D : \mathcal{L}(G) \longrightarrow \mathbb{F}^n, \quad f \longmapsto \big(f(Q_1), \ldots, f(Q_n)\big), \text{ resp.}$$
$$res_D : \Omega(G-D) \longrightarrow \mathbb{F}^n, \quad \omega \longmapsto \big(\mathrm{res}_{Q_1}(\omega), \ldots, \mathrm{res}_{Q_n}(\omega)\big).$$

In practice, there are two main difficulties when looking for an *effective method* to compute the generator matrices of such codes: Given a plane (singular) model C of $\widetilde{C}$, how to compute the places of C and how to compute a basis for the linear system $\mathcal{L}(G)$ (cf. below), resp. for the vector space of rational one-forms

$$\Omega(G-D) = \big\{\omega \in \Omega(X)^* \,\big|\, (\omega) + G - D \geq 0\big\} \cup \{0\}.$$

One possible solution, making use of blowing-up theory (to compute the places of C) and of the (classical) Brill-Noether algorithm (for the computation of a basis of $\mathcal{L}(G)$) is presented in (Le Brigand and Risler; Haché and Le Brigand). In the following, we should like to point to the modified approach of Campillo and Farrán (1998), using Hamburger-Noether expansions instead of blowing-up theory, and to present in some detail the resulting algorithm as implemented in the computer algebra system SINGULAR (Greuel, Pfister and Schönemann).

2.1. PRELIMINARIES, NOTATIONS

Throughout the following, let $\mathbb{F}$ be a finite field and $\overline{\mathbb{F}}$ an algebraic closure of $\mathbb{F}$. Moreover, let $C \subset \mathbb{P}^2(\overline{\mathbb{F}})$ be an *absolutely irreducible*, reduced, projective plane curve given by a homogeneous form of degree d, $F \in \mathbb{F}[X,Y,Z]_d$.

A point $P \in C$ is called *rational* if its coordinates are in $\mathbb{F}$. More generally, by a *closed point* $[P] \in C$ we denote the formal sum of a point (defined over $\overline{\mathbb{F}}$) with its conjugates. If there is no risk of confusion, we sometimes write $P \in C$ to denote the closed point $[P]$. Note that closed points are invariant under the action of the Galois group $\mathrm{Gal}(\overline{\mathbb{F}}/\mathbb{F})$.

We denote by $n : \widetilde{C} \to C$ the normalization and by

$$N := i \circ n : \ \widetilde{C} \to \mathbb{P}^2(\overline{\mathbb{F}})$$

the parametrization of C (with $i : C \hookrightarrow \mathbb{P}^2(\overline{\mathbb{F}})$ being the inclusion).

The points of $\widetilde{C}$ are called *places of* C. Again, a place is called *rational* if its coordinates are in $\mathbb{F}$, and by a *closed place* we denote the formal sum of a place (defined over $\overline{\mathbb{F}}$) with its conjugates. Note that each smooth (rational, resp. closed) point $P \in C$ corresponds to a unique (rational, resp. closed) place $n^{-1}(P) \in \widetilde{C}$. If P is a singular point of C then each local branch of C at P corresponds to a unique place of C. Hence, the set of places of C can be identified with the set consisting of the non-singular points of C and all tuples $\big(P;(C_i,P)\big)$, P a singular point of C and (C_i,P) a local branch of C at P.

A *(rational) divisor* D on $\widetilde{C}$ is a finite, weighted, formal sum of (closed) places Q of C, $D = \sum n_Q Q$ with integer coefficients $n_Q =: \mathrm{ord}_Q(D)$. The divisor D is called *effective* if there are no negative n_Q. Moreover, we introduce the *degree* of the divisor D, $\deg D := \sum n_Q$ and the *support* of D, $\mathrm{supp}\, D := \{Q \mid n_Q \neq 0\}$.

To each element g in the function field $\mathbb{F}(\widetilde{C})$ of $\widetilde{C}$ one associates the *principal divisor* $(g) := \sum \mathrm{ord}_Q(g) \cdot Q$. Note that (g) has degree 0, by the residue theorem.

Finally, the *linear system* associated to a divisor D is defined to be

$$\mathcal{L}(D) := \big\{ g \in \mathbb{F}(\widetilde{C}) \mid (g) \geq -D \big\} \cup \{0\},$$

where the order relation is defined by

$$D \geq D' \ :\Longleftrightarrow \ \mathrm{ord}_Q(D) \geq \mathrm{ord}_Q(D') \text{ for all places } Q \in \widetilde{C}.$$

2.2. SYMBOLIC HAMBURGER-NOETHER EXPRESSIONS

We recall the definition of Hamburger-Noether expansions (HNE), resp. symbolic Hamburger-Noether expressions, for the branches of a reduced plane curve singularity. They can be regarded as the analogue of Puiseux expansions when working over a field of positive characteristic (cf. the discussion in (Campillo), 2.1). As being the case for Puiseux expansions, many invariants of plane curve singularities (such as the multiplicity sequence, the δ-invariant, the intersection multiplicities of the branches, etc.), can be computed directly from the corresponding system of HNE.

Let $P \in C$ be a point and x, y local parameters at P. Moreover, let the germ (C, P) be given by a local equation $f \in \mathbb{F}[[x,y]]$ with irreducible decomposition $f = f_1 \cdot \ldots \cdot f_r \in \overline{\mathbb{F}}[[x,y]]$. Finally, let's suppose that x is a *transversal parameter* for (C, P), that is, the order of $f(0,y)$ is equal to the order of $f(x,y)$.

DEFINITION 2.1. A *Hamburger-Noether expansion (HNE)* of C at P for the local branch given by f_ν (defined over some finite algebraic extension $\mathbb{F} \subset \mathbb{F}_{\text{ext}}$) is a finite sequence of equations

$$\begin{aligned}
z_{-1} &= a_{0,1}z_0 + a_{0,2}z_0^2 + \cdots + a_{0,h_0}z_0^{h_0} + z_0^{h_0}z_1 \\
z_0 &= \qquad a_{1,2}z_1^2 + \cdots + a_{1,h_1}z_1^{h_1} + z_1^{h_1}z_2 \\
&\vdots \\
z_{i-1} &= \qquad a_{i,2}z_i^2 + \cdots + a_{i,h_i}z_i^{h_i} + z_i^{h_i}z_{i+1} \\
&\vdots \\
z_{s-2} &= \qquad a_{s-1,2}z_{s-1}^2 + \cdots + a_{s-1,h_{s-1}}z_{s-1}^{h_{s-1}} + z_{s-1}^{h_{s-1}}z_s \\
z_{s-1} &= \qquad a_{s,2}z_s^2 + a_{s,3}z_s^3 + \cdots\cdots
\end{aligned} \tag{2.1}$$

where s is a non-negative integer, $a_{j,i} \in \mathbb{F}_{\text{ext}}$ and h_j, $j = 1, \ldots, s-1$, are positive integers, such that $f_\nu(z_0(t), z_{-1}(t)) = 0$ in $\mathbb{F}_{\text{ext}}[[z_s]]$.

If the local equation of (C, P) is polynomial in x, y, i.e., $f \in \mathbb{F}[x,y]$, then the last (infinite) row of (2.1) can be replaced, equivalently, by a (finite) implicit equation

$$g(z_s, z_{s-1}) = 0, \qquad g \in \mathbb{F}_{\text{ext}}[x,y], \quad \frac{\partial g}{\partial z_{s-1}}(0,0) \neq 0.$$

The resulting system is called a *symbolic Hamburger-Noether expression* (sHNE) for the branch.

Any HNE leads to a primitive parametrization $\varphi : \mathbb{F}[[x,y]] \to \mathbb{F}_{\text{ext}}[[t]]$ of the branch (setting $t := z_s$ and mapping $x \mapsto z_0(z_s)$, $y \mapsto z_{-1}(z_s)$). It can be computed from a sHNE up to an arbitrary finite degree in t.

REMARK 2.2. There exist constructive algorithms to compute a system of sHNE's (resp. HNE's up to a given degree) for the branches of a reduced plane curve singularity (cf. (Campillo) for the irreducible case, resp. (Rybowicz) for the reducible case). A modification of the latter algorithm is implemented in the computer algebra system SINGULAR (Lamm; Greuel, Pfister and Schönemann).

To perform the algorithm one does not need any knowledge about the irreducible factorization of f in $\overline{\mathbb{F}}[[x,y]]$. Moreover, in the reducible case, the probably necessary (finite) field extension is not performed a priori, but by successive (primitive) extensions introduced exactly when needed for computations (factorization).

REMARK 2.3. Given a Hamburger-Noether expansion (2.1) for a local branch (C_v, P), the corresponding *multiplicity sequence* $m_1, \dots, m_n$ can be easily read off:

$$\underbrace{n_0,\dots,n_0}_{h_0 \text{ times}},\underbrace{n_1,\dots,n_1}_{h_1 \text{ times}},\dots\dots,\underbrace{n_{s-1},\dots,n_{s-1}}_{h_{s-1} \text{ times}},1,\dots,1$$

where the n_j are defined recursively by setting $n_s := 1$ and, for $j = s, \dots, 1$,

$$n_{j-1} := \begin{cases} n_j h_j + n_{j+1} & \text{if } a_{j,\ell} = 0 \text{ for all } 2 \leq \ell \leq h_j, \\ n_j \ell & \text{otherwise, with } \ell \geq 2 \text{ minimal s.th. } a_{j,\ell} \neq 0. \end{cases}$$

In particular, we can compute the δ-*invariant* of the branch directly from the Hamburger-Noether expansion, since

$$\delta(C_v, P) = \sum_{i=1}^{n} \frac{m_i(m_i - 1)}{2}.$$

Moreover, if $\varphi : \mathbb{F}[[x,y]] \to \mathbb{F}_{\text{ext}}[[t]]$ is the primitive parametrization of (C_v, P) obtained from a Hamburger-Noether expansion, then the intersection multiplicity of (C_v, P) with the plane curve germ (C', P) given by $g \in \overline{\mathbb{F}}[[x,y]]$ can be computed as $i_P(C_v, C') = \operatorname{ord}_t g\big(\varphi(x), \varphi(y)\big)$.

2.3. ADJOINT CURVES AND BRILL-NOETHER RESIDUE THEOREM

This section is devoted to the "core" of the presented algorithm, the Brill-Noether residue theorem. In an intuitive formulation it states that "the intersection divisors of C with all adjoint curves of a fixed degree m form a complete linear system, up to being shifted by the adjoint divisor".

Let P be a closed point of C. We introduce the local (resp. semilocal) rings $\mathcal{O} := \mathcal{O}_{C,P}$, resp. $\overline{\mathcal{O}} := (n_* \mathcal{O}_{\tilde{C}})_P \cong \prod_{i=1}^{r} \mathcal{O}_{\tilde{C},Q_i}$, where $Q_1, \dots, Q_r$ denote the closed places of C over P. Note that $\mathcal{O}$ is a subring of $\overline{\mathcal{O}}$. The *conductor*

$$\mathcal{C}_P := \mathcal{C}_{\overline{\mathcal{O}}/\mathcal{O}} := \{\phi \in \overline{\mathcal{O}} \mid \phi \cdot \overline{\mathcal{O}} \subset \mathcal{O}\}$$

is an ideal both in $\mathcal{O}$ and in $\overline{\mathcal{O}}$. It defines a divisor $\mathcal{A}_P$ on $\tilde{C}$ whose support is $\{Q_1, \dots, Q_r\}$, the set of all places of C over the singular point P. Recall that such a place corresponds to a unique branch of (C,P). By abuse of notation, we sometimes do not distinguish between the place and the corresponding branch.

DEFINITION 2.4. We call the divisor $\mathcal{A} := \sum_P \mathcal{A}_P$ the *adjunction divisor* (or the *divisor of double points*) of C. Its support is the set of all places over singular points of C.

REMARK 2.5. The adjunction divisor $\mathcal{A} = \sum_Q d_Q Q$ is rational, that is, conjugate branches Q, Q' satisfy $d_Q = d_{Q'}$. We have $d_Q = 2 \cdot \delta(C,P)$ if Q is a place over an *irreducible* plane curve singularity (C,P), resp.

$$d_Q = 2 \cdot \delta(C_i, P) + \sum_{j \neq i} i_P(C_i, C_j) \tag{2.2}$$

if Q corresponds to the local branch (C_i, P) of (C,P) (δ denoting the δ-invariant and i_P the local intersection multiplicity at P). Alternatively, one can use the *Dedekind formula* to compute the multiplicities d_Q: let $\varphi_Q : \mathbb{F}[[x,y]] \to \overline{\mathbb{F}}[[t]]$ be a primitive parametrization of the branch Q then

$$d_Q = \operatorname{ord}_t \left(\frac{f_y(\varphi_Q(x), \varphi_Q(y))}{\frac{d}{dt}\varphi_Q(x)} \right) = \operatorname{ord}_t \left(\frac{f_x(\varphi_Q(x), \varphi_Q(y))}{\frac{d}{dt}\varphi_Q(y)} \right)$$

if the respective expressions are finite (notice that either $\frac{d}{dt}\varphi_Q(x)$ or $\frac{d}{dt}\varphi_Q(y)$ does not vanish identically).

Notation. Let $H \in \mathbb{F}[X,Y,Z]_m$ be a homogeneous form of degree m such that F does not divide H. Then we denote by N^*H the intersection divisor on $\widetilde{C}$ cut out by the (preimage under N of the) plane curve defined by H.

DEFINITION 2.6. Let D be a rational divisor on $\mathbb{P}^2(\mathbb{F})$ such that C is not contained in the support of D. We call D an *adjoint divisor* of C iff the pull-back divisor satisfies $N^*D \geq \mathcal{A}$.

Let $H \in \mathbb{F}[X,Y,Z]_m$, $F \not| H$, such that $N^*H \geq \mathcal{A}$. Then we call H an *adjoint form* and the plane curve defined by H an *adjoint curve* of C.

Note that D is an adjoint divisor iff the intersection multiplicity of D with every local branch Q of C is at least d_Q.

PROPOSITION 2.7 (BRILL-NOETHER RESIDUE THEOREM). *Let $C \overset{i}{\hookrightarrow} \mathbb{P}^2(\mathbb{F})$ be a reduced absolutely irreducible plane projective curve given by the homogeneous polynomial $F \in \mathbb{F}[X,Y,Z]_d$. Moreover, let $n : \widetilde{C} \to C$ be the normalization and D a rational divisor on $\widetilde{C}$.*

Finally, let $H_0 \in \mathbb{F}[X,Y,Z]_m$ be an adjoint form of degree $m > 0$ such that $N^*H_0 \geq \mathcal{A} + D$. Then we can identify

$$\mathcal{L}(D) \equiv \left\{ \left[\frac{H}{H_0}\right] \in \mathbb{F}(C) \;\middle|\; \begin{array}{c} H \in \mathbb{F}[X,Y,Z]_m,\ F \not| H, \\ N^*H \geq N^*H_0 - D \end{array} \right\} \cup \{0\}$$

under the isomorphism $\mathbb{F}(\widetilde{C}) \cong \mathbb{F}(C)$ induced by N.

This proposition is an immediate corollary of

THEOREM 2.8 (M. NOETHER). *Let $G,H \in \mathbb{F}[X,Y,Z]$ be homogeneous forms such that $N^*H \geq \mathcal{A} + N^*G$. Then there exist homogeneous forms $A,B \in \mathbb{F}[X,Y,Z]$ such that $H = AF + BG$.*

For a complete proof we refer to (Van der Waerden), pp. 215ff, resp. (Le Brigand and Risler), Prop. 4.1.

REMARK 2.9. Haché (1996) has shown that a form $H_0 \in \mathbb{F}[X,Y,Z]_m$ as in the Brill-Noether residue theorem exists whenever

$$m > \max\left\{ d-1, \frac{d-3}{2} + \frac{\deg(\mathcal{A} + D_+)}{d} \right\}, \tag{2.3}$$

where D_+ denotes the effective part of the divisor $D = D_+ - D_-$.

2.4. COMPUTATIONAL ASPECTS

Let C be an absolutely irreducible plane projective curve defined over the prime field $\mathbb{F} = \mathbb{F}_p$ (given by a homogeneous form $F \in \mathbb{F}[X,Y,Z]_d$).

2.4.1. *Computing the places of C*

A place Q of C is represented by a triple, consisting of

- the closed point $[P] \in C$ corresponding to Q,
- the *degree* k_Q of the place (that is, the minimal degree of a field extension defining Q) and
- a symbolic Hamburger-Noether expression HN for the local branch corresponding to Q (defined over a primitive algebraic field extension $\mathbb{F} \subset \mathbb{F}(a)$ of degree k_Q).

Recall that $[P] \in C$ denotes the formal sum of a point $P \in C$ with its conjugates. Affine closed points will be represented by a defining (triangular) ideal $I = \langle \phi, \psi \rangle$ in $\mathbb{F}[x,y]$, while closed points at infinity are usually stored in form of a homogeneous polynomial $\Phi \in \mathbb{F}[X,Y]$ (the defining prime factor of $F(X,Y,0)$).

Note that the conjugates of a place Q are given by the triples $([P], k_Q, HN')$, where HN' runs through the conjugates of HN. Hence, when computing the closed places of C, we can restrict ourselves to computing one representing place for each. We apply the following algorithm:

INPUT: Squarefree homogeneous polynomial $F \in \mathbb{F}[X,Y,Z]_d$, degree bound $k \in \mathbb{N}$.

OUTPUT: List L of all closed singular places and all closed non-singular places up to degree k of the plane curve C defined by F.

1. *Affine singular points.* Let $f(x,y) := F(x,y,1)$ and $I := \langle f, f_x, f_y \rangle$ the Tjurina ideal of f. Compute a *triangular system* for I, that is, a system of triangular bases T_i such that $V(I) = V(T_1) \cup \ldots \cup V(T_s)$.
Here, by a *triangular basis* one denotes a reduced lexicographical Gröbner basis of the form $T = \{\phi, \psi\}$ with $\phi \in \mathbb{F}[y]$ a monic polynomial in y and $\psi = y^b + \sum_{i<b} \psi_i(x) y^i \in \mathbb{F}[x,y]$. Triangular systems can be computed effectively, basically by two different methods, one due to Lazard (1992) (see also Della Dorca, Dicrescenzo and Duval), the other due to Möller (1993). Choose any of these methods to compute a triangular system for I, $T_i = \{\phi_i, \psi_i\}$, $i = 1, \ldots, s$. For each i,
   - compute a prime factorization of ϕ_i in $\mathbb{F}[y]$,
   $$\phi_i = \phi_{i,1} \cdot \ldots \cdot \phi_{i,r_i} \in \mathbb{F}[y];$$
   - for $j = 1, \ldots, r_i$, let $\mathbb{F}(a_{i,j}) = \mathbb{F}[y]/\langle \phi_{i,j} \rangle$ be the primitive field extension defined by the irreducible polynomial $\phi_{i,j}$. Compute a prime factorization of ψ_i in $\mathbb{F}(a_{i,j})[x,y]$,
   $$\psi_i = \psi_{i,1} \cdot \ldots \cdot \psi_{i,s_i} \in \mathbb{F}(a_{i,j})[x,y].$$

   Finally, the closed affine singular points are given by the set of ideals
   $$\mathrm{SING}_{\mathrm{aff}} := \left\{ \langle \phi_{i,j}, \overline{\psi}_{i,k} \rangle \subset \mathbb{F}[x,y] \;\middle|\; \begin{array}{c} j = 1, \ldots, r_i,\, k = 1, \ldots, s_i, \\ i = 1, \ldots, s \end{array} \right\},$$
   where $\overline{\psi}_{i,j}$ is the image of $\psi_{i,j}$ when substituting the parameter $a_{i,j}$ by y.

2. *Points at infinity.* Let $f_\infty(x) := F(x,1,0)$ and compute a prime factorization of the polynomial $f_\infty \in \mathbb{F}[x]$,
   $$f_\infty = f_{\infty,1} \cdot \ldots \cdot f_{\infty,d'} \in \mathbb{F}[x], \qquad d' \leq d. \tag{2.4}$$
   Let $a_j \in \overline{\mathbb{F}}$ be a root of $f_{\infty,j}$ and define
   $$\mathrm{PTS}_\infty := \{ [(a_j:1:0)] \mid j = 1, \ldots, d' \},$$
   where $[(a_j:1:0)]$ denotes the formal sum of the point $(a_j:1:0)$ (defined over $\overline{\mathbb{F}}$) with its conjugates. (It is represented by $f_{\infty,j}$.)
   We denote by $\mathrm{SING}_\infty \subset \mathrm{PTS}_\infty$ the subset of closed singular points. To check whether a point $[(a_j:1:0)]$ is singular or not, one has to check whether $F_X(a_j,1,0) = F_Z(a_j,1,0) = 0$ (these computations can be performed over the finite field extension $\mathbb{F}(a_j) = \mathbb{F}[x]/\langle f_{\infty,j} \rangle$).
   Finally, consider the (closed) point $P = (1:0:0)$: if $F(1,0,0) = 0$ then P has to be added to PTS_∞; if, additionally, $F_Y(1,0,0)$ and $F_Z(1,0,0)$ vanish then it has to be added to SING_∞, too.
   The sets PTS_∞ (resp. SING_∞) are the sets of closed (singular) points at infinity.

3. *Affine singular places.* To each closed affine singular point $[P]$ given by a (triangular) ideal $\langle\phi,\psi\rangle \in \mathrm{SING}_{\mathrm{aff}}$ we compute the corresponding places in form of a system of symbolic Hamburger-Noether expressions for the respective germ (C,P) (defined over $\overline{\mathbb{F}}$). More precisely, a closed place over $[P]$ is the formal sum of a place Q described by one of the computed sHNE with its conjugates. The computation of the sHNE has to be performed in the local ring $\mathbb{F}(a)[x,y]_P$ where $\mathbb{F} \subset \mathbb{F}(a)$ is a primitive field extension (of degree $k_P = \deg_y(\phi)\deg_x(\psi)$) such that ϕ,ψ decompose into linear factors. Note that during the computation of a sHNE further field extensions might be necessary.

4. *Singular places at infinity.* To each closed singular point $[P]$ in SING_∞ we compute a system of sHNE for the local germ (C,P) (defined over $\overline{\mathbb{F}}$). To be precise, if $P=(a_j:1:0)$ then we compute a system of sHNE for $F(x+a_j,1,z)$ in $\mathbb{F}(a_j)[x,z]_{\langle x,z\rangle}$; if $P=(1:0:0)$ then the system of sHNE is computed for $F(1,y,z)$ in $\mathbb{F}[y,z]_{\langle y,z\rangle}$.

5. *Non-singular affine closed points up to degree k.* For each $1 \leq \ell \leq k$ do the following:

   - let $q := p^\ell$ and set $I := \langle f, x^q - x, y^q - y\rangle$.
   - Proceed as in Step 1 to obtain a set of (triangular) ideals $\mathrm{PTS}_{\mathrm{aff}}$ corresponding to the set of closed points defined over $\mathbb{F}_q$.
   - For all non-singular $[P] \in \mathrm{PTS}_{\mathrm{aff}}$ (given by $\langle\phi,\psi\rangle \subset \mathbb{F}[x,y]$) compute the degree $k_P = \deg_y(\phi)\deg_x(\psi)$). If $k_P = \ell$ then compute the corresponding closed place (that is, a sHNE for the germ (C,P)) and add it to the list L of closed places.

REMARK 2.10. It is interesting to notice that triangular sets have mainly been used for numerical purpose, since they allow a fast and stable numerical solving of polynomial systems (cf. (Lazard; Möller; Greuel, 2000b)), and this has been the reason for implementing it in SINGULAR. Several experiments have shown that they behave also superior against other methods to represent closed points over finite fields.

2.4.2. *Computing the adjunction divisor*

For any (closed) singular place we determine the multiplicity $d_{[Q]} := \mathrm{ord}_Q(\mathcal{A})$, alternatively, by the formula (2.2) or the Dedekind formula (cf. Remarks 2.5 and 2.3). Note that to compute a local intersection number $i_P(f,g)$ (as appearing in both formulas), we can proceed inductively, computing the primitive parametrization $\varphi : \mathbb{F}[[x,y]] \to \overline{\mathbb{F}}[[t]]$ of g up to degree k, until $\mathrm{ord}_t(\varphi(x) \bmod t^k, \varphi(y) \bmod t^k)$ is less than $k/\mathrm{ord}(f)$.

2.4.3. *Computing the divisors N^*H*
The intersection divisors N^*H can be computed by the following algorithm:

INPUT: $F \in \mathbb{F}[X,Y,Z]_d$, L list of closed places $[Q]$ of C (defined by $F=0$), $H \in \mathbb{F}[X,Y,Z]_m$.

OUTPUT: Extended list of closed places $[Q]$ and list of integers $m_{[Q]}$ such that $N^*H = \Sigma_{[Q]} m_{[Q]}[Q]$.

1. *Affine Intersection.* Let $h(x,y) := H(x,y,1)$, $f(x,y) := F(x,y,1)$ and consider $I := \langle h, f\rangle$. Proceed as in Step 1 of the algorithm in Section 2.4.1 to obtain a set of (triangular) ideals corresponding to the set of closed points in $V(I)$. For each closed point $[P]$ in $V(I)$ do the following:
   - check for each closed place $[Q]$ in L whether $[Q]$ lies above $[P]$ (that is, the first entry of the triple representing $[Q]$ has to be $[P]$). If this is the case, compute the multiplicity $m_{[Q]} = \mathrm{ord}_t\, h(\varphi_Q(x), \varphi_Q(y))$, where φ_Q is the primitive parametrization obtained from the sHNE (third entry) of $[Q]$.
   - If there is no such $[Q]$ in L then compute the sHNE for the (smooth) germ (C,P), add the resulting place $[Q]$ to the list L and proceed as before to obtain $m_{[Q]}$.
2. *Intersection at infinity.* Let $h_\infty(x) := H(x,1,0)$ and compute a prime factorization of the polynomial $h_\infty \in \mathbb{F}[x]$,

$$h_\infty = h_{\infty,1} \cdot \ldots \cdot h_{\infty,d''} \in \mathbb{F}[x], \qquad d'' \le d.$$

   Each factor that appeared also in the prime factorization (2.4) of f_∞ corresponds to a closed point $[(a_j:1:0)]$ in the intersection of C with the plane curve defined by H. For the corresponding closed places $[Q]$ we compute

$$m_{[Q]} = \mathrm{ord}_t\, H(\varphi_Q(x)+a_j,\, 1,\, \varphi_Q(z)),$$

   where $\varphi_Q : \mathbb{F}[[x,z]] \to \overline{\mathbb{F}}[[t]]$ is the primitive parametrization obtained from the sHNE of Q (cf. Step 4 of the algorithm in Section 2.4.1).
   Finally, if $(1:0:0) \in \mathrm{PTS}_\infty$ and $H(1,0,0) = 0$ then we compute for the corresponding closed places $[Q]$ the multiplicities

$$m_{[Q]} = \mathrm{ord}_t\, H(1, \varphi_Q(y), \varphi_Q(z)).$$

2.4.4. *Computing $\mathbb{F}$-bases of adjoint forms*
Let $\mathrm{Monom}(m) = \{G_1, \ldots, G_M\}$ denote the monomial basis for $\mathbb{F}[X,Y,Z]_m$ over $\mathbb{F}$, resp. the corresponding vector of monomials. We represent a homogeneous form $H \in \mathbb{F}[X,Y,Z]_m$ by the vector $v \in \mathbb{F}^n$ such that $H = \mathrm{Monom}(m)^T \cdot v$. To compute $\mathbb{F}$-bases of adjoint forms as needed by the Brill-Noether algorithm, we apply the following algorithm:

INPUT: $F \in \mathbb{F}[X,Y,Z]_d$, L list of closed places of C (defined by $F=0$),
$m > d$ a positive integer,
non-negative integers $d_{[Q]}$, $[Q] \in L$, s.th. $\mathcal{A} = \sum_{[Q]} d_{[Q]}[Q]$,
non-negative integers $n_{[Q]}$, $[Q] \in L$, s.th. $E = \sum_{[Q]} n_{[Q]}[Q]$.

OUTPUT: $\mathbb{F}$-basis of a subspace of

$$V := \left\{ H \in \mathbb{F}[X,Y,Z]_m \,\middle|\, F \mid H \text{ or } N^*H \geq \mathcal{A} + E \right\}$$

complementary to $W := F \cdot \mathbb{F}[X,Y,Z]_{m-d}$ (given in form of a matrix of coefficients w.r.t. $\mathrm{Monom}(m)$).

1. *The subspace $W \subset V$ of forms divisible by F.* Compute $W \in \mathrm{Mat}(\ell \times M, \mathbb{F})$ such that $W \cdot \mathrm{Monom}(m) = F \cdot \mathrm{Monom}(m-d)$.
   Note that $\ell = \binom{m-d+2}{2}$, $M = \binom{m+2}{2}$.
2. *The space V.* For each closed place $[Q]$ with $m_{[Q]} := d_{[Q]} + n_{[Q]} > 0$ do the following:
   – compute a matrix $A_Q = (a_{ij})_{i,j} \in \mathrm{Mat}(m_{[Q]} \times M, \overline{\mathbb{F}})$ such that

   $$G_j\big(\varphi_Q(x), \varphi_Q(y)\big) = \sum_{i=0}^{m_{[Q]}-1} a_{ij} t^i, \quad j = 1, \ldots, M,$$

   where φ_Q is the primitive parametrization as computed from the sHNE of Q (up to degree $m_{[Q]}$).
   – Let k_Q be the degree of the place Q and let $A_Q^{(i)}$ denote the image of A_Q after applying i times the Frobenius map over $\mathbb{F}$. Then compute

   $$A_{[Q]} := \text{row-red NF} \begin{pmatrix} A_Q \\ A_Q^{(1)} \\ \vdots \\ A_Q^{(k_Q-1)} \end{pmatrix} \in \mathrm{Mat}(k_Q m_{[Q]} \times M, \mathbb{F}).$$

   Finally concatenate $A_{[Q]}$ to obtain $A \in \mathrm{Mat}(K \times M, \mathbb{F})$ with $K = \sum_{[Q]} k_Q m_{[Q]}$ and compute V as the kernel of A, that is

   $$V := \mathrm{Ker}(A) \in \mathrm{Mat}(k \times M, \mathbb{F}), \qquad A \cdot V^T = 0.$$

3. *Compute a complement of W in V.* This can be done, for instance, by using the `lift` command in SINGULAR.

2.4.5. *Computing a basis for $\mathcal{L}(G)$*

Let G be any rational divisor on $\widetilde{C}$. The algorithms of the preceding sections finally allow to compute an $\mathbb{F}$-basis of $\mathcal{L}(G)$ by the *Brill-Noether Algorithm* (cf. Prop. 2.7):

INPUT: $F \in \mathbb{F}[X,Y,Z]_d$, L list of closed places of $C = \{F=0\}$, non-negative integers $d_{[Q]}$, $[Q] \in L$, s.th. $\mathcal{A} = \sum_{[Q]} d_{[Q]} [Q]$, integers $g_{[Q]}$, $[Q] \in L$, s.th. $G = \sum_{[Q]} g_{[Q]} [Q]$.

OUTPUT: Vector space basis of $\mathcal{L}(G)$ (in terms of rational functions on C).

1. *Choose m sufficiently large* (for instance, according to (2.3), above).
2. *Compute* $H_0 \in \mathbb{F}[X,Y,Z]_m$, $F \nmid H_0$, *such that* $N^*H_0 \geq \mathcal{A} + G_+$. To do so, we compute an $\mathbb{F}$-basis for a vector subspace of

$$V_0 := \left\{ H \in \mathbb{F}[X,Y,Z]_m \;\middle|\; F\,|\,H \text{ or } N^*H \geq \mathcal{A} + G_+ \right\},$$

complementary to $W := F \cdot \mathbb{F}[X,Y,Z]_{m-d}$ (cf. Section 2.4.4) and choose any element H_0 of this basis (for instance, with the minimal number of monomials).
3. *Compute the effective divisor* $R := N^*H_0 - G - \mathcal{A}$ (cf. Section 2.4.3).
4. *Compute an* $\mathbb{F}$*-basis* $H_1, \ldots, H_s$ *of a vector subspace of*

$$V := \left\{ H \in \mathbb{F}[X,Y,Z]_m \;\middle|\; F\,|\,H \textit{ or } N^*H \geq \mathcal{A} + R \right\},$$

complementary to $W = F \cdot \mathbb{F}[X,Y,Z]_{m-d}$ (cf. Section 2.4.4).
5. *Return the set of rational functions* $\mathcal{B} := \left\{ \frac{H_1}{H_0}, \ldots, \frac{H_s}{H_0} \right\}$.

2.5. EXAMPLE

The above algorithms are implemented in the library `brnoeth.lib` of SINGULAR, together with procedures for coding and decoding and are distributed with version 2.0. To compute an example, we first have to load the library.

```
LIB "brnoeth.lib";
```

Let C be the absolutely irreducible plane projective curve given by the affine equation $y^2 + y + x^5 \in \mathbb{F}_2[x,y]$. We compute all places up to degree 4, by

```
ring r=2,(x,y),dp;
poly f=y2+y+x5;
list CURVE=Adj_div(f);
==> The genus of the curve is 2.
CURVE=NSplaces(3,CURVE); // places up to degree 4=1+3
```

We can consider the curve as being defined over $\mathbb{F}_{16}$ in order to get many *rational* places.

```
CURVE=extcurve(4,CURVE);
==> Total number of rational places : NrRatPl = 33
```

The degree of the computed (conjugacy classes of) places is displayed by

```
list L=CURVE[3]; L;
==>  [1]: 1,1    [2]: 1,2    [3]: 1,3
==>  [4]: 2,1
==>  [5]: 3,1    [6]: 3,2
==>  [7]: 4,1    [8]: 4,2    [9]: 4,3  [10]: 4,4
==> [11]: 4,5  [12]: 4,6  [13]: 4,7
```

In particular, besides the $33 = 3*1+1*2+7*4$ rational places over $\mathbb{F}_{16}$ there are 2 closed places $[Q_5], [Q_6]$ of degree 3. The adjunction divisor is given by $8Q_1$, where Q_1 is the unique (rational) point on $\widetilde{C}$ mapping to the singular point $(0:1:0)$. This can be read off as follows:

```
CURVE[4];      // the mult's d_Q at L[1],L[2],...
               // (zeroes omitted)
==> 8
def r1=CURVE[5][1][1];
setring r1;
POINTS[1];     // coordinates of the base point of L[1]
==> [1]:  0  [2]:  1   [3]:  0
PARAMETRIZATIONS[1];  // parametrization of L[1]
==> [1]:  _[1]=t3+t8
          _[2]=t5+t15
     // exact up to order:
     [2]:  8,10
```

We construct the evaluating AG-code $C_L(G,D,\widetilde{C})$ where all rational points of $\widetilde{C}$ appear in the support of D and $G = [Q_5]+[Q_6]$.

```
intvec G=0,0,0,0,1,1,0,0,0,0,0,0,0;
intvec D=1..33;
def R=CURVE[1][4];
setring R;
matrix CODE=AGcode_L(G,D,CURVE);
```

The echelon form of the resulting 5×33-matrix is

$$\left(\begin{array}{cccccccccccccccccc}
1 & 0 & 0 & 0 & 0 & 0 & a & a^8 & a^{12} & a^{12} & 0 & a & a^8 & a & a^5 & a^5 & a^9 & a^5 \\
0 & 1 & 0 & 0 & 1 & 0 & 0 & a^8 & a^6 & a^7 & a^5 & a^{10} & a^4 & 1 & a^{12} & a^2 & a^2 & 1 \\
0 & 0 & 1 & 0 & 1 & 0 & a^7 & 0 & a^{13} & a^9 & a^5 & a^6 & a^5 & a^{11} & a^{11} & a^{14} & a^{14} & a^6 \\
0 & 0 & 0 & 1 & 1 & 0 & a^4 & a^{13} & a^6 & a^6 & 0 & a^4 & a^{13} & a^{13} & a^{13} & a^5 & a^8 & a^5 \\
0 & 0 & 0 & 0 & 0 & 1 & a^7 & a^6 & a^4 & a^4 & 1 & a^7 & a^6 & 1 & a^7 & a^6 & a^4 & a^6
\end{array}\right.$$

$$\left.\begin{matrix} a^9 & a & a^5 & a^7 & a & a^7 & a & a^3 & a^{13} & a^4 & a^{12} & a^4 & a^{12} & a^3 & a^{13} \\ a^5 & a^8 & a^6 & a^4 & a^{14} & a^2 & a^{12} & a^4 & a^4 & a^{11} & a^5 & a^8 & a^{12} & a^{11} & a^{13} \\ a^7 & a^9 & a^{13} & a^{14} & a^6 & a^{11} & a^9 & a^5 & a^{13} & a^5 & a^7 & a^{13} & a & a^7 & a^4 \\ a^8 & a^{13} & a^{13} & a^3 & a^7 & a^3 & a^7 & a^{10} & a^{11} & a^{14} & a & a^{14} & a & a^{10} & a^{11} \\ a^4 & 1 & a^7 & a^3 & a^{13} & a^3 & a^{13} & a^7 & a^6 & a^4 & 1 & a^4 & 1 & a^7 & a^6 \end{matrix}\right) .$$

Note that the constructed code $C_L(G,D,\widetilde{C})$ has block length 33, dimension 5 and designed distance $27 = 33 - \deg G$. On the other hand, the first row corresponds to a word of weight 27, whence the designed distance coincides with the minimal distance. As a result, we get that $C_L(G,D,\widetilde{C})$ is a $[33,5,27]$-code. Note that the parameters, that is, the information rate $R = 5/33$ and the relative minimum distance $\delta = 27/33$, lie above the Gilbert-Varshamov bound.

3. V-filtration and spectral numbers of hypersurface singularities

3.1. INTRODUCTION

The *Gauß-Manin connection* $\mathscr{G}$ is a *regular* $\mathscr{D}$-module associated to an isolated hypersurface singularity (Pham). The *V-filtration* on $\mathscr{G}$ is defined by the $\mathscr{D}$-module structure. One can describe $\mathscr{G}$ in terms of integrals of holomorphic differential forms over vanishing cycles (Brieskorn). Classes of these differential forms in the *Brieskorn lattice* $\mathscr{H}''$ can be considered as elements of $\mathscr{G}$. The *V-filtration* on $\mathscr{H}''$ reflects the embedding of $\mathscr{H}''$ in $\mathscr{G}$ and determines the *singularity spectrum* which is an important invariant of the singularity.

E. Brieskorn (1970) gave an algorithm to compute the *complex monodromy* based on the $\mathscr{D}$-module structure which is implemented in the computer algebra system SINGULAR (Greuel, Pfister and Schönemann, 2001) in the library `mondromy.lib` (Schulze, 1999). In many respects, the *microlocal structure* of $\mathscr{G}$ and $\mathscr{H}''$ (Pham) seems to be more natural.

After a brief introduction to the theory of the *Gauß-Manin connection*, we describe how to use this structure for computing in $\mathscr{G}$ and give an explicit algorithm to compute the *V-filtration* on $\mathscr{H}''$. This also leads to a much more efficient algorithm to compute the *complex monodromy* and the singularity spectrum of an arbitrary isolated hypersurface singularity. All algorithms are implemented in the SINGULAR library `gaussman.lib` (Schulze, 2001) and are distributed with version 2.0.

For more theoretical background on this section see (Schulze and Steenbrink).

3.2. MILNOR FIBRATION

Let $f : X \to T$ be a *Milnor representative* (Milnor, 1968) of an isolated hypersurface singularity $f : (\mathbb{C}^{n+1}, 0) \to (\mathbb{C}, 0)$ with Milnor number μ. Then

$$f^{-1}(T') \cap X =: X' \xrightarrow{f} T' := T \setminus \{0\}$$

is a C^∞ fibre bundle with fibres $X_t := f^{-1}(t)$, $t \in T'$, of homotopy type of a bouquet of μ n-spheres.

The cohomology bundle $H^n := \bigcup_{t\in T'} H^n(X_t, \mathbb{C})$ is a flat complex vector bundle on T'. Hence, there is a natural flat connection on the sheaf $\mathscr{H}^n$ of holomorphic sections in H^n with covariant derivative $\partial_t : \mathscr{H}^n \to \mathscr{H}^n$. It induces a differential operator ∂_t on $(i_*\mathscr{H}^n)_0$ where $i : T' \hookrightarrow T$ denotes the inclusion.

Let $u : T^\infty \to T$, $\tau \mapsto \exp(2\pi i\tau)$, be the *universal covering* of T' and consider the *canonical Milnor fibre* $X^\infty := X' \times_{T'} T^\infty$. The natural maps $X_{u(\tau)} \cong X^\infty_\tau \hookrightarrow X^\infty$, $\tau \in T^\infty$, are homotopy equivalences. Hence, $H^n(X^\infty, \mathbb{C})$ can be considered as the space of global flat multivalued sections in H^n and as a trivial complex vector bundle on T^∞.

3.3. GAUß-MANIN CONNECTION

There is a natural action of the fundamental group $\Pi_1(T', t)$, $t \in T'$, on the space $H^n(X_t, \mathbb{C}) \cong H^n(X^\infty, \mathbb{C})$. A positively oriented generator of $\Pi_1(T', t)$ operates via the *monodromy operator* M defined by

$$(Ms)(\tau) := s(\tau + 1)$$

for $s \in H^n(X^\infty, \mathbb{C})$ and $\tau \in T^\infty$. Let $M = M_s M_u$ be the decomposition of M into the semisimple part M_s and the unipotent part M_u, and set $N := \log M_u$. By the *monodromy theorem* (Brieskorn; Katz), the eigenvalues of M_s are roots of unity and $N^{n+1} = 0$. Let

$$H^n(X^\infty, \mathbb{C}) \cong \bigoplus_\lambda H^n(X^\infty, \mathbb{C})_\lambda$$

be the decomposition of $H^n(X^\infty, \mathbb{C})$ into the generalized eigenspaces of M, namely $H^n(X^\infty, \mathbb{C})_\lambda := \operatorname{Ker}(M_s - \lambda)$, $\lambda = \exp(-2\pi i\alpha)$, $\alpha \in \mathbb{Q}$, and let $M_\lambda := M|_{H^n(X^\infty, \mathbb{C})_\lambda}$. For $A \in H^n(X^\infty, \mathbb{C})_\lambda$, $\lambda = \exp(-2\pi i\alpha)$,

$$s(A, \alpha)(t) := t^{\alpha - \frac{N}{2\pi i}} A(t) = t^\alpha \exp\left(-\frac{N}{2\pi i} \log t\right) A(t),$$

is monodromy invariant and defines a holomorphic section in H^n. The sections $i_*s(A, \alpha)$ span a ∂_t-invariant, finitely generated, free $\mathscr{O}_T[t^{-1}]$-submodule $\mathscr{G}$ of rank μ of $i_*\mathscr{H}^n$. Note that the direct image sheaf $i_*\mathscr{H}$ is in general not finitely generated. The *Gauß-Manin connection* is the *regular* $\mathbb{C}\{t\}[\partial_t]$-module $\mathscr{G}_0$, the stalk of $\mathscr{G}$ at 0 (Brieskorn; Pham).

3.4. V-FILTRATION

Since $t^{\alpha - \frac{N}{2\pi i}}$ is invertible, the map given by $\psi_\alpha(A) := (i_*s(A, \alpha))_0$ defines an inclusion $\psi_\alpha : H^n(X^\infty, \mathbb{C}) \hookrightarrow \mathscr{G}_0$ satisfying the relations $\partial_t \psi_\alpha = \psi_{\alpha-1}(\alpha - \frac{N}{2\pi i})$ and

$t\psi_\alpha = \psi_{\alpha+1}$, by definition of $s(A,\alpha)$. This implies that $(t\partial_t - \alpha)\psi_\alpha = \psi_\alpha(-\frac{N}{2\pi i})$, $\exp(-2\pi i t\partial_t)\psi_\alpha = \psi_\alpha M_\lambda$, and the image

$$C_\alpha := \operatorname{im}\psi_\alpha = \operatorname{Ker}(t\partial_t - \alpha)^{n+1}$$

of ψ_α is the generalized α-eigenspace of $t\partial_t$. Moreover, $T : C_\alpha \to C_{\alpha+1}$ is bijective, and $\partial_t : C_\alpha \to C_{\alpha-1}$ is bijective for $\alpha \neq 0$. The *V-filtration* V on $\mathcal{G}_0$ is defined by

$$V^\alpha := V^\alpha \mathcal{G}_0 := \sum_{\alpha \le \beta} \mathbb{C}\{t\} \cdot C_\beta, \quad V^{>\alpha} := V^{>\alpha}\mathcal{G}_0 := \sum_{\alpha < \beta} \mathbb{C}\{t\} \cdot C_\beta.$$

V^α and $V^{>\alpha}$ are free $\mathbb{C}\{t\}$-modules of rank μ with $V^\alpha / V^{>\alpha} \cong C_\alpha$.

3.5. BRIESKORN LATTICE

The *Brieskorn lattice* (Brieskorn, 1970) $\mathcal{H}'' := f_*\Omega_X^{n+1}/df \wedge d(f_*\Omega_X^{n-1})$ is a free $\mathcal{O}_T$-module of rank μ (Sebastiani, 1970). The *Gelfand-Leray form*

$$s([\omega])(t) := \left[\frac{\omega}{df}\Big|_{X_t}\right] = \left[\left(\frac{\omega}{f-t}\right)\right]$$

defines a map $s : \mathcal{H}'' \to i_*\mathcal{H}^n$ with image in $\mathcal{G}$, which induces an isomorphism $\mathcal{H}''\big|_{T'} \cong \mathcal{H}^n$ and satisfies $\partial_t s([df \wedge \eta]) = s([d\eta])$ by the *Leray residue formula* (Brieskorn). Since $\mathcal{H}_0'' = \Omega_{X,0}^{n+1}/df \wedge d\Omega_{X,0}^{n-1}$ is a torsion free $\mathbb{C}\{t\}$-module, the map $s : \mathcal{H}_0'' \hookrightarrow \mathcal{G}_0$ is an inclusion, and we identify $\mathcal{H}_0''$ with its image in $\mathcal{G}_0$. By a result of (Malgrange), we have $\mathcal{H}_0'' \subset V^{>-1}$.

3.6. MICROLOCAL STRUCTURE

The ring of *microdifferential operators* with constant coefficients

$$\mathbb{C}\{\{\partial_t^{-1}\}\} := \left\{ \sum_{i \ge 0} a_i \partial_t^{-i} \in \mathbb{C}[[\partial_t^{-1}]] \;\middle|\; \sum_{i \ge 0} \frac{a_i}{i!} t^i \in \mathbb{C}\{t\} \right\}$$

is a discrete valuation ring and $\mathbb{C}\{t\}$ is a free $\mathbb{C}\{\{\partial_t^{-1}\}\}$-module of rank 1. For $\alpha > -1$, $\partial_t t : C_\alpha \to C_\alpha$ is bijective and hence $t^{\partial_t t|C_\alpha}$ is a $\mathbb{C}\{t\}$-automorphism of $\mathbb{C}\{t\} \cdot C_\alpha$ mapping the trivial $\mathbb{C}\{t\}[\partial_t t]$-structure to that of $\mathbb{C}\{t\} \cdot C_\alpha$. Hence the module $\mathbb{C}\{t\} \cdot C_\alpha$ is free of rank $\dim_\mathbb{C} C_\alpha$ over $\mathbb{C}\{\{\partial_t^{-1}\}\}$.

In particular, V^α, resp. $V^{>\alpha}$, is a free $\mathbb{C}\{\{\partial_t^{-1}\}\}$-module of rank μ for $\alpha > -1$, resp. $\alpha \ge -1$. Hence, $\mathcal{G}_0$ is a μ-dimensional vector space over the quotient field $\mathbb{C}(\partial_t^{-1}) := \mathbb{C}\{\{\partial_t^{-1}\}\}[\partial_t]$. Since $\partial_t^{-1}\mathcal{H}_0'' \subset \mathcal{H}_0''$ and $\mathcal{H}_0'' \subset V^{>-1}$, $\mathcal{H}_0''$ is a free $\mathbb{C}\{\{\partial_t^{-1}\}\}$-module of rank μ.

3.7. SINGULARITY SPECTRUM

The *Hodge filtration* F on $\mathcal{G}_0$ is defined by $F_k := F_k\mathcal{G}_0 := \partial_t^k \mathcal{H}''$. The *singularity spectrum* $\mathrm{Sp} : \mathbb{Q} \to \mathbb{N}$, defined by

$$\mathrm{Sp}(\alpha) := \dim_{\mathbb{C}} \mathrm{Gr}_V^{\alpha} \mathrm{Gr}_0^F \mathcal{G}_0$$

reflects the embedding of $\mathcal{H}_0''$ in $\mathcal{G}_0$ and satisfies the symmetry relation $\mathrm{Sp}(\alpha) = \mathrm{Sp}(n-1-\alpha)$. Since $\mathcal{H}_0'' \subset V^{>-1}$, this implies $V^{>-1} \supset \mathcal{H}_0'' \supset V^{n-1}$ or, equivalently, $\mathrm{Sp}(\alpha) = 0$ for $\alpha \leq -1$ or $\alpha \geq n$.

3.8. ALGORITHM

We abbreviate $\Omega := \Omega_{X,0}$, $\mathcal{H}'' := \mathcal{H}_0''$, $\mathcal{G} := \mathcal{G}_0$, $s := \partial_t^{-1}$, and we consider the rings $\mathbb{C}\{t\}$ and $\mathbb{C}\{\{s\}\}$. Then $[s^{-2}t, s] = 1$ and, hence, $t = s^2\partial_s$.

3.8.1. *Idea*

Since $\partial_t t = s^{-1}t = s^{-1}s^2\partial_s = s\partial_s$, the

$$\mathcal{H}_k'' := \sum_{j=0}^{k} (\partial_t t)^j \mathcal{H}'' = \sum_{j=0}^{k} (s\partial_s)^j \mathcal{H}''.$$

are $\mathbb{C}\{t\}$-lattices and $\mathbb{C}\{\{s\}\}$-lattices. Since $\mathcal{G}$ is regular, the *saturation*

$$\mathcal{H}_\infty'' := \sum_{j=0}^{\infty} (\partial_t t)^j \mathcal{H}''$$

of $\mathcal{H}''$ is a $\mathbb{C}\{t\}$-lattice and, hence, $k_\infty := \min\{k \geq 0 \mid \mathcal{H}_k'' = \mathcal{H}_\infty''\}$ is a finite number. For any $K \geq n+1$ the inclusions $V^{>-1} \supset \mathcal{H}'' \supset V^{n-1}$ imply inclusions

$$V^{>-1} \supset \mathcal{H}_\infty'' \supset \mathcal{H}'' \supset \partial_t^{-1}\mathcal{H}'' \supset V^n \supset \partial_t^{-K} V^{>-1} \supset \partial_t^{-K}\mathcal{H}_\infty''$$

and $\mathcal{H}_\infty''$ and $\partial_t^{-K}\mathcal{H}_\infty''$ are $\partial_t t$-invariant. Hence, $\partial_t t$ and $t\partial_t$ induce endomorphisms $\overline{\partial_t t}, \overline{t\partial_t} \in \mathrm{End}_{\mathbb{C}}\left(\mathcal{H}_\infty''/\partial_t^{-K}\mathcal{H}_\infty''\right)$ such that the V-filtration $V_{\overline{t\partial_t}} = V_{\overline{\partial_t t}}^{\bullet+1}$ defined by $\overline{t\partial_t}$ on $\mathcal{H}_\infty''/\partial_t^{-K}\mathcal{H}_\infty''$ induces the V-filtration on the subquotient $\mathcal{H}''/\partial_t^{-1}\mathcal{H}'' = \mathrm{Gr}_0^F \mathcal{G}_0$.

3.8.2. *Computation*

By the *finite determinacy theorem*, we may assume that $f \in \mathbb{C}[\mathbf{x}]$, $\mathbf{x} := x_0, \ldots, x_n$, is a polynomial. Since $\mathbb{C}[\mathbf{x}]_{(\mathbf{x})} \subset \mathbb{C}\{\mathbf{x}\}$ is faithfully flat and all data will be defined over $\mathbb{C}[\mathbf{x}]_{(\mathbf{x})}$, we may replace $\mathbb{C}\{\mathbf{x}\}$ by $\mathbb{C}[\mathbf{x}]_{(\mathbf{x})}$ and, similarly, $\mathbb{C}\{t\}$ by $\mathbb{C}[t]_{(t)}$ and $\mathbb{C}\{\{s\}\}$ by $\mathbb{C}[s]_{(s)}$ for the computation. With the additional assumption $f \in \mathbb{Q}[\mathbf{x}]$, all data will be defined over $\mathbb{Q}$, and we can apply methods of computer algebra.

Using standard basis methods for *local rings*, $\mathbb{C}[\mathbf{x}]_{\langle\mathbf{x}\rangle}$ one can compute a monomial $\mathbb{C}$-basis $\mathbf{m} = (m_1,\ldots,m_\mu)^T$ of

$$\Omega_f := \Omega^{n+1}/df\wedge\Omega^n \cong \mathbb{C}\{\mathbf{x}\}/(\partial_{\mathbf{x}}f).$$

Since $\mathscr{H}''/s\mathscr{H}'' \cong \Omega_f$, $\mathbf{m}$ represents a $\mathbb{C}\{\{s\}\}$-basis of $\mathscr{H}''$ and a $\mathbb{C}(s)$-basis of $\mathscr{G}$ by *Nakayama's lemma*.

The matrix A of t with respect to $\mathbf{m}$ is defined by $t\mathbf{m} =: A\mathbf{m}$. Since $t = s^2\partial_s$, we obtain for $g \in \mathbb{C}(s)^\mu$

$$tg\mathbf{m} = \big(gA + s^2\partial_s(g)\big)\mathbf{m}.$$

So the action of t in terms of the $\mathbb{C}(s)$-basis $\mathbf{m}$ is determined by the matrix A by the above formula.

A *reduced normalform* with respect to a *local monomial ordering* allows to compute the projection to the first summand in

$$\Omega_f \oplus df\wedge\Omega^n \longrightarrow\!\!\!\!\!\rightarrow \Omega_f \oplus df\wedge\Omega^n/df\wedge d\Omega^{n-1} \cong \mathscr{H}''/s\mathscr{H}'' \oplus s\mathscr{H}''.$$

Since $t[\omega] = [f\omega]$ and $s^{-1}[df\wedge\eta] = \partial_t[df\wedge\eta] = [d\eta]$, the matrix A of t with respect to $\mathbf{m}$ can be computed up to arbitrarily high order.

The basis representation H''_k of $\mathscr{H}''_k$ with respect to $\mathbf{m}$ defined by $\mathscr{H}''_k =: H''_k\mathbf{m}$ can be computed inductively by

$$\begin{aligned}
\delta H_0 &:= H_0 := \mathbb{C}\{\{s\}\}^\mu,\\
\delta H_{k+1} &:= \mathrm{jet}_{-1}\big(s^{-1}\delta H_k \mathrm{jet}_k(A) + s\partial_s\delta H_k\big),\\
H_{k+1} &= H_k + \delta H_{k+1}.
\end{aligned}$$

Using standard basis methods, one can check if $H_k = H_{k+1}$ and compute a $\mathbb{C}\{\{s\}\}$-basis $\mathbf{m}'$ of $H_{k_\infty} =: H_\infty$ with

$$\delta(\mathbf{m}') := \max\big\{\mathrm{ord}\big(m'_{i_1,j_1}\big) - \mathrm{ord}\big(m'_{i_2,j_2}\big) \,\big|\, m'_{i_1,j_1} \neq 0 \neq m'_{i_2,j_2}\big\} \leq k_\infty.$$

Then the matrix A' of t with respect to the $\mathbb{C}\{\{s\}\}$-basis $\mathbf{m}'\mathbf{m}$ of $\mathscr{H}''_\infty$ is defined by the formula $\mathbf{m}'A + s^2\partial_s\mathbf{m}' =: A'\mathbf{m}'$, and $\mathrm{jet}_k(A') = \mathrm{jet}_k(A'_{\leq k})$ for $A'_{\leq k}$ defined by

$$\mathbf{m}'\mathrm{jet}_{k+\delta(\mathbf{m}')}(A) + s^2\partial_s\mathbf{m}' =: A'_{\leq k}\mathbf{m}'.$$

Hence, the basis representation of $\overline{\partial_t t} \in \mathrm{End}_{\mathbb{C}}\big(\mathscr{H}''_\infty/\partial_t^{-K}\mathscr{H}''_\infty\big)$ with respect to $\mathbf{m}'\mathbf{m}$ is

$$\overline{s^{-1}A' + s\partial_s} = \overline{s^{-1}A'_{\leq K} + s\partial_s} \in \mathrm{End}_{\mathbb{C}}\big(\mathbb{C}\{\{s\}\}^\mu/s^K\mathbb{C}\{\{s\}\}^\mu\big).$$

The basis representation H' of $\mathscr{H}''$ with respect to $\mathbf{m}'\mathbf{m}$ is defined by $H_0 =: H'\mathbf{m}'$, and $V^{\bullet+1}_{\overline{s^{-1}A'_{\leq K}+s\partial_s}}(H'/sH')\mathbf{m}'$ is the basis representation of $V(\mathscr{H}''/s\mathscr{H}'')$ with respect to **m**. The matrix of $\overline{s^{-1}A'_{\leq K}+s\partial_s}$ with respect to the canonical $\mathbb{C}$-basis

$$\begin{pmatrix} 1 & s & s^2 & & s^{K-1} & \\ \ddots & \ddots & \ddots & \cdots & \ddots & \\ & 1 & s & s^2 & & s^{K-1} \end{pmatrix}^t$$

of $\mathbb{C}\{\{s\}\}^\mu/s^K\mathbb{C}\{\{s\}\}^\mu$ is given by the block matrix

$$\begin{pmatrix} A'_1 & A'_2 & A'_3 & A'_4 & \cdots & A'_K \\ & A'_1+1 & A'_2 & A'_3 & \cdots & A'_{K-1} \\ & & A'_1+2 & A'_2 & \cdots & A'_{K-2} \\ & & & A'_1+3 & \ddots & \vdots \\ & & & & \ddots & A'_2 \\ & & & & & A'_1+K-1 \end{pmatrix}$$

where $A' = \sum_{k\geq 0} A'_k s^k$. Since the eigenvalues of A'_1 are rational, they can be computed using univariate factorization over the rational numbers. Then the V-filtration $V^{\bullet+1}_{\overline{s^{-1}A'_{\leq K}+s\partial_s}}$ can be computed using methods of linear algebra.

3.8.3. *Summary*

Here are the main steps of the algorithm:

1. Compute a $\mathbb{C}\{\{\partial_t^{-1}\}\}$-basis **m** of $\mathscr{H}''$.
2. For increasing k compute successively the lattices $\mathscr{H}''_k$ in terms of **m**, and t in terms of **m** up to order k until $k = k_\infty$ and $\mathscr{H}''_k$ is the saturation $\mathscr{H}''_\infty$ of $\mathscr{H}''$.
3. Compute a $\mathbb{C}\{\{\partial_t^{-1}\}\}$-basis $\mathbf{m}'\mathbf{m}$ of $\mathscr{H}''_\infty$.
4. Compute t in terms of **m** up to order $\delta(\mathbf{m}') + n + 1$.
5. Compute t in terms of $\mathbf{m}'\mathbf{m}$ up to order $K := n+1$.
6. Compute $\mathscr{H}''$ in terms of $\mathbf{m}'\mathbf{m}$.
7. Compute the V-filtration on $\mathscr{H}''_\infty/\partial_t^{-K}\mathscr{H}''_\infty$ in terms of $\mathbf{m}'\mathbf{m}$.
8. Compute the induced V-filtration on $\mathscr{H}''/\partial_t^{-1}\mathscr{H}''$ in terms of $\mathbf{m}'\mathbf{m}$.

3.8.4. *Example*

The SINGULAR library `gaussman.lib` (Schulze, 2001) contains an implementation of the algorithm. We use it to compute an example. First, we have to load the library:

```
LIB "gaussman.lib";
```

We define the ring $R := \mathbb{Q}[x,y]_{(x,y)}$ and $f = x^5+x^2y^2+y^5 \in R$:

```
ring R=0,(x,y),ds;
poly f=x5+x2y2+y5;
```

Finally, we compute the V-filtration:

```
vfiltration(f);
==> [1]:  _[1]=-1/2  _[2]=-3/10  _[3]=-1/10  _[4]=0
==>       _[5]=1/10  _[6]=3/10   _[7]=1/2
==> [2]:  1,2,2,1,2,2,1
==> [3]:  [1]:  _[1]=gen(11)
==>       [2]:  _[1]=gen(10)  _[2]=gen(6)
==>       [3]:  _[1]=gen(9)   _[2]=gen(4)
==>       [4]:  _[1]=gen(5)
==>       [5]:  _[1]=gen(8)   _[2]=gen(3)
==>       [6]:  _[1]=gen(7)   _[2]=gen(2)
==>       [7]:  _[1]=gen(1)
==> [4]:  _[1]=y5  _[2]=y4  _[3]=y3  _[4]=y2  _[5]=xy
==>       _[6]=y   _[7]=x4  _[8]=x3  _[9]=x2  _[10]=x
==>      _[11]=1
==> [5]:  _[1]=2x2y+5y4  _[2]=2xy2+5x4  _[3]=x5-y5
==>       _[4]=y6
```

The result is a list with 5 entries: The first contains the spectral numbers, the second the corresponding multiplicities, the third $\mathbb{C}$-bases of the graded parts of the V-filtration on Ω_f in terms of the monomial $\mathbb{C}$-basis in the fourth entry, and the fifth a standard basis of the Jacobian ideal. A monomial $x^\alpha y^\beta$ in the fourth entry is considered as $x^\alpha y^\beta dx \wedge dy \in \Omega_f$.

As an application of the implementation, the third author could verify a conjecture of Hertling (2000) about the variance of the spectral numbers for all isolated hypersurface singularities of Milnor number ≤ 16.

References

Brieskorn, E. (1970) Die Monodromie der isolierten Singularitäten von Hyperflächen, *Manuscr. Math.* **2**, 103–161.

Campillo, A. (1980) *Algebroid curves in positive characteristic*, SLN 813, Springer-Verlag.

Campillo, A., and Farrán, J.I. (1998) Construction of AG codes from symbolic Hamburger-Noether expressions of plane curves, Preprint Univ. Valladolid.

Corso, A., Huneke, C., and Vasconcelos, W. (1998) On the integral closure of ideals, *Manuscripta math.* **95**, 331–347.

Decker, W., Greuel, G.-M., de Jong, T., and Pfister, G. (1999) The Normalization: a new Algorithm, Implementation and Comparisons, *Progress in Mathematics* **173**, 177–185.

De Jong, T. (1998) An algorithm for computing the integral closure, *J. of Symbolic Computation* **26**, 273–277.

Della Dorca, J., Dicrescenzo, C., and Duval, D. (1985) About a new method for computing in algebraic number fields, in *EUROCAL 1985*, LNCS 204, pp. 289–290.

Eisenbud D., Huneke, C., and Vasconcelos, W. (1992) Direct Methods for Primary Decomposition, *Invent. Math.* **110**, 207–235.

Farrán, J.I., and Lossen, C. (2001) `brnoeth.lib`. A SINGULAR 2.0 library for computing AG codes and Weierstraß semigroups.

Gianni, P., and Trager, B. Integral closure of Noetherian rings, Preprint, to appear.

Gianni, P., Trager, B., and Zacharias, G. (1988) Gröbner Bases and Primary Decomposition of Polynomial Ideals, *J. Symbolic Computation* **6**, 149–167.

Grauert H., and Remmert, R. (1971) *Analytische Stellenalgebren*, Springer, Berlin.

Greuel, G.-M. (2000a) Computer Algebra and Algebraic Geometry - Achievements and Perspectives, *J. Symbolic Computation* **30**, 3, 253–290.

Greuel, G.-M. (2000b) Applications of Computer Algebra to Algebraic Geometry, Singularity Theory and Symbolic-Numerical Solving, to appear in *Proceedings of the 3ECM*, Barcelona.

Greuel, G.-M., and Pfister, G. (2001a) `normal.lib`. A SINGULAR 2.0 library for computing the normalization of affine rings.

Greuel, G.-M., and Pfister, G. (2001b) *A* SINGULAR *Introduction to Commutative Algebra*, Book in preparation.

Greuel, G.-M., Pfister, G., and Schönemann, H. (2001) SINGULAR 2.0. A Computer Algebra System for Polynomial Computations. Centre for Computer Algebra, University of Kaiserslautern. `http://www.singular.uni-kl.de`.

Haché, G. (1996) Construction effective des codes géométriques, Ph.D. Thesis, Univ. Paris 6.

Haché, G., and Le Brigand, D. (1995) Effective Construction of Algebraic Geometry Codes, *IEEE Trans. Inf. Theory* **41**, no. 6, 1615–1628.

Hertling, C. (1998) Brieskorn lattices and Torelli type theorems for cubics in $\mathbb{P}^3$ and for Brieskorn-Pham singularities with coprime exponents, *Progress in Math.* **162**, 167–194.

Hertling, C. (2000) Variance of the spectral numbers, Preprint, math.CV/0007187.

Hirsch, T. (2000a) Die Berechnung des ganzen Abschlusses eines Ideals mit Hilfe seiner Rees-Algebra, Diplomarbeit, Univ. Kaiserslautern.

Hirsch, T. (2000b) `reesclos.lib`, A SINGULAR 2.0 library for computing the integral closure of an ideal.

Katz, N.M. (1970) The regularity theorem in algebraic geometry, *Actes Congrès intern. math.* **1**, 437–443.

Krick, T., and Logar, A. (1991) An algorithm for the computation of the radical of an ideal in the ring of polynomials, *AAECC 9*, Springer LNCS 539, pp. 195–205.

Lamm, M. (1999) Hamburger-Noether-Entwicklung von Kurvensingularitäten, Diplomarbeit, Univ. Kaiserslautern.

Lazard, D. (1992) Solving Zero–dimensional Algebraic Systems, *J. Symbolic Computation* **13**, 117–131.

Le Brigand, D., and Risler, J.J. (1988) Algorithme de Brill-Noether et codes de Goppa, *Bull. Soc. math. France* **116**, 231–253.

Malgrange, B. (1974) Intégrales asymptotiques et monodromie, *Ann. sci. Ec. Norm. Sup.* **4**, 405–430.

Milnor, J. (1968) *Singular Points on Complex Hypersurfaces*, Princeton Univ. Press.

Möller, H.M. (1993) On decomposing systems of polynomial equations with finitely many solutions, *Appl. Algebra Eng. Commun. Comput.* **4**, 217–230.

Pham, F. (1979) *Singularités des systèmes de Gauss-Manin*, Birkhäuser.

Pretzel, O. (1998) *Codes and Algebraic Curves*, Clarendon, Oxford.

Rybowicz, M. (1990) Sur le calcul des places et des anneaux d'entiers d'un corps de fonctions algébriques, Thèse d'état, Univ. de Limoges.

Schulze, M. (1999) Computation of the monodromy of an isolated hypersurface singularity, Diplomarbeit, Univ. Kaiserslautern.

Schulze, M. (2001) `gaussman.lib`, A SINGULAR 2.0 library for computing the Gauß-Manin connection of isolated hypersurface singularities.

Schulze, M., and Steenbrink, J.H.M. (2001) Computing Hodge-theoretic invariants of singularities, to appear in *Proceedings of New Developments in Singularity Theory*, Kluwer.

Sebastiani, M. (1970) Preuve d'une conjecture de Brieskorn, *Manuscr. Math.* **2**, 301–308.

Stolzenberg, G. (1968) Constructive normalization of an algebraic variety, *Bull. AMS.* **74**, 595–599.

Tsfasman, M.A., and Vlăduţ, S.G. (1991) *Algebraic-Geometric Codes*, Kluwer Acad. Publ., Amsterdam.

Tsfasman, M.A., Vlăduţ, S.G., and Zink, T. (1982) Modular curves, Shimura curves, and Goppa codes, better than Varshamov-Gilbert bound, *Math. Nachr.* **109**, 21–28.

Van der Waerden, B.L. (1973) *Einführung in die algebraische Geometrie*, Springer-Verlag.

Vasconcelos, W. (1991) Computing the integral closure of an affine domain, *Proc. AMS* **113** (3), 633–639.

Vasconcelos, W. (1994) *Arithmetic of blowup algebras*, LMS Lecture Notes 195, Cambridge Univ. Press.

Vasconcelos, W. (1997) *Computational Methods in Commutative Algebra and Algebraic Geometry*, Springer-Verlag.

COUNTING POINTS ON CALABI-YAU THREEFOLDS

Some computational aspects

K. HULEK and J. SPANDAW
University of Hannover, Germany

Abstract. We discuss a number of examples of rigid Calabi-Yau varieties for which one can prove modularity. In this situation the number N_p of $\mathbb{F}_p$-rational points of the Calabi-Yau is (via the Lefschetz fixed point formula) related to the Fourier coefficient a_p of some modular form. In all cases which we discuss it turns out that it is much faster to compute a_p than N_p.

Key words: Calabi-Yau varieties, L-function, modularity, diophantine equations, modular forms.

Mathematics Subject Classification (2000): 14J32, 14G10, 14K10, 11D45, 11G18, 11G25.

1. Introduction

Let $f_1,\ldots,f_k \in \mathbb{Q}[x_0,\ldots,x_N]$ be homogenous polynominals and consider the projective variety

$$X = \{(x_0 : \ldots : x_N) \in \mathbb{P}^N_{\mathbb{Q}};\ f_1(x_0,\ldots,x_N) = \ldots = f_k(x_0,\ldots,x_N) = 0\}.$$

After clearing denominators, we can assume that $f_1,\ldots,f_k \in \mathbb{Z}[x_0,\ldots,x_N]$. For each prime p we define

$$N_p := \#X(\mathbb{F}_p),$$

i.e. N_p is the number of points of X considered as an algebraic variety over the field $\mathbb{F}_p$. One can try and calculate N_p by means of a computer. The time needed to do this depends on the equations f_i, but in any case the difficulty grows immensely as p increases.

Let $\bar{X}_p = X(\mathbb{F}_p) \times_{\mathbb{F}_p} \bar{\mathbb{F}}_p$. Then there exists the Frobenius morphism $F_p : \bar{X}_p \to \bar{X}_p$ whose fixed point set is exactly $X(\mathbb{F}_p)$. We shall assume at this point that $\bar{X}_p$ is smooth. If $l \neq p$ is a different prime, then by the Lefschetz fixed point formula

$$N_p = \sum_{i=0}^{2\dim X} (-1)^i \mathrm{Tr}(F_p^*; H^i_{\acute{e}t}(\bar{X}_p, \mathbb{Q}_l)). \tag{1}$$

C. Ciliberto et al. (eds.),
Applications of Algebraic Geometry to Coding Theory, Physics and Computation, **195–205.**

The theme of this talk is that it is sometimes easier to control the right hand side of this formula than to compute N_p. In exceptional case one can compute the right hand side by computing the coefficients of some modular form or by computing N_p of a different and possibly easier to handle variety. We shall explain this in examples.

We would like to stress here that the theoretical background is well known to specialists. Our only claim to a genuine contribution is our contribution to the joint paper (Hulek et al.) with B. van Geemen and D. van Straten in which we treat the case of the Barth-Nieto quintic and its relatives. It seemed to us worthwile, however, to advertise the computational aspects of the Weil conjectures to a wider audience which includes mathematicians who are interested in applications of algebraic geometry.

Acknowledgement. We are grateful to E. Schellhammer who helped us with the running of the programmes and to Ch. Meyer for information on his programmes.

2. Counting points on Calabi-Yau varieties

Let X be a smooth projective n-fold which is smooth and defined over the integers. The Frobenius morphism $F_p : \bar{X}_p \to \bar{X}_p$ defines endomorphisms

$$F^*_{i,p} : H^i_{ét}(X, \mathbb{Q}_l) \to H^i_{ét}(X, \mathbb{Q}_l).$$

(In order to simplify the notation we simply write $H^i_{ét}(X) = H^i_{ét}(X, \mathbb{Q}_l)$ for the l-adic cohomology.) Let

$$P_i(t) = P_{i,p}(t) = \det(1 - tF^*_{i,p}).$$

The *zeta-function* for the prime p is then defined by

$$Z_p(t) = \frac{P_1(t)P_3(t)\dots P_{2n-1}(t)}{P_0(t)P_2(t)\dots P_{2n}(t)}.$$

This is a priori a rational function with coefficients in $\mathbb{Q}_l$, but one can show that it is in fact contained in $\mathbb{Q}(t)$. One has $P_0(t) = 1 - t$ and $P_{2n}(t) = 1 - p^n t$. An important theorem of Deligne says that the $P_i(t)$ have integer coefficients and moreover

$$P_i(t) = \prod_j (1 - \alpha_{ij} t)$$

where the α_{ij} are algebraic integers with $|\alpha_{ij}| = p^{i/2}$. The most interesting part of the cohomology is the midde cohomology $H^n_{ét}(X)$. For each prime p we define the Euler factor of this prime by

$$L_p(H^n_{ét}(X), s) = \frac{1}{P_{n,p}(p^{-s})}$$

and the L-function by

$$L(H^n_{ét}(X),s) = \prod_p \frac{1}{P_{n,p}(p^{-s})}.$$

We shall now specialize to the case where X is a rigid Calabi-Yau threefold. By a Calabi-Yau threefold we mean a projective threefold X with $K_X = \mathcal{O}_X$ and $q(X) = h^1(\mathcal{O}_X) = 0$. By Serre duality it follows that also $h^2(\mathcal{O}_X) = 0$. The Calabi-Yau 3-fold X is rigid if $H^1(X,T_X) = H^1(X,\Omega^2_X) = 0$, i.e. if and only if $h^{12} = h^{21} = 0$. In this case $H^3_{ét}(X,\mathbb{Q}_l)$ is 2-dimensional. Note that the absolute Galois group $\mathrm{Gal}(\bar{\mathbb{Q}}/\mathbb{Q})$ acts on the l-adic cohomology. In particular the action on $H^3_{ét}(X,\mathbb{Q}_l)$ defines a 2-dimensional representation of $\mathrm{Gal}(\bar{\mathbb{Q}}/\mathbb{Q})$.

The determinant of $F^*_{3,p}$ is known to be p^3. This is a consequence of Poincaré duality (see also (Meyer, 2000, Lemma 4.4).) Hence

$$P_{3,p}(t) = 1 - a_p t + p^3 t^2 \text{ where } a_p = \mathrm{Tr}(F^*_{3,p})$$

and the L-function is of the form

$$L(H^3_{ét}(X),s) = \prod_p \frac{1}{1 - a_p p^{-s} + p^3 p^{-2s}} = \sum_{k=1}^{\infty} a_k k^{-s}.$$

For a prime p the coefficient $a_p = \mathrm{Tr}(F^*_{3,p})$ whereas for general k the coefficient a_k is determined by the coefficients a_p corresponding to the prime divisors p of k.

CONJECTURE (FONTAINE AND MAZUR). *The L-function $L(H^3_{ét}(X),s)$ is (up to the factors associated to bad primes) the Mellin transform of a modular form f.*

For a discussion of this see (Fontaine and Mazur, Conjecture 3). The Mellin transform of a modular form $f(q) = \sum_n a_n q^n$ is defined as $\mathrm{Mell}(f) = \sum_n a_n n^{-s}$.

On the other hand the numbers a_p are closely related to the numbers N_p. In our case $H^1_{ét}(X) = H^5_{ét}(X) = 0$. Using Poincaré duality for $H^2_{ét}(X)$ and $H^4_{ét}(X)$ formula (1) becomes

$$N_p = 1 + (1+p)\mathrm{Tr}(F^*_{2,p}) + p^3 - a_p. \tag{2}$$

In some cases it is possible to determine $\mathrm{Tr}(F^*_{2,p})$ This is, for example, the case if $H^2(X,\mathbb{Z})$ is spanned by divisors which are defined over $\mathbb{Z}$. Then all eigenvalues of $\mathrm{Tr}(F^*_{2,p})$ are equal to p, hence $\mathrm{Tr}(F^*_{2,p}) = b_2(X)p$. But even if this is not the case it is sometimes possible to compute $\mathrm{Tr}(F^*_{2,p})$ without too many difficulties (for such an example see Remark 3.3, resp. the proof of Theorem 1 in (Hulek et al.)).

Now, if one can prove that the L-function $L(H^3_{ét}(X),s)$ is modular, i.e. is the Mellin transform of a modular form f, then computing N_p is under the above conditions equivalent to computing the coefficients a_p. Indeed, the latter can be much easier. Fortunately there is a technique which one can use to prove the modularity of $L(H^3_{ét}(X),s)$. The main point is a theorem of Serre based on work of Faltings and recast by Livné. To explain this we work with the prime $l = 2$. We have already remarked that the action of Frobenius defines a 2-dimensional representation $\rho_1 : \mathrm{Gal}(\bar{\mathbb{Q}}/\mathbb{Q}) \to \mathrm{Aut}(H^3_{ét}(X,\mathbb{Q}_2)) \cong \mathrm{GL}(2,\mathbb{Q}_l)$. On the other hand, if f is a newform of weight $2k$, then by a theorem of Deligne (1971) one can associate to f a piece of the l-adic cohomology $H^{k+1}(X^{(k)},\mathbb{Q}_l)$ where $X^{(k)}$ is the k-fold fibre product over C, of a universal elliptic curve X over a modular curve C. Which curve C one has to take depends on the modular group for which f is a modular form. In this way one can associate to f another 2-dimensional representation $\rho_2 : \mathrm{Gal}(\bar{\mathbb{Q}}/\mathbb{Q}) \to \mathrm{GL}(2,\mathbb{Q}_2)$. The crucial theorem of Faltings, Serre and Livné is now the following

THEOREM 2.1. *Let ρ_1,ρ_2 be two continuous 2-dimensional 2-adic representations of $Gal(\bar{\mathbb{Q}}/\mathbb{Q})$ unramified outside a finite set S of prime numbers. Let $\mathbb{Q}_S$ be the compositum of all quadratic extensions of $\mathbb{Q}$ which are unramified outside S. Let T be a set of primes, disjoint from S, such that $Gal(\mathbb{Q}_S/\mathbb{Q}) = \{F_p|_{\mathbb{Q}_S}; p \in T\}$ where F_p again denotes the Frobenius homomorphism. Suppose that*

(a) $Tr\rho_1(F_p) = Tr\rho_2(F_p)$ *for all* $p \in T$,
(b) $\det\rho_1(F_p) = \det\rho_2(F_p)$ *for all* $p \in T$,
(c) $Tr\rho_1 \equiv Tr\rho_2 \equiv 0 \pmod 2$ *and* $\det\rho_1 \equiv \det\rho_2 \pmod 2$.

Then ρ_1 and ρ_2 have isomorphic semisimplifications, hence $L(\rho_1,s) = L(\rho_2,s)$. In particular the good Euler factors of ρ_1 and ρ_2 coincide.

PROOF. See Theorem 4.3 in (Livné). □

In particular examples condition (c) is often not difficult to check and in this case this theorem says that by checking finitely many numbers a_p one can conclude the equality of almost all numbers a_p! This theorem has been used in a number of cases in exactly this way (see (Saito and Yui); (Verrill); (Yui)).

In this situation one can replace the computation of the numbers N_p by the computation of the Fourier coefficients of a modular form. Another possible application is when one has two varieties which have the same L-function. Then one can choose the variety which is computationally easier to deduce the numbers N_p for the other variety.

3. The Barth-Nieto quintic and its relatives

We consider the following three varieties.

(1) The *Barth-Nieto quintic* N is given by the equations

$$(N) \quad \begin{aligned} x_0 + \cdots + x_5 &= 0 \\ \frac{1}{x_0} + \ldots + \frac{1}{x_5} &= 0. \end{aligned}$$

This defines a variety $N \subset \mathbb{P}^5$ contained in the hyperplane given by the first equation. Hence N is a quintic threefold and it is singular along 20 lines and has 10 isolated A_1-singularities. Barth and Nieto have shown that N has a smooth Calabi-Yau model Y with Euler number $e(Y) = 100$. Their construction shows that Y is defined over $\mathbb{Z}$ and it can be checked that Y has good reduction for $p \geq 5$.

(2) Let $\tilde{N}$ be the pullback of N under the double cover of $\mathbb{P}^5$ branched along the union $S = \cup S_k$ of the 6 coordinate hyperplanes $S_k = \{x_k = 0\}$. In affine coordinates $(x_0 = 1)$ the variety $\tilde{N}$ is given by the equations

$$(\tilde{N}) \quad \begin{aligned} & y^2 = x_1 \dots x_5 \\ & 1 + x_1 + \dots + x_5 = 0 \\ & x_1 \dots x_5 + x_2 \dots x_5 + x_1 x_3 \dots x_5 + \dots + x_1 \dots x_4 = 0. \end{aligned}$$

The variety $\tilde{N}$ is singular and has a desingularization $\tilde{Y}$ which is defined over the integers and which has the property that there exists a morphism $\tilde{Y} \to Z$ which contracts 20 quadrics to lines. The variety Z is a Calabi-Yau variety with Euler number $e(Z) = 80$. Since the morphism $\tilde{Y} \to Z$ comes from Mori theory we do not know whether Z is defined over the integers. The map $\tilde{Y} \to Z$ induces an isomorphism $H^3(Z) \cong H^3(\tilde{Y})$ and we shall work with the l-adic cohomology group $H^3_{\acute{e}t}(\tilde{Y}, \mathbb{Q}_l)$. Again $\tilde{Y}$ has good reduction for $p \geq 5$. It is also worth noting that $\tilde{N}$ (and hence also $\tilde{Y}$ and Z) are birationally equivalent to the moduli of (1,3)-polarized abelian surfaces with a level-2 structure.

(3) The universal elliptic curve with a point of order 6 is given by the pencil of cubics

$$t_0(x_0 + x_1 + x_2)(x_1x_2 + x_2x_0 + x_0x_1) = t_1 x_0 x_1 x_2.$$

Let W be the product of this pencil with itself over the base $\mathbb{P}^1 = \mathbb{P}^1(t_0, t_1)$. Then $W \subset \mathbb{P}^2 \times \mathbb{P}^1 \times \mathbb{P}^2$ and using the affine coordinate t it is given by the equations

$$(W) \quad \begin{aligned} (x_0 + x_1 + x_2)\left(\frac{1}{x_0} + \frac{1}{x_1} + \frac{1}{x_2}\right) &= t \\ (y_0 + y_1 + y_2)\left(\frac{1}{y_0} + \frac{1}{y_1} + \frac{1}{y_2}\right) &= t \end{aligned}$$

W is also singular and it is well known that it has a desingularisation $\widehat{W}$ which is a rigid Calabi-Yau variety.

THEOREM 3.1. *The varieties Y and Z are rigid Calabi-Yau threefolds.*

PROOF. For the technical details we refer the reader to (Hulek et al.). The basic idea is a method which was pioneered by B. van Geemen and which also uses a counting argument. One can show for Y that the Néron-Severi group is generated by divisors which are defined over the integers. Hence all eigenvalues of $F^*_{2,p}$ are equal to p. We know that $e(Y) = 100$. Let $a = h^{1,2}(Y) = h^{2,1}(Y)$. Then $h^{1,1}(Y) = h^{2,2}(Y) = 50 + a$. Moreover one can check easily that $\det F^*_{3,p} = p^3$. Hence by formula (2)

$$a_p = \mathrm{Tr}(F^*_{3,p}) = 1 + (a+50)(p+p^2) + p^3 - N_p.$$

By the Riemann hypothesis

$$|1 + (a+50)(p+p^2) + p^3 - N_p| \leq b_3(Y)p^{3/2} = (2a+2)p^{3/2}.$$

Computing $N_{13}(Y) = 11260$ by a computer then shows $a = 0$.

The situation for Z is more complicated. We first note that it is enough to prove $h^{1,2}(\tilde{Y}) = h^{2,1}(\tilde{Y}) = 0$. In this case it is no longer true that all eigenvalues of $F^*_{2,p}$ are equal to p. One can, however, show (Hulek et al., Proposition 2.21) that this is still true for $p \equiv 1 \pmod 4$. Then the same argument goes through where we can again work with the prime $p = 13$. □

REMARK. *We know that* $Tr(F^*_{2,p}) = 60p$, *if* $p \equiv 1 \pmod 4$ *and conjecture that* $Tr(F^*_{2,p}) = 40p$ *if* $p \equiv 3 \pmod 4$. *We have checked this for all primes* $p \leq 59$. *In any case we know that all eigenvalues are* $\pm p$.

The varieties described above are modular in the sense that their L-functions are (up to possibly the bad primes) the Mellin transform of a modular form. Let

$$\Gamma_0(6) = \left\{ \begin{pmatrix} a & b \\ c & d \end{pmatrix} \in SL(2,\mathbb{Z});\ c \equiv 0 \pmod 6 \right\}$$

and

$$\Gamma_1(6) = \left\{ \begin{pmatrix} a & b \\ c & d \end{pmatrix} \in \Gamma_0(6);\ a \equiv 1 \pmod 6 \right\}.$$

Note that both groups have the same image in PSL$(2,\mathbb{Z})$. Hence the corresponding modular curves $X_0(6)$ and $X_1(6)$ are isomorphic. They parametrize elliptic curves with a subgroup of order 6, resp. elliptic curves with a point of order 6. The space $S_4(\Gamma_0(6)) = S_4(\Gamma_1(6))$ of cusps forms of weight 4 has dimension 1. The form

$$\begin{aligned} f(q) &= [\eta(q)\eta(q^2)\eta(q^3)\eta(q^6)]^2 \\ &= q\prod_{n=1}^{\infty}(1-q^n)^2(1-q^{2n})^2(1-q^{3n})^2(1-q^{6n})^2 \\ &= q - 2q^2 - 3q^3 + 4q^4 + 6q^5 + 6q^6 - 16q^7 \\ &\quad -8q^8 + 9q^9 - 12q^{10} + 12q^{11} - 12q^{12} + 38q^{13} + \ldots \\ &= \sum b_n q^n \end{aligned}$$

is the normalized generator of this space. Here

$$\eta(q) = q^{1/24}\prod_{n=1}^{\infty}(1-q^n)$$

is the Dedekind η-function. The L-function of f is the Mellin transform

$$L(f,s) = \sum_{n=1}^{\infty} b_n n^{-s}.$$

Since $b_n = O(n^{3/2})$ for $n \to \infty$ it converges for $Re(s) > 5/2$. It has an analytic continuation to an entire function. Furthermore, there is a functional equation relating $L(f,s)$ and $L(f,4-s)$. Since f is a Hecke eigenform its L-function is an Euler product

$$L(f,s) = \prod_{p \text{ prime}} L_p(f,s)$$

with Euler factors

$$L_p(f,s) = \frac{1}{1 - b_p p^{-s} + p^3 \cdot p^{-2s}}, \quad \text{for } p \geq 5$$

and $L_p(f,s) = (1 + p \cdot p^{-s})^{-1}$ for $p < 5$. Recall also that by Deligne (1971) $L(f,s) = L(\rho_f,s)$ where $\rho_f : Gal(\bar{\mathbb{Q}}/\mathbb{Q}) \to Aut(H^3_{ét}(\widehat{W},\mathbb{Q}_2)) \cong GL(2,\mathbb{Q}_2)$ and where $\widehat{W}$ is a small resolution of the second fibre product of the universal elliptic curve $S_1(6) \to X_1(6)$ over the base.

THEOREM 3.2. *The varieties Y,Z and $\widehat{W}$ are modular. More precisely for the L-functions $L(H^3_{ét}(Y),s) \overset{\circ}{=} L(H^3_{ét}(Z),s) \overset{\circ}{=} L(H^3_{ét}(\widehat{W}),s) \overset{\circ}{=} L(f,s)$ where $\overset{\circ}{=}$ means that the Euler factors for $p \geq 5$ coincide.*

PROOF. The modularity of $\widehat{W}$ goes, of course, back to Deligne (1971). A proof for Y and Z can be found in (Hulek et al., Theorem 3.2). See also the historic remarks here. The crucial point is the application of Theorem 2.1. For this one can take the following set of primes $T = \{5,7,11,13,17,19,23,73\}$. Using slightly more theory Meyer (2000) showed that it already suffices to compute a_p and N_p for p in $T' = \{5,7,11,13,17,19,23\}$. □

By a conjecture of Tate this result should imply the existence of correspondences between these varieties inducing an isomorphism of the middle cohomology groups. This is easy for Y and Z since Z is (birationally) a 2:1 cover of Y. In (Hulek et al., Theorem 4.1) we found an explicit birational equivalence, which is defined over $\mathbb{Z}$, between Y and Z. We also found (Hulek et al., Theorem 4.3) a birational equivalence to Verrill's Calabi-Yau variety V (see (Verrill, 2000)) which has the same L-series and is given by the equation

$$\text{(V)} \qquad (1+x+xy+xyz)(1+z+yz+xyz) = \frac{(t+1)^2}{t}xyz.$$

We now have 5 series of numbers, namely $\#W(p), \#Y(p), \#Z(p), \#V(p)$ and a_p and knowing one of these numbers determines the others (in the case of $Z(p)$ one has to assume $p \equiv 1 \pmod 4$), although one can probably prove with some more effort that this assumption is unnecessary). The computation of the numbers $\#Y(p)$ etc. falls into two parts of which one is theoretical and the other uses a computer. We shall explain this here for Y, the other cases are similar. The variety N given by the equations (N) is singular along 20 lines and 10 isolated nodes, namely the "Segre" points given by the orbit of (1:1:1: -1:-1:-1) under the permutation group S_6. One obtains the non-singular model Y from N as follows. One first blows up $\mathbb{P}^4 = \{x_0 + \cdots + x_5 = 0\}$ in the 15 points $P_{klmn} = \{x_k = x_l = x_m = x_n = 0\}$ and then in the strict transforms of the 20 lines $L_{klm} = \{x_k = x_l = x_m = 0\}$. Finally one replaces the 10 nodes by $\mathbb{P}^1$'s. All of this can be done over the integers. We have already introduced the hyperplanes $S_k = \{x_k = 0\}$. Note that

$$S_k \cap N = \bigcup_{l=0}^{5} F_{kl} \quad \text{where } F_{kl} = \{x_k = x_l = 0\}.$$

Let

$$U = N \setminus \cup F_{kl}.$$

The resolution $Y \to N$ affects U only in the last step where we replace the 10 nodes by $\mathbb{P}^1$'s. Since these $\mathbb{P}^1$'s are defined over the integers we have to add $10p$ to the number of points. To compute the number $\#U(p)$ before this last step in the resolution we use a computer. What happens outside U can be controlled by hand. Blowing up the points P_{klmn} introduces Cayley cubics, i.e. a $\mathbb{P}^2$ blown up in 6 points. The exceptional locus which results from blowing up the strict transforms $L_{klm}^{(1)}$ of the lines L_{klm} is a union of quadrics and the strict transforms of the planes F_{kl} are again Cayley cubics. In each of these case we can count the number of points modulo p by hand for all primes p.

We used a Maple programme to compute the number of points on the various varieties. All computations were done on a Duron Processor 700 MHz with 64 kB RAM. The variety W is easier to handle than the other varieties, since it is the product (over the base) of a pencil of plane cubics with itself. To compute the number of points on Z we count the number of points in $U(p)$ such that $u = x_1 \ldots x_5$ is a square modulo p. This is the case if and only if $u^q \equiv 1 \pmod p$ where $q = (p-1)/2$. We give two times for each of the calculations. The first column gives for each variety the time needed (in seconds) using a naive programme which simply runs through all possibilities. The second programme makes use of the symmetries of the equations. In either case the time needed is of order $O(n^3)$ for W and of order $O(n^4)$ for the other varieties. Our use of the symmetries gives us roughly a factor of 2 for W and V and a factor of 20 for the other varieties, but is still not optimal. Meyer (2000) has developed a more subtle approach for the variety Y. He gains the following factors where the primes are given in brackets $46(p=37), 43(p=47), 39(p=59), 33(p=67)$ and $26(p=97)$. Another way to speed up the computations is to write a C++ programme instead of using Maple. The improvement is considerable, see the times given in (Meyer).

The computation of the Fourier expansion of f can be done in more than one way, at least in this case. The naive approach is to make use of the fact that the form f has a product expansion and to simply expand it. This is still faster than counting points on any of the varieties with the exception of W which needs roughly the same time. This method has the disadvantage that one soon encounters integers which produce an error message in MAPLE because they are too large. On the other hand there is the package HECKE developed by (W. A. Stein). This programme enables one to calculate a basis of a space of modular forms e.g. for the groups $\Gamma_0(N)$ for given level, weight and character. The problem is reduced to computing a basis for the space of newforms. These spaces are spanned by eigenforms with respect to the Hecke operator. Using modular symbols and theoretical work of Manin (1972) the computation of the coefficients of the Fourier expansion of a basis consisting of eigenform can thus be reduced to a linear algebra problem. Note that in our case the form f is a newform and the space of cusp forms $S_4(\Gamma_0(6))$ has dimension 1.

Comparing counting points and the computation of the Fourier coefficients one should be aware of the following difference. Counting points is done for each prime p separately, whereas the programmes computing the Fourier coefficients produce the numbers a_p simultaneously up to a given prime. Hence we produce two tables. In the first table we give the times needed to compute the numbers $N(p)$ for the various varieties for a given prime p. In the second table we compare the times needed to compute all numbers $N(p)$ and the Fourier coefficients a_p up to a fixed prime.

The final result is that the computation of the Fourier coefficients a_p using HECKE is much faster than any of the counting methods. The fastest counting method known to us is Meyer's C++ programme making use of the symmetries. We have compared this to the programme HECKE by computing the numbers $\#Y(p)$ up to $p = 211$, resp. the coefficients a_p. Counting points takes 14 min. 30 seconds, whereas computing the Fourier coefficients takes 7 seconds.

References

Barth, W., and Nieto, I. (1994) Abelian surfaces of type $(1,3)$ and quartic surfaces with 16 skew lines, *J. Alg. Geom.* **3**, 173–222.

Deligne, P. (1971) Formes modulaires et représentations l-adiques, *Sem. Bourbaki* 355 (1968/69), Lect. Notes Math. 179.

Fontaine, J.M., and Mazur, M. (1995) Geometric Galois representations, in J. Coates, S.T. Yau (eds.), *Elliptic curves, Modular Forms and Fermat's Last Theorem* (Hong Kong, 1993), International Press, pp. 41–78.

Hulek, K., Spandaw, J., van Geemen, B., and van Straten, D. (2001) The modularity of the Barth-Nieto quintic and its relatives, to appear in *Advances in Geometry.*

Livné, R. (1987) Cubic exponential sums and Galois representations, in K. Ribet (ed.), *Current trends in arithmetical algebraic geometry* (Arcata, Calif. 1985), Contemp. Math. 67, Amer. Math. Soc., Providence R.I., pp. 247–261.

Manin, J. I. (1972) Parabolic points and zeta functions of modular curves, *Izv. Akad. Nauk SSSR Ser. Mat.* **36** (1972), 19–66.

Meyer, Ch. (2000) Die L-Reihen einiger symmetrischer Quintiken, Diplomarbeit, Mainz.

Saito, M.-H., and Yui, N., The modularity conjecture for rigid Calabi-Yau threefolds over $\mathbb{Q}$, preprint, math.AG/0009041.

Stein, W.A., Modular forms database, `http://modular.fas.harvard.edu/`.

Verrill, H. (2000) The L-series of certain rigid Calabi-Yau threefolds, *J. Number Theory* **81**, 509–542.

Yui, N. (to appear) Arithmetic of certain Calabi-Yau varieties and mirror symmetry, *IAS/Park City Proceedings on Arithmetic Geometry, AMS.*

TABLE I.

prime	W		Y		Z		V	
5	.0	.0	.0	.0	.0	.0	.0	.0
7	.0	.0	.1	.0	.1	.0	.1	.0
11	.0	.0	.5	.0	.6	.1	.4	.2
13	.0	.0	1.1	.1	1.2	.1	.8	.4
17	.0	.1	3.7	.2	3.9	.3	2.5	1.4
19	.1	.1	6.0	.4	6.4	.4	4.1	2.4
23	.2	.1	13.6	.8	14.4	.9	9.5	5.2
29	.4	.3	36.6	2.0	38.5	2.1	26.0	13.9
31	.5	.3	48.7	2.6	51.3	2.8	34.8	18.4
37	.9	.5	102.5	5.2	108.8	5.7	74.1	38.8
41	1.2	.7	157.8	8.0	168.8	8.5	114.9	60.0
43	1.4	.8	192.5	9.6	207.5	10.4	140.9	73.6
47	1.9	1.1	279.4	13.8	303.4	14.8	204.5	106.9
53	2.8	1.5	461.4	22.6	506.8	24.1	338.8	177.0
59	3.9	2.1	761.3	34.9	828.1	37.5	560.8	282.2
61	4.5	2.4	892.8	40.7	954.7	43.5	641.7	334.1
67	6.2	3.2	1320.1	60.9	1411.2	65.7	950.3	500.4
71	7.4	4.0	1683.5	77.6	1813.1	83.2	1214.6	628.4
73	7.9	4.4	1892.3	86.9	2023.0	93.6	1362.0	716.3
79	10.3	5.5	2622.1	120.5	2811.6	129.1	1889.0	993.6
83	12.0	6.5	3218.4	146.8	3448.6	157.8	2301.0	1214.1
89	14.9	8.1	4245.9	194.8	4578.8	209.6	3096.2	1615.4
97	19.7	10.5	6135.6	276.1	6573.7	297.0	4404.6	2300.1
101	22.4	12.3	7242.1	325.0	7526.9	349.6	5140.1	2721.3
103		13.0		352.4		379.5		2940.9
107		14.6		412.1		443.5		3442.6
109		15.5		443.3		480.1		3699.3
113		17.4		511.8		554.1		4211.8

TABLE II.

prime	W		Y		Z		V		Cusp	Hecke
5	.0	.0	.0	.0	.0	.0	.0	.0	.0	
7	.0	.0	.1	.0	.1	.0	.1	.0	.0	
11	.0	.0	.6	.0	.6	.1	.4	.2	.1	
13	.0	.0	1.7	.1	1.8	.2	1.2	.7	.1	
17	.1	.1	5.4	.3	5.7	.4	3.8	2.1	.2	.0
19	.2	.2	11.3	.7	12.1	.8	7.9	4.5	.2	
23	.4	.3	24.9	1.5	26.5	1.7	17.3	9.7	.4	
29	.8	.5	61.5	3.5	65.0	3.8	43.3	23.5	.9	
31	1.3	.8	110.2	6.1	116.3	6.6	78.1	41.9	1.2	
37	2.2	1.3	212.7	11.3	225.0	12.3	152.2	80.7	2.1	
41	3.4	2.0	370.5	19.3	393.8	20.7	267.0	140.8	3.0	
43	4.8	2.8	563.0	28.9	601.3	31.1	407.4	214.4	3.6	.3
47	6.7	3.8	842.3	42.7	904.7	45.9	612.4	321.3	5.5	
53	9.4	5.3	1303.8	65.3	1411.6	70.0	951.2	498.3	8.0	
59	13.3	7.5	2065.0	100.2	2239.7	107.5	1512.0	780.5	11.6	
61	17.8	9.8	2957.8	140.9	3194.3	151.1	2153.7	1114.7	12.6	
67	24.0	13.1	4277.9	201.9	4605.5	216.8	3104.0	1615.1	18.5	
71	31.3	17.0	5961.4	279.5	6418.7	300.0	4318.6	2243.5	24.2	
73	39.2	21.4	7853.7	366.3	8441.6	393.6	5680.5	2959.8	28.3	1.4
79	49.5	26.9	10475.8	486.8	11253.2	522.6	7569.5	3953.4	ERROR	
83	61.5	33.5	13694.2	633.7	14701.8	680.4	9870.5	5167.5		
89	76.3	41.6	17940.1	828.5	19280.6	890.1	12966.8	6782.9		
97	96.0	52.0	24075.8	1104.6	25854.3	1187.1	17371.3	9082.9		
101	118.4	64.3	31317.9	1429.6	33381.3	1536.2	22511.4	11804.2		2.5
103		77.2		1782.0		1916.2		14745.1		2.9
107		91.8		2194.0		2359.6		18187.7		3.0
109		107.3		2637.3		2839.8		21886.9		3.2
113		124.6		3149.1		3393.8		26098.8		3.6

SUBVARIETIES OF ABELIAN VARIETIES

E. IZADI*
University of Georgia, Athens, U.S.A.

Abstract. We discuss various constructions which allow one to embed a principally polarized abelian variety in the jacobian of a curve. Each of these gives representatives of multiples of the minimal cohomology class for curves which in turn produce subvarieties of higher dimension representing multiples of the minimal class. We then discuss the problem of producing curves representing multiples of the minimal class via deformation-theoretic methods.

Key words: subvarieties of abelian varieties, curves in abelian varieties, jacobians, minimal cohomology classes, Prym varieties, Prym-Tjurin varieties, deformations of curves in jacobians.

Mathematics Subject Classification (2000): *Primary* 14K12, 14C25; *Secondary* 14B10, 14H40.

Introduction

Let (A,Θ) be a principally polarized abelian variety (*ppav*) of dimension g over the field $\mathbb{C}$ of complex numbers. This means that Θ is an ample divisor on A, well-determined up to translation, with $h^0(A,\Theta) := \dim H^0(A,\Theta) = 1$. Let $[\Theta] \in H^2(A,\mathbb{Z})$ be the cohomology class of the theta divisor Θ. Then the cohomology class $[\Theta]^{g-e}$ is divisible by $(g-e)!$. The class $\frac{[\Theta]^{g-e}}{(g-e)!}$ is not divisible and it is called the dimension e minimal cohomology class in (A,Θ). This class is positive in the sense that some multiple of it can be represented by an algebraic subvariety (for instance $[\Theta]^{g-e}$ is the class of a complete intersection of $g-e$ general translates of Θ) and, furthermore, any subvariety whose class is a multiple of $\frac{[\Theta]^{g-e}}{(g-e)!}$ is non-degenerate, i.e., generates A as a group. We are interested in the representability of multiples of the minimal classes by algebraic subvarieties of A. We begin by discussing two special cases.

* This material is based upon work partially supported by the National Science Foundation under Grant No. DMS-0071795. Any opinions, findings and conclusions or recomendations expressed in this material are those of the author and do not necessarily reflect the views of the National Science Foundation (NSF).

C. Ciliberto et al. (eds.),
Applications of Algebraic Geometry to Coding Theory, Physics and Computation, 207–214.

1. Jacobians

Let C be a smooth, complete, irreducible curve of genus g over the complex numbers. The jacobian $JC = Pic^0 C$ of C is the connected component of its Picard group parametrizing degree 0 invertible sheaves. For any nonnegative integer e, the choice of an invertible sheaf $\mathcal{L}$ of degree e on C gives a morphism

$$\begin{aligned} \phi_{\mathcal{L}} : C^{(e)} &\longrightarrow JC \\ D_e &\longmapsto \mathcal{O}_C(D_e) \otimes \mathcal{L}^{-1} \end{aligned}$$

where $C^{(e)}$ is the e-th symmetric power of C. For $e \geq g$, such a morphism is surjective. When $e = g-1$, the image of $C^{(g-1)}$ by $\phi_{\mathcal{L}}$ is a theta divisor on JC which we will denote by Θ_C (always well-determined up to translation). For any e between 1 and g the image of $\phi_{\mathcal{L}}$ has class $\frac{[\Theta]^{g-e}}{(g-e)!}$. If $e = 1$, the map $\phi_{\mathcal{L}}$ is an embedding and its image is called an Abel curve. By a theorem of Matsusaka (1959), the dimension 1 minimal class is represented by an algebraic curve C in (A, Θ) if and only if (A, Θ) is the polarized jacobian (JC, Θ_C) of C. The higher-dimensional analogue of this theorem has the following counterexample: by a result of Clemens and Griffiths (1972), the intermediate jacobian of a smooth cubic threefold in $\mathbb{P}^4$ is not the jacobian of a curve but it contains a surface (the image of the Fano variety of lines in the cubic threefold) whose cohomology class is the dimension 2 minimal class. Debarre (1995) has shown that for any e strictly between 1 and $g-1$ jacobians form an irreducible component of the family of ppav in which the dimension e minimal class is represented by an algebraic subvariety.

As the above suggests, not every ppav is a jacobian. In fact, the moduli space of ppav of dimension g has dimension $\frac{g(g+1)}{2}$ whereas the moduli space of curves of genus $g \geq 2$ has dimension $3g-3$. As soon as $g \geq 4$, we have $\frac{g(g+1)}{2} > 3g-3$ and so not all ppav of dimension g are jacobians. The question then becomes how else can one parametrize ppav? The first step of a generalization of the notion of jacobian is the construction of a Prym variety which we describe below.

2. Pryms

Suppose that C is a smooth, complete and irreducible curve of genus $g+1$ with an étale double cover $\pi : \widetilde{C} \to C$. Then $\widetilde{C}$ has genus $\tilde{g} := 2g+1$. Let $\sigma : \widetilde{C} \to \widetilde{C}$ be the involution of the cover π. The involution σ acts on the jacobian $J\widetilde{C}$ and the Prym variety P of π is an abelian variety of dimension g defined by

$$P := im(\sigma - 1) \subset J\widetilde{C}.$$

The principal polarization of $J\widetilde{C}$ induces twice a principal polarization on P which we will denote Ξ.

A priori, there are two ways of obtaining subvarieties of a Prym variety: by projection and intersection. Since P is a quotient of $J\widetilde{C}$ via $\sigma - 1$, we can take the images of subvarieties of $J\widetilde{C}$ in P. For $e = 1$, we can take the images in P of Abel curves in $J\widetilde{C}$. We obtain in this way embeddings of $\widetilde{C}$ in P whose images are called Prym-embedded curves. The class of a Prym-embedded curve is $2\frac{[\Xi]^{g-1}}{(g-1)!}$. Welters (1987) classified all curves of class twice the minimal class in a ppav. The list is short but it is not limited to Prym-embedded curves. Therefore the analogue of Matsusaka's theorem is false for curves representing twice the minimal class. For $e > 1$, using Pontrjagin product, it is easy to see that the projection in P of the image of $\phi_{\mathcal{L}}$ has class $2^e\frac{[\Xi]^{g-e}}{(g-e)!}$ (provided it has dimension e).

Secondly, since the Prym variety is also a subvariety of $J\widetilde{C}$, we can intersect the images of the symmetric powers of $\widetilde{C}$ with it. Since $\Theta_{\widetilde{C}}$ induces twice Ξ on P, for $\mathcal{L}$ of degree e, the intersection of the image of $\phi_{\mathcal{L}}$ with P has class $2^{\tilde{g}-e}\frac{[\Xi]^{\tilde{g}-e}}{(\tilde{g}-e)!}$, provided that the intersection is proper. There is at least one case in which one can do better: there are a finite number of theta divisors in $J\widetilde{C}$ whose intersection with P is 2Ξ as a divisor, see for instance (Mumford, 1974). In particular, we have a nice way of parametrizing the theta divisor of the ppav (P, Ξ).

It can easily happen that the intersection $\phi_{\mathcal{L}}(\widetilde{C}^{(e)}) \cap P$ is not proper. In such a case the cohomolgy class of the resulting subvariety needs to be determined by other means. Such subvarieties appear in the work of Recillas (1974); Donagi (1981); Clemens and Griffiths (1972); Beauville (1982). Following Beauville, we shall call them special subvarieties. A different way of defining a special subvariety of a Prym variety which allows one to compute its cohomology class is as follows (Recillas; Donagi; Clemens and Griffiths; Beauville, 1982). Let g^r_e be a *complete* linear system of dimension r and degree e on C. Let L be the corresponding invertible sheaf on C and let $\mathcal{L}$ be an invertible sheaf on $\widetilde{C}$ whose norm is $Nm(\mathcal{L}) = L$ (i.e., if $\mathcal{L} \cong \mathcal{O}_{\widetilde{C}}(D)$, then $L \cong \mathcal{O}_C(\pi_* D)$). Consider g^r_e as a subvariety of $C^{(e)}$, isomorphic to $\mathbb{P}^r$. Assuming that g^r_e contains reduced divisors, the inverse image of g^r_e in $\widetilde{C}^{(e)}$ is reduced. It splits as the union of two connected components whose images in $J\widetilde{C}$ by $\phi_{\mathcal{L}}$ are contained in P and the translate P' of P such that $P \cup P'$ is the kernel of the Norm $Nm : J\widetilde{C} \to JC$. Therefore, after translating one of these subvarieties, we obtain two subvarieties of P. They are isomorphic if e is odd but not if e even. They both have cohomology class $2^{e-2r-1}\frac{[\Xi]^{g-r}}{(g-r)!}$ at least when $1 \leq e \leq 2g+1$ and $e > 2r$ (Beauville, 1982).

To see that these special subvarieties are indeed the nonproper intersections that we mentioned above, one needs to note that the fibers of the map $J\widetilde{C} \to JC$ are translates of $P \cup P'$. The special subvarieties are intersections of these fibers with images of maps $\phi_{\mathcal{L}}$. Note that when the g^r_e is nonspecial, i.e., $h^1(g^r_e) = 0$, the special subvariety is in fact a proper intersection and Beauville's cohomology class is equal (as it should be) to the cohomology class of the proper intersection above. As we shall see below, looking at special subvarieties as such nonproper intersections allows us to define them for arbitrary ppav.

For $g \leq 5$, all ppav are Prym varieties (in the generalized sense of Beauville (1977)). For $g \geq 5$ Prym varieties of dimension g depend on the same number of moduli as curves of genus $g+1$, meaning $3g$ moduli. Therefore, for $g \geq 6$, a general ppav is not a Prym variety. So we need to find a different way to parametrize a ppav.

3. Prym-Tjurin varieties

Again, one would want to use a construction involving curves. Higher degree coverings $\widetilde{C} \to C$ do not yield general ppav because the dimension of the Prym variety (defined as the connected component of $\mathcal{O}_{\widetilde{C}}$ of the kernel of the Norm (or pushforward on divisors) $J\widetilde{C} \to JC$) is too high and therefore the families of ppav that one would obtain are too small, their dimensions being the dimensions of the moduli spaces for the bottom curves C. Looking back at a Prym variety, we note that it was defined as a special type of abelian subvariety of a jacobian. An abelian subvariety A of a jacobian (JX, Θ_X) such that Θ_X induces m-times a principal polarization Θ on A is called a Prym-Tjurin variety. Welters (1987) has proved that all ppav are Prym-Tjurin varieties.

To say that Θ_X induces m-times Θ on A is equivalent to saying that the class of the image of an Abel curve in A is m-times the minimal class for curves (Welters, 1987). Here we are taking the image of an Abel curve by the composition

$$JX \xrightarrow{\cong} \hat{JX} \longrightarrow \hat{A} \xrightarrow{\cong} A$$

where $\hat{}$ denotes the dual abelian variety, the first and the last map are induced by the polarizations Θ_X and Θ respectively, and the middle map is the transpose of the embedding of A in JX.

Therefore, finding a structure of Prym-Tjurin variety on a ppav is equivalent to finding a reduced and irreducible curve $\overline{X}$ in A representing m-times the minimal class and such that A embeds in the jacobian of the normalization X of $\overline{X}$. Given such a structure, we can find subvarieties of A as in the case of Prym varieties: by projection and intersection. Since the class of the image of an Abel curve is m-times the minimal class, Pontrjagin product shows that the projection in P of the image of $\phi_{\mathcal{L}}$ has class $m^e \frac{[\Theta]^{g-e}}{(g-e)!}$ (provided it has dimension e). Since Θ_X induces m-times Θ on A, for $\mathcal{L}$ of degree e, the intersection of the image of $\phi_{\mathcal{L}}$ with P has class $m^{g_X-e} \frac{[\Theta]^{g_X-e}}{(g_X-e)!}$, provided that the intersection is proper (g_X is the genus of X). However, unlike Prym varieties, it is not clear whether one can find translates of Θ_X whose intersection with A is m-times a theta divisor. Kanev (1987) has shown that this is possible under a restrictive hypothesis which we explain below.

Any abelian subvariety of JX is the image of an endomorphism of JX (which is not unique). The datum of an endomorphism of JX is equivalent to the datum of a correspondence, i.e., a divisor in $X \times X$, up to addition and subtraction of fibers of the two projections. This is best seen as follows. Start with a divisor $D \subset X \times X$. To D one can associate an endomorphism of JX in the following way

$$\begin{aligned} \psi_D: \quad JX &\longrightarrow JX \\ \mathcal{O}_X(E) &\longmapsto \mathcal{O}_X(p_{2*}((p_1^*E)\cdot D)) \end{aligned}$$

where p_1 and p_2 are the two projections $X \times X \to X$. If D is linearly equivalent to a sum of fibers of p_1 and p_2, then ψ_D is the zero endomorphism. If we exchange the roles of p_1 and p_2 in the above definition then ψ_D is replaced by its image under the Rosati involution. The correspondence D is said to be symmetric if there are (not necessarily effective) divisors a and b on X such that $D - D^t$ is linearly equivalent to $p_1^*(a) + p_2^*(b)$, where D^t is the transpose of D, i.e., the image of D under the involution exchanging the two factors of $X \times X$. So D is symmetric if and only if ψ_D is fixed by the Rosati involution. We shall assume that this is the case. This is not restrictive since any abelian subvariety of JX is always the image of an endomorphism fixed by the Rosati involution, see e.g. (Welters, 1987).

Kanev (1987) has shown that if the endomorphism can be represented by a symmetric fixed-point-free correspondence D (i.e., the support of D does not intersect the diagonal of $X \times X$), then one can find theta divisors Θ_X such that $\Theta_X|_A = m\Theta$ as divisors. Furthermore, fixing an invertible sheaf $\mathcal{L}_0 \in Pic^{g_X-1}X$, an invertible sheaf $\mathcal{L} \in A \subset JX$ is on Θ if and only if $h^0(\mathcal{L} \otimes \mathcal{L}_0) \geq m$ and $\mathcal{L} \notin \Theta$ if and only if $h^0(\mathcal{L} \otimes \mathcal{L}_0) = 0$. This gives a nice parametrization of Θ and even allows one to analyze the singularities of Θ. It is not known however, whether every ppav is a Prym-Tjurin variety for a (symmetric) fixed-point-free correspondence. In addition, two correspondences could induce the same endomorphism of JX while one is fixed-point-free and the other is not. In general it is difficult to determine whether a given endomorphism can be induced by a fixed-point-free correspondence.

As we noted above, we can generalize the notion of special subvarieties to Prym-Tjurin varieties by defining them to be non-proper intersections of A with images of symmetric powers of X. It would be interesting to compute the cohomology classes of these special subvarieties and see whether the analogue of Beauville's formula holds, meaning, the cohomology class of a special subvariety of dimension r is $m^{e-2r-1}\frac{[\Theta]^{g-r}}{(g-r)!}$.

Welters (1987) showed that every principally polarized abelian variety is a Prym-Tjurin variety. Birkenhake and Lange (1992, Corollary 2.4, p. 374)[1] showed that every principally polarized abelian variety is a Prym-Tjurin variety for an integer $m \leq 3^g(g-1)!$

[1] Their proof uses 3-theta divisors. Using the fact that a general 2-theta divisor is smooth, the exact same proof would give $m \leq 2^g(g-1)!$. For abelian varieties with a smooth theta divisor, the same proof would give $m \leq (g-1)!$. One needs the Lefschetz hyperplane theorem which also works for mildly singular theta divisors, see e.g. Goresky and MacPherson (1988), Chapter 2).

4. Deforming curves

The question is to find the smallest integer m for which $m\frac{[\Theta]^{g-1}}{(g-1)!}$ can be represented by an algebraic curve. This naturally defines a stratification of the moduli space $\mathcal{A}_g$ of ppav. Using results of Kanev, Debarre (1992) shows that if (A,Θ) is the Prym-Tjurin variety for a symmetric fixed-point-free correspondence, then either $Sing(\Theta)$ is empty or its dimension is at least $g-2m-2$. Since the theta divisor of a general ppav is smooth, this suggests that, for a general ppav A, the smallest integer m for which there is a curve of class $m\frac{[\Theta]^{g-1}}{(g-1)!}$ in A which in addition gives A a structure of Prym-Tjurin variety should be at least $\frac{g-1}{2}$. It is unlikely however that this bound is effective. Debarre (1994) has proved that the smallest integer m for which $m\frac{[\Theta]^{g-1}}{(g-1)!}$ is the class of an algebraic curve is at least $\sqrt{\frac{g}{8}}-\frac{1}{4}$ if (A,Θ) is general.

The difficulty is to produce curves in ppav in nontrivial ways. One approach that we have considered is to deform curves in jacobians of curves out of the jacobian locus. More precisely, let C be a curve of genus g with a g^1_d (a pencil of degree d). Define

$$X_e(g^1_d) := \{D_e : \exists D \in C^{(d-e)} \text{ such that } D_e + D \in g^1_d\} \subset C^{(e)}$$

(for the precise scheme-theoretical definition see Izadi (2000), when $e=2$, and Izadi (2001) for $e>2$). If $d \geq e+1$, the restriction of a given morphism $\phi_{\mathcal{L}}$ to $X_e(g^1_d)$ is nonconstant and so we can map $X_e(g^1_d)$ to JC. The cohomology class of the image of $X_e(g^1_d)$ in JC is m-times the minimal class with $m=\binom{d-2}{e-1}$. Given a one-parameter infinitesimal deformation of the jacobian of C out of the jacobian locus $\mathcal{J}_g$ we ask when the curve $X_e(g^1_d)$ deforms with it. Infinitesimal deformations of JC are parametrized by $H^1(T_{JC})$ where T_{JC} is the tangent sheaf of JC. The principal polarization Θ_C provides an isomorphism between $H^1(T_{JC})$ and the second tensor power $H^1(\mathcal{O}_C)^{\otimes 2}$. Under this isomorphism the globally unobstructed deformations of the pair (JC,Θ_C) are identified with the symmetric square $S^2H^1(\mathcal{O}_C)$. Therefore any quadric in the canonical space of C defines a linear form on the space of these infinitesimal deformations. When we say that an infinitesimal deformation $\eta \in S^2H^1(\mathcal{O}_C)$ is in the annihilator of a quadric, we mean that it is in the kernel of the corresponding linear form. We prove the following in (Izadi, 2000)

THEOREM 1. *Suppose C nonhyperelliptic and $d \geq 4$. If the curve $X_2(g^1_d)$ deforms out of $\mathcal{J}_g$ then*

1. *either $d=4$*
2. *or $d=5$, $h^0(g^1_5)=3$ and C has genus 5 or genus 4 and only one g^1_3.*

In the case $g=5$ if $X_2(g^1_5)$ deforms in a direction $\eta \in S^2H^1(C_C)$ out of $\mathcal{J}_5$, then η is in the annihilator of the quadric $\bigcup_{D\in g^1_5}\langle D\rangle$.

Here $\langle D\rangle$ denotes the span of the divisor D in the canonical space of C. For $d=3$ the image of $X_2(g_3^1)$ in JC is an Abel curve and so by the result of Matsusaka (1959), the curve $X_2(g_3^1)$ cannot deform out of $\mathcal{J}_g$. For $d=4$, it follows from the theory of Prym varieties that $X_2(g_4^1)$ deforms out of $\mathcal{J}_g$ (into the locus of Prym varieties): in fact $X_2(g_4^1)$ is a Prym-embedded curve (Recillas). For $d=5$, $h^0(g_d^1)=3$ and $g=4$ (with only one g_3^1) or $g=5$ we believe that $X_2(g_5^1)$ deforms out of $\mathcal{J}_g$ but we do not have a proof of this. An interesting question is what are these deformations of (JC,Θ_C) into which $X_2(g_5^1)$ deforms. Can one describe them in a concrete geometric way?

For $e>2$, the analogous result would be the following. The curve $X_e(g_d^1)$ deforms out of $\mathcal{J}_g$ if and only if

- either $e=h^0(g_d^1)$ and $d=2e$
- or $e=h^0(g_d^1)-1$ and $d=2e+1$.

We expect this to be true most of the time. There could, however, be special pairs (C,g_d^1) for which the curve $X_e(g_d^1)$ deforms out of the jacobian locus but g_d^1 does not verify the above conditions. For instance, so far my calculations (Izadi, 2001) seem to indicate that if there is a divisor $D\in X$ with $h^0(D)\geq 2$, then X might deform in directions η whose images in the projectivization $\mathbb{P}(S^2H^1(\mathcal{O}_C))$ are in the span of the image of $\bigcup_{D'\in|D|}\langle D'\rangle$. Finally, note that a standard Brill-Noether calculation shows that for general curves of genus ≥ 7, the smallest d for which they can have g_d^1's satisfying $d=2h^0(g_d^1)$ is $d=2g-4$. In such a case the class of $X_e(g_d^1)$ is m-times the minimal class with $m=\binom{2g-6}{g-3}$ which is then what we would find for a general ppav. We address the case $e>2$ in (Izadi, 2001).

5. The genus

The cohomology class is one discrete invariant that one can associate to a curve in a ppav. Another discrete invariant is the genus of the curve. We refer the reader to the nice paper by Bardelli, Ciliberto and Verra (1995) for a discussion of this.

References

Bardelli, F., Ciliberto, C., and Verra, A. (1995) Curves of minimal genus on a general abelian variety, *Compositio Math.* **96**(2), 115–147.

Beauville, A. (1977) Prym varieties and the Schottky problem, *Inventiones Math.* **41**, 149–196.

Beauville, A. (1982) Sous-variétés spéciales des variétés de Prym, *Compositio Math.* **45**, Fasc. 3, 357–383.

Birkenhake, C. and Lange, H. (1992) *Complex abelian varieties*, Grundlehren der Mathematischen Wissenschaften 302, Springer-Verlag, Berlin.

Clemens, H. and Griffiths, Ph. (1972) The intermediate jacobian of the cubic threefold, *Annals of Math.* **95**, 281–356.

Debarre, O. (1992) Vers une stratification de l'espace des modules des variétés abeliennes principalement polarisées, in *Complex algebraic varieties (Bayreuth 1990)*, Lecture Notes in Math. 1507, Springer-Verlag, Berlin, pp. 71–86.

Debarre, O. (1994) Degrees of curves in abelian varieties, *Bull. Soc. Math. France* **122**(3), 343–361.
Debarre, O. (1995) Minimal cohomology classes and jacobians, *J. Algebraic Geometry* **4**(2), 321–335.
Donagi, R. (1981) The tetragonal construction, *Bull. of the A.M.S.* **4**(2), 181–185.
Goresky, M., and MacPherson, R. (1988) *Stratified Morse Theory*, Springer-Verlag, New York.
Izadi, E. (2000) Deforming curves representing multiples of the minimal class in jacobians to non-jacobians I, preprint math.AG/0103204.
Izadi, E. (2001) Deforming curves representing multiples of the minimal class in jacobians to non-jacobians II, in preparation.
Kanev, V. (1987) Principal polarizations of Prym-Tjurin varieties, *Compos. Math.* **64**, 243–270.
Matsusaka, T. (1959) On a characterization of a jacobian variety, *Memo. Coll. Sci. Univ. Kyoto* **32**, 1–19.
Mumford, D. (1974) Prym varieties I, in L. Ahlfors, I. Kra, B. Maskit and L. Niremberg (eds.), *Contributions to Analysis*, Academic Press, pp. 325–350.
Recillas, S. (1974) Jacobians of curves with a g^1_4 are Prym varieties of trigonal curves, *Bol. Soc. Math. Mexicana* **19**, 9–13.
Welters, G. E. (1987) Curves of twice the minimal class on principally polarized abelian varieties, *Nederl. Akad. Wetensch. Indag. Math.* **49**(1), 87–109.

CHARACTERISTIC VARIETIES OF ALGEBRAIC CURVES

A. LIBGOBER
University of Illinois at Chicago, U.S.A.

Abstract. We study an invariant of plane algebraic curves with several components. Such invariant, called here a characterisitic variety, is a collection of subtori in the group of characters of the fundamental group of the complement to the curve. This invariant is a generalization of one variable Alexander polynomial. The paper discusses the basic properties of characterisitic varieties and their calculation in terms of position of the singularities of the curve in the plane.

Key words: plane algebraic curves, singularities, fundamental groups of the complements.

Mathematics Subject Classification (2000): 14E20, 14H30, 14H50.

Introduction

A procedure for calculation of fundamental groups for the complements to algebraic curves in complex projective plane was found by Zariski (1971) and van Kampen (1933). Their methods yielded several important calculations and results on the fundamental groups of the complements (cf. for example (Libgober, 1983) for references). However, only limited information was obtained about their algebraic structure or what actually affects the complexity of these fundamental groups. This paper is a result of attempts to find alternative ways for calculating the fundamental groups of the complements or at least some invariants of these groups. The invariants of the fundamental groups, which we consider here, are certain subvarieties of complex tori $\mathbb{C}^{*r}$. They were called characteristic varieties in (Libgober, 1992). These subvarieties are unions of translated subtori, as follows from recent work of (Arapura). We calculate these subtori in terms of local type of singularities and dimensions of linear systems which we attach to the configuration of singularities of the curve.

These characteristic varieties can be defined as follows. Let $C = \bigcup_{1 \leq i \leq r} C_i$ be an algebraic curve in $\mathbb{C}^2$ and $\pi_1 = \pi_1(\mathbb{C}^2 - C)$ be the fundamental group of its complement. Then π_1/π_1' is isomorphic to $\mathbb{Z}^r$ and acts on π_1'/π_1'' by conjugation. This makes π_1'/π_1'' into a module over the group ring of π_1/π_1'. The latter is just the ring of Laurent polynomials $\mathbb{Z}[t_1, t_1^{-1}, \ldots, t_r, t_r^{-1}]$. After tensoring with $\mathbb{C}$, we

C. Ciliberto et al. (eds.),
***Applications of Algebraic Geometry to Coding Theory, Physics and Computation*, 215–254.**

obtain a $\mathbb{C}[\pi_1/\pi_1']$-module $\pi_1'/\pi_1''\otimes\mathbb{C}$. The support of its i-th exterior power is a subvariety of the torus $\mathrm{Spec}\mathbb{Z}[t_1,t_1^{-1},\ldots,t_r,t_r^{-1}]$ called the i-th characteristic variety of C: $\mathrm{Char}_i(C)$, cf. (Libgober, 1992).

This invariant of the fundamental group can be used to calculate the homology of abelian covers of $\mathbb{C}^2-C$ (Libgober, 1992) and covers of $\mathbb{P}^2$ branched over the projective closure of C (Sakuma). The above construction of characteristic varieties can be, of course, carried out for any topological space with $H_1=\mathbb{Z}^r$ and several known results can be recasted using them. For example, the modules π_1'/π_1'' were widely studied in the context of the complements to links in spheres (cf. e.g. (Hillman) and references there; for the case of algebraic links cf. (Sabbah)). In this case, Char_1 is the set of zeros of the multi-variable Alexander polynomial. We shall see, however, that the characteristic varieties of algebraic curves rarely have codimension equal to one and hence cannot be described using single polynomial. The varieties $\mathrm{Char}_i(C)$ coincide with the cohomology support loci for local systems of rank 1 considered in (Arapura). The homology of the Milnor fiber of the function obtained by homogenizing a defining equation of C (i.e. the Milnor fiber of the cone over the projective closure of C) can be found from the characteristic varieties of the latter. These Milnor fibers earlier were considered in the case when C is an arrangement i.e. when all components of C are lines, cf. (Cohen and Suciu).

For an irreducible C the characteristic varieties are subsets of $\mathbb{C}^*$ i.e. collections of complex numbers. Those are the roots of the Alexander polynomial of C (Libgober, 1982) and the results of this paper are equivalent to the results of (Libgober, 1983).

Our calculation of Char_i based on the following observations. Firstly, by Arapura Theorem the characteristic varieties are translated tori and hence can be described by simple discrete data. Secondly, Sakuma's formula (cf. (1.3.2.2)), relating the homology of abelian branched over C cover to the characteristic varieties of C, can be used to calculate such data for essential components (cf. (1.4.3)) of characteristic varieties completely from the information about the homology of *all* abelian covers with the branching locus C. Thirdly, these abelian covers can be realized as complete intersections and one can use the theory of adjoints (cf. (1.5)) to calculate the homology of these covers (generalizing the calculations in the case of hypersurfaces cf. (Zariski; Libgober, 1983 and 1986)); complete intersections were used by (Ishida) (cf. also (Zuo)) in similar context for calculations in the case of abelian covers of $\mathbb{P}^2$ branched along some arrangements). We associate with each singular point a collection of polytopes in the unit cube $\mathcal{U}$ in $\mathbb{R}^r$ union of which is $\mathcal{U}$ and call them the local polytopes of quasiadjunction (cf. sect. 2.4.1). Moreover, every such polytope defines the ideal in the local ring of the singular point. The collection of local polytopes defines new partition of the unit cube which is a refinement of partitions corresponding to all singular points of C and reflecting the global information about singularities of C (cf. sect. 2.6). We call the polytopes of this partition the global polytopes of quasiadjunction. In the set of faces of global polytopes of quasiadjnction we single out a subset of contributing ones. To each contributing face δ corresponds the linear system $H^0(\mathbb{P}^2, \mathcal{J}_\delta(\deg C - 3 - l(\delta))$ where the ideal sheaf $\mathcal{J}_\delta \subset \mathcal{O}_{\mathbb{P}^2}$ and the integer $l(\delta)$ are determined by the face δ. The components of $\mathrm{Char}_i C$ correspond to contributing faces of global polytopes of quasiadjunction for which $\dim H^1(\mathbb{P}^2, \mathcal{J}_\delta(\deg C - 3 - l(\delta))) = i$. The main result of the paper is Theorem 3.1 where the equations for translated tori are given explicitly in terms of corresponding faces of polytopes of quasiadjunction (cf. sect. 3).

The procedure for calculating the characteristic varieties, though involving possibly large calculations, is entirely algorithmic. For example, suppose that the curve C is an arrangement of r lines. (cf. sect. 3.3 for several examples of explicite calculations for such arrangements including Ceva's and Hesse's arrangements). Then any component of the characteristic variety $\mathrm{Char}_i(C)$, having positive dimension either belongs to a component of characteristic variety of a sub-arrangement (i.e. is inessential and can be found by applying this algorithm to a sub-arrangement) or is a connected component of a subgroup corresponding to a collection $\mathcal{S}$ of vertices in arrangement having multiplicity greater than 2 (more detailed calculation allows to pick the component as well, cf. th.3.1). A collection $\mathcal{S}$ yields a component of characteristic variety if it satisfies the following conditions.

a) Certain system of linear homogeneous equations attached to $\mathcal{S}$ has a non zero solution. This system is constructed as follows: the unknowns x_i are in one to one correspondence with $r(\mathcal{S})$ lines of arrangement containing points from $\mathcal{S}$. The equations are in one to one correspondence with the elements of $\mathcal{S}$. Left hand side of each equation is the sum of variables corresponding to the lines through a point of $\mathcal{S}$ and the right hand side is a positive integer (not exceeding the number of lines in the arrangements).

b) The solution set of the system in a) belongs to a hyperplane $\sum_{i=1}^{r(\mathcal{S})} x_i = l(\mathcal{S})$.

c) Let $I_\mathcal{S}$ be the ideal sheaf with $\mathrm{Supp}\, \mathcal{O}_{\mathbb{P}^2}/I_\mathcal{S} = \mathcal{S}$ which stalk at $P \in \mathcal{S}$ is $\mathcal{M}_P^{m-2-\rho(P)}$ where $\mathcal{M}_P$ is the maximal ideal of P, m is the number of lines in $\mathcal{S}$ containing P and $\rho(P)$ is the right hand side of the equaition in the system from a) corresponding to P. Then $\dim H^1(\mathbb{P}^2, I_\mathcal{S}(r-3-l(\mathcal{S}))) = i \neq 0$.

Moreover if a)-c) are satisfied then the corresponding to $\mathcal{S}$ subgroup is the set of solutions of $\exp(L_P) = 1$ where $L_P = \rho(P)$ are the equations of the system mentioned in a). Selection of particular connected component follows from more technical description in Theorem 3.1

This algorithm yields complete description of characteristic tori of positive dimension (and essential torsion points). An interesting problem which left unanswered here is the realization problem: which tori (or collections of tori) can appear as characteristic tori of an algebraic curve with fixed degrees of components and given local type of singularities. Some information in this direction however is provided in section 4. Finally in the case of line arrangements (or equivalently the case of fundamental groups of the complements to arbitrary arrangements) the characteristic varieties give new sufficient conditions (resonance conditions) for Aomoto complex of an arrangement to be quasi-isomorphic to the corresponding twisted DeRham complex (in many situations less restrictive than previously used, cf. (Esnault, Schechtman and Viehweg). We describe in a new way the space of "resonant" Aomoto complexes on given arrangement i.e. those with the cohomology different from the cohomology of generic Aomoto complexes (Th. 5.4.1; this space was considered in (Falk)). Vice versa, this relation between the space of resonant Aomoto complexes and characteristic varieties shows that components of characteristic varieties which are subgroups of the group of characters are *combinatorial* invariants of arrangements. Moreover, Aomoto complexes provide another algorithm for calculating these components of characteristic varieties of the fundamental groups of the complements to arrangements.

The announcement of these results is presented in (Libgober, 1999). This work was supported by NSF grants DMS-9803623, DMS-9872025 and Mittag Leffler Institute. I am also grateful to S. Yuzvinsky for very interesting correspondence regarding the material in section 5 and to J. Cogolludo for useful discussions of the manuscript.

1. Preliminaries

1.1. SETTING

Let $\overline{C} = \bigcup \bar{C}_i (i=1,\ldots,r)$ be a reduced algebraic curve in $\mathbb{P}^2$ where $\bar{C}_i (i=1,\ldots,r)$ are the irreducible components of $\overline{C}$. We shall denote by d_i the degree of the component $\bar{C}_i$. Let L_∞ be a line in $\mathbb{P}^2$ which we shall view as the line at infinity. We shall be concerned with the fundamental groups of the complements to $\overline{C}$ in $\mathbb{P}^2$ and in $\mathbb{C}^2 = \mathbb{P}^2 - L_\infty$. Let $C = \bigcup C_i$ be the affine portion of $\overline{C}$. The homology of these complements are the following (Libgober, 1982):

$$H_1(\mathbb{C}^2 - C, \mathbb{Z}) = \mathbb{Z}^r, \ H_1(\mathbb{P}^2 - \overline{C}, \mathbb{Z}) = \mathbb{Z}^r/(d_1,\ldots,d_r) \tag{1.1.1}$$

Generators of these homology groups are represented by the classes of the loops γ_i, each of which is the boundary of a small 2-disk intersecting C_i (resp. $\bar{C}_i$) transversally at a non singular point.

For the fundamental groups we have the exact sequence:

$$\pi_1(\mathbb{C}^2 - C) \to \pi_1(\mathbb{P}^2 - \overline{C}) \to 1 \tag{1.1.2}$$

If the line L_∞ is transversal to $\overline{C}$, then the kernel of the surjection (1.1.2) is isomorphic to $\mathbb{Z}$ and belongs to the center of $\pi_1(\mathbb{C}^2 - C)$, cf. (Libgober, 1994). In general, the fundamental group of the affine portion of the complement to $\overline{C}$ in $\mathbb{P}^2$

depends on position of L_∞ relative to $\overline{C}$. Throughout the paper we assume that L_∞ is transversal to $\overline{C}$.

1.2. CHARACTERISTIC VARIETIES OF ALGEBRAIC CURVES

1.2.1. Let R be a commutative Noetherian ring and M be a finitely generated R-module. Let $\Phi : R^m \to R^n$ be such that $M = \operatorname{Coker}\Phi$. Recall that the k-th Fitting ideal of M is the ideal generated by $(n-k+1)\times(n-k+1)$ minors of the matrix of Φ (clearly depending only on M rather than on Φ). The k-th characteristic variety M is the reduced sub-scheme of $\operatorname{Spec} R$ defined by $F_k(M)$.

If $R = \mathbb{C}[H]$ where H is a free abelian group then R can be identified with the ring of Laurent polynomials and $\operatorname{Spec} R$ is a complex torus. In particular each k-th characteristic variety of an R-module is a subvariety $V_k(M)$ of $(\mathbb{C}^*)^{\operatorname{rk} H}$.

If $\operatorname{Ann} \bigwedge^k M \subset R$ is the annihilator of the k-th exterior power of M then (Buchsbaum and Eisenbud, Cor. 1.3): $(\operatorname{Ann} \bigwedge^k M)^t \subseteq F_k(M) \subseteq \operatorname{Ann} \bigwedge^k M$ for some integer t. In particular, if $\operatorname{Supp}(M) \subset \operatorname{Spec}(R)$ is the set of prime ideals in R containing $\operatorname{Ann}(M)$ (alternatively $\{\wp \in \operatorname{Spec} R | M \otimes R/\wp R \neq 0\}$, cf. (Serre), p.3), then $\operatorname{Supp}(\bigwedge^k M) = \operatorname{Supp}(R/F_k(M))$ is the k-th characteristic variety of M.

Note the following:

LEMMA 1.2.1. *Let* $0 \to M' \to M \to M'' \to 0$. *Then* $V_1(M) = V_1(M') \cup V_1(M'')$ *and for* $k \geq 2$: $V_k(M'') \subset V_k(M) \subset V_k(M'') \cup V_{k-1}(M'') \cap V_1(M')$.

The first equality is Prop. 4(a) in (Serre). The second follows from the first and the exact sequence: $\Lambda^{k-1} M'' \otimes M' \to \Lambda^k(M) \to \Lambda^k(M'') \to 0$, since $\operatorname{Supp}(A \otimes B) = \operatorname{Supp}(A) \cap \operatorname{Supp}(B)$ for any R-modules of finite type (Serre, Prop. 4(c)).

1.2.2. Let G be a finitely generated, finitely presented group such that $H_1(G,\mathbb{Z}) = G/G' = \mathbb{Z}^r$ (for example $G = \pi_1(\mathbb{C}^2 - C)$ where $C = \bigcup C_i$ is a plane curve as in 1.1; another class of examples which was studied in detail is given by link groups (Hillman)). Then $G'/G'' \otimes \mathbb{C}$ can be viewed as $H_1(\tilde{X},\mathbb{C})$ where X is a topological space with $\pi_1(X) = G$ and $\tilde{X}$ is the universal abelian cover of X. The group $G/G' = H_1(X,\mathbb{Z})$ acts as the group of deck transformations on $\tilde{X}$ and hence $G'/G'' \otimes \mathbb{C}$ has a structure of a $\mathbb{C}[G/G']$-module. We shall denote the i-th characteristic variety of this module as $V_i(G)$ (or $V_i(C)$ if $G = \pi_1(\mathbb{C}^2 - C)$) and call it the i-th characteristic variety of G (resp. C). The *depth* of a component V is the integer $i = \max\{j | V \subset V_j(G)\}$. We shall see that if a component has depth i and dimension $\rho > 0$ and contains identity, then $i = \rho - 1$, cf. footnote 1, p. 224.

1.2.2.1. If $G = F_r$ is a free group on r-generators then $G'/G'' = H_1(\widetilde{\bigvee_r S^1},\mathbb{Z})$, where $\widetilde{\bigvee_r S^1}$ is the universal abelian cover of the wedge of r circles. It fits into the exact sequence:

$$0 \to H_1\left(\widetilde{\bigvee_r S^1},\mathbb{C}\right) \to \mathbb{C}[\mathbb{Z}^r]^r \to I \to 0$$

with I denoting the augmentation ideal of the group ring of $\mathbb{Z}^r$. (As an universal abelian cover of $\widetilde{\bigvee_r S^1}$ one can take the subset of $\mathbb{R}^r$ of points having at least $r-1$ integer coordinates with the action of $\mathbb{Z}^r$ given by translations; unit vectors of the standard basis provide identification of 1-chains with $\mathbb{C}[\mathbb{Z}^r]^r$ while the module of 0-chains is identified with $\mathbb{C}[\mathbb{Z}^r]$). Thus $H_1(\widetilde{\bigvee_r S^1}, \mathbb{C})$ is the cokernel of the map $\Lambda^{\binom{r}{3}}\mathbb{C}[\mathbb{Z}^r]^r \to \Lambda^{\binom{r}{2}}\mathbb{C}[\mathbb{Z}^r]^r$ in the Koszul resolution corresponding to the $(x_1 - 1), \dots, (x_r - 1)$. This implies that $V_i(F_r) = \mathbb{C}^{*r}$ for $0 < i \leq r-1$ and $V_i(F_r) = (1, \dots, 1)$ for $r \leq i \leq \binom{r}{2}$ (cf. also (1.4.1) below).

If G is the fundamental group of a link in a 3-sphere S^3 with r components, then the first determinantal ideal is generated by $(t_1 - 1), \dots, (t_r - 1)$ and certain principal ideal. A generator $\Delta(t_1, \dots, t_r)$ of the latter is called the Alexander polynomial. Alexander polynomial satisfies $\Delta(1, \dots, 1) = 0$ and hence $V_1(G)$ is the hypersurface $\Delta(t_1, \dots, t_r) = 0$. Extensive calculations of the Alexander polynomials of links can be found in (Sumners and Woods). In particular, if G is the fundamental group of the complement to the Hopf link in S^3 with r components then $V_1(G)$ is the set of zeros of $t_1 \cdot t_2 \cdots t_r - 1$. Moreover $V_1(G) = \cdots = V_{r-1}(G)$, cf. p. 165 in (Libgober, 1994). From a presentation of G using Fox calculus one can calculate a presentation of $\pi_1'/\pi_1'' \otimes \mathbb{C}$ as a $\mathbb{C}([H])$-module and hence the characteristic varieties of π_1, cf. (Hironaka) and (Cohen and Suciu) for examples of such calculations.

1.2.3. Let $T(L_\infty)$ be a small tubular neighborhood of L_∞ in $\mathbb{P}^2$. If $\partial T(L_\infty)$ is its boundary then $\pi_1(\partial T(L_\infty) - \partial T(L_\infty) \cap C) \to \pi_1(\mathbb{P}^2 - L_\infty \cup \overline{C})$ is a surjection. Lemma 1.2.1 implies that the characteristic variety of C is a subset of the torus $t_1^{d_1} \cdots t_r^{d_r} = 1$ since $C \cap \partial T(L_\infty) \subset \partial T(L_\infty) = S^3$ is the Hopf link with $d_1 + \cdots + d_r$ components with d_i components of the link belonging to C_i and hence corresponding to t_i for each $1 \leq i \leq r$, cf. (Libgober, 1992). In fact the characteristic varieties of an affine curve can be determined from the projectivization as follows.

It is a corollary of (1.1.1) that $\mathbb{T}_a = \operatorname{Spec} \mathbb{C}[H_1(\mathbb{C}^2 - C)]$ is the torus of dimension r and that $\mathbb{T}_p = \operatorname{Spec} \mathbb{C}[H_1(\mathbb{P}^2 - \overline{C})]$ is the sub-scheme of zeros of $t_1^{d_1} \cdots t_r^{d_r} - 1$ in $\mathbb{T}_a$. We denote by $E : \mathbb{T}_p \to \mathbb{T}_a$ the corresponding embedding. On the other hand, the construction of (1.2.1) and (1.2.2) yields subvarieties $V_i(\overline{C})_p$ in $\mathbb{T}_p$.

PROPOSITION 1.2.3. *The characteristic variety of projective and affine curves satisfy:*

$$V_i(C) = E(V_i(\overline{C})_p). \tag{1.2.3.1}$$

PROOF. It follows from the isomorphism:

$$\pi_1'(\mathbb{P}^2 - \overline{C})/\pi_1''(\mathbb{P}^2 - \overline{C}) = \pi_1'(\mathbb{C}^2 - C)/\pi_1''(\mathbb{C}^2 - C) \tag{1.2.3.2}$$

equivariant with respect to the action of $H_1(\mathbb{C}^2 - C)$. This isomorphism is a consequence of (1.1.2) because in the latter the left map induces isomorphism on commutators. Indeed, the kernel of surjection (1.1.2) is isomorphic to $\mathbb{Z}$ (cf. (1.1)) and does not intersect $\pi_1'(\mathbb{C}^2 - C)$ because it injects into $H_1(\mathbb{C}^2 - C)$, cf. also (Libgober, 1994). □

1.3. ABELIAN COVERS

1.3.1. Let $m_1, \ldots, m_r$ be positive integers and $h_{m_1,\ldots,m_r} : H_1(\mathbb{C}^2 - C, \mathbb{Z}) \to \mathbb{Z}/m_1\mathbb{Z} \oplus \cdots \oplus \mathbb{Z}/m_r\mathbb{Z}$ be the surjection $\gamma_i \to \gamma_i \bmod m_i$. The kernel of the homomorphism $\pi_1(\mathbb{C}^2 - C) \to \mathbb{Z}/m_1\mathbb{Z} \oplus \cdots \oplus \mathbb{Z}/m_r\mathbb{Z}$, which is the composition of the abelianization $ab : \pi_1(\mathbb{C}^2 - C) \to H_1(\mathbb{C}^2 - C)$ and $h_{m_1,\ldots,m_r}$, defines an unbranched cover of $\mathbb{C}^2 - C$. We shall denote it as $(\mathbb{C}^2 - C)^{\sim}_{m_1,\ldots,m_r}$. This is a quasi-projective algebraic variety defining a birational class of projective surfaces $\overline{(\mathbb{C}^2 - C)^{\sim}_{m_1,\ldots,m_r}}$. Birational invariants of surfaces in this class (in particular the first Betti number of a non singular model) depend only on C and the homomorphism $h_{m_1,\ldots,m_r}$.

If $h_{m_1,\ldots,m_r}(d_1\gamma_1 + \cdots + d_r\gamma_r) = 0$, then the corresponding branched covering of $\mathbb{C}^2$ is a restriction of the covering of $\mathbb{P}^2$ unbranched over the line at infinity. It can be easily checked that the first Betti numbers of those two branched coverings are the same, since we assume (cf. (1.1)) that the line at infinity is transversal to C, cf. (Libgober, 1982).

A model (singular, in general) for a surface birational to $(\mathbb{C}^2 - C)^{\sim}_{m_1,\ldots,m_r}$ can be constructed as follows. Let $f_i(u,x,y) = 0$ be an equation of the component C_i $(i = 1, \ldots, r)$. Let $V_{m_1,\ldots,m_r}$ be a complete intersection on $\mathbb{P}^{r+2}$ (coordinates of which we shall denote $z_1, \ldots, z_r, u, x, y$) given by the equations

$$z_1^{m_1} = u^{m_1 - d_1} f_1(u,x,y), \ldots, z_r^{m_r} = u^{m_r - d_r} f_r(u,x,y) \tag{1.3.1.1}$$

Projection from the subspace given by $u = x = y = 0$ onto the plane $z_1 = \cdots = z_r = 0$ (i.e. $(z_1, \ldots, z_r, u, x, y) \to (u,x,y)$), when restricted on the preimage in $V_{m_1,\ldots,m_r}$ of $\mathbb{C}^2 - C$, is unbranched cover of $\mathbb{C}^2 - C$ corresponding to $\mathrm{Ker}(h_{m_1,\ldots,m_r} \circ ab)$.

1.3.2. The first Betti number of *unbranched* cover $(\mathbb{C}^2 - C)^{\sim}_{m_1,\ldots,m_r}$ can be found in terms of the characteristic varieties of C as follows, cf. (Libgober, 1992). For $P \in \mathbb{C}^{*r}$ let $f(P,C) = \max\{i | P \in V_i(C)\}$. Then

$$b_1((\mathbb{C}^2 - C)^{\sim}_{m_1,\ldots,m_r}) = r + \sum_{\substack{\omega_i^{m_i} = 1 \\ (\omega_{m_1},\ldots,\omega_{m_r}) \neq (1,\ldots,1)}} f((\omega_{m_1}, \ldots, \omega_{m_r}), C). \tag{1.3.2.1}$$

The first Betti number of a resolution of branched cover of $\mathbb{P}^2$ (i.e. of $V_{m_1,\ldots,m_r}$) can be calculated using the characteristic varieties of curves formed by components of C (Sakuma). Let $\tilde{V}_{m_1,\ldots,m_r}$ be such a resolution. For a torsion point of $\omega = (\omega_1, \ldots, \omega_r), \omega_i^{m_i} = 1$ in the torus $\mathbb{C}^{*r}$ let $C_\omega = \bigcup_{i|\omega_i \neq 1} C_i$. Then the first Betti number of $\tilde{V}_{m_1,\ldots,m_r}$ equals:

$$\sum_{\omega} \max\{i \,|\, \omega \in \mathrm{Char}_i(C_\omega)\}. \tag{1.3.2.2}$$

More precisely, if χ_ω is the character of $\pi_1(\mathbb{C}^2 - C)$ such that $\chi_\omega(\gamma_i) = \omega_i$ and for a character χ of the Galois group $\mathrm{Gal}(\tilde{V}_{m_1,\dots,m_r}/\mathbb{P}^2)$ we put:

$$H_{1,\chi}(\tilde{V}_{m_1,\dots,m_r}) = \{x \in H_1(\tilde{V}_{m_1,\dots,m_r}) | g(x) = \chi(g)\cdot x, \\ \forall g \in \mathrm{Gal}(\tilde{V}_{m_1,\dots,m_r}/\mathbb{P}^2)\} \tag{1.3.2.3}$$

then

$$\dim H_{1,\chi_\omega} = \max\{i | \omega \in \mathrm{Char}_i(C_\omega)\}. \tag{1.3.2.4}$$

1.3.3. *A bound on the growth of Betti number*

PROPOSITION 1.3.3. *Let $b_1(\bar{C},n)$ (resp. $b_1(C,n)$) be the first Betti number of the cover of $\mathbb{P}^2$ (resp. $\mathbb{C}^2 - C$) branched over $L_\infty \cup \bar{C}$ (resp. unbranched) and corresponding to the surjection $h_{n,\dots,n} : \pi_1(\mathbb{P}^2 - L_\infty \cup \bar{C}) \to (\mathbb{Z}/n\mathbb{Z})^r$ (given by evaluation modulo n of the linking numbers of loops with the components of C modulo n). Then $b_1(\bar{C},n) \le \bar{C}_1 \cdot n^{r-1}$. (resp. $b_1(C,n) \le C_1 \cdot n^{r-1}$) for some constants $C_1, \bar{C}_1$ independent of n.*

PROOF. This follows from the Sakuma's formula (1.3.2.2) (resp. (1.3.2.1)) and the obvious remark that the number of n-torsion points on a torus of dimension l grows as n^l since $\dim(\mathrm{Char}_i(\pi_1(\mathbb{C}^2 - C)'/\pi_1(\mathbb{C}^2 - C)'') \le r - 1$ by 1.2.3. □

1.3.4. *Characteristic varieties and the homology of Milnor fibers*

The polynomial $f_1(u,x,y)\cdots f_r(u,x,y)$ (which set of zeros in $\mathbb{P}^2$ is $\bar{C}$) defines a cone in $\mathbb{C}^3$ having a non isolated singularity, provided C is singular. The Milnor fiber M_c of this singularity (cf. (Cohen and Suciu) in the case when $\deg f_i = 1, \forall i$) is diffeomorphic to an affine hypersurface given by the equation: $f_1 \cdots f_r = c$, $c \ne 0$. Quotient of the latter by the action of the cyclic group $\mathbb{Z}/d\mathbb{Z}$ ($d = \sum_i d_i, d_i = \deg f_i$) acting via $(u,x,y) \to (\omega_d u, \omega_d x, \omega_d y), \omega_d^d = 1$ is $\mathbb{P}^2 - \bar{C}$. In other words, the Milnor fiber is the cyclic cover $p: M_c \to \mathbb{P}^2 - \bar{C}$ corresponding to the homomorphism sending $\gamma_i \to 1 \bmod d$. The exact sequence of the pair $(M_c, p^{-1}(\mathbb{P}^2 - \bar{C} \cup L_\infty))$ shows that $\mathrm{rk}\, H_1(M_c) = \mathrm{rk}\, H_1(p^{-1}(\mathbb{P}^2 - \bar{C} \cup L_\infty) - 1$, since we assume that $\bar{C}$ is transversal to L_∞, cf. (Libgober, 1982). Hence it follows from (1.3.2.1) that

$$\mathrm{rk}\, H_1(M_c) = r - 1 + \sum_{i=1}^{d-1} f((\omega_d^i, \dots, \omega_d^i), C). \tag{1.3.4.1}$$

1.4. CHARACTERISTIC VARIETIES AND SUPPORT LOCI FOR RANK ONE LOCAL SYSTEMS

1.4.1. Let again G be a group such that $G/G' = \mathbb{Z}^r$. If X is a topological space with $\pi_1(X) = G$ then the local systems of rank one on X correspond to the points

$\mathrm{Hom}(G,\mathbb{C}^*)$ (Steenrod). The latter has a natural identification with $H^1(X,\mathbb{C}^*)$. Each γ_i, corresponding to a component C_i of C (cf. 1.1), defines the homomorphism $t_i : \mathrm{Hom}(G,\mathbb{C}^*) \to \mathbb{C}^*$ given by $t_i(\chi) = \chi(\gamma_i), \chi \in \mathrm{Hom}(G,\mathbb{C}^*)$. Therefore t_i's provide an identification of $\mathrm{Hom}(G,\mathbb{C}^*)$ with $\mathbb{C}^{*r}$.

The homology groups $H_i(X,\rho)$ of X with coefficients in a local system corresponding to a homomorphism $\rho : \pi_1(X) \to H_1(X,\mathbb{Z}) \to \mathbb{C}^*$ are the homology of the complex $C_i(\tilde{X}) \otimes_{H_1(X,\mathbb{Z})} \mathbb{C}$ where $\mathbb{C}$ is equipped with the structure of module over $\mathbb{Z}[H_1(X,\mathbb{Z})]$ using ρ. If $\rho \neq 1$ then

$$H_1(\tilde{X},\mathbb{C}) \otimes_{\mathbb{C}[H_1(X,\mathbb{Z})]} \mathbb{C} = H_1(X,\rho). \tag{1.4.1.1}$$

This follows, for example, from the exact sequence of the low degree terms in the spectral sequence corresponding to the action of $H_1(X,\mathbb{Z})$ on the universal abelian cover $\tilde{X}$: $H_p(H_1(X,\mathbb{Z}),H_q(\tilde{X})_\rho) \Rightarrow H_{p+q}(X,\rho)$ (here $H_q(\tilde{X})_\rho$ denotes the homology of the complex $C_i(\tilde{X}) \otimes_{\mathbb{Z}} \mathbb{C}$ with the action of $H_1(X,\mathbb{Z})$ given by $g(e \otimes \alpha) = g \cdot e \otimes \rho(g^{-1})\alpha, g \in H_1(X,\mathbb{Z}), e \in C_i(\tilde{X}), \alpha \in \mathbb{C}$ i.e. the usual homology $H_q(\tilde{X},\mathbb{C})$ with the action of $H_1(X,\mathbb{Z})$ changed by the character ρ, cf. Ch. XVI, Th. 8.4 in (Cartan and Eilenberg). This exact sequence is:

$$\begin{aligned} H_2(X,\rho) \to H_2(H_1(X,\mathbb{Z}),\rho) \to (H_1(\tilde{X})_\rho)_{H_1(X,\mathbb{Z})} \to \\ \to H_1(X,\rho) \to H_1(H_1(X,\mathbb{Z}),\rho) \to 0, \end{aligned}$$

cf. Ch. XVI, (4a) in (Cartan and Eilenberg). For $\rho \neq 1$ we have $H_i(H_1(X,\mathbb{Z}),\rho) = 0$, so we obtain (1.4.1.1). For $\rho = 1$, an argument similar to sect. 1 in (Libgober, 1992) yields that $\dim H_1(\tilde{X},\mathbb{C}) \otimes_{\mathbb{C}[H_1(X,\mathbb{Z})]} \mathbb{C}$ is the dimension of the kernel of the map $\cup_X : \Lambda^2 H^1(X,\mathbb{C}) \to H^2(X,\mathbb{C})$ given by the cup product. From the definition of Fitting ideals (cf. 1.2.1) it follows that for $\rho \neq 1$ one has:

$$V_i(X) = \{\rho \in \mathrm{Hom}(G,\mathbb{C}^*) | H_1(X,\rho) \geq i\} \tag{1.4.1.2}$$

and that $\rho = 1$ belongs to $V_{\dim \mathrm{Ker} \cup_X}$, cf. Prop. 1.1 in (Libgober, 1992).

For example if $G = F_r$ then $\dim H_0(F_r,\rho)$ is 0, if ρ is non trivial, and 1 otherwise. Using $e(F_r,\rho) = r-1$ we obtain that $\dim H^1(F_r,\rho)$ is $r-1$, if ρ is non trivial, and otherwise is r. Since $\dim \mathrm{Ker} \cup_{F_r} = \binom{r}{2}$ we recover the description of the characteristic varieties for F_r mentioned in (1.2.2.1).

1.4.2. *Structure of characteristic varieties*

We will need the following theorem of D. Arapura (1997) which generalizes the results of C. Simpson to quasi-projective case.

Let $\bar{X}$ be a Kähler manifold with $H^1(\bar{X},\mathbb{C}) = 0$, D a normal crossings divisor and $X = \bar{X} - D$. Then, for each characteristic variety V, there exist a finite number of torsion characters $\rho_i \in \mathrm{Hom}(G,\mathbb{C}^*)$, a finite number of unitary characters ρ'_j and surjective maps onto (quasiprojective) curves $f_i : X \to C_i$ such that

$$V(X) = \bigcup_i \rho_i f^* H^1(C_i,\mathbb{C}^*) \cup \bigcup \rho'_j. \tag{1.4.2.1}$$

A consequence of 1.4.2.1 for curves in $\mathbb{C}^2$ is that the components of positive dimensions of their characteristic varieties are subtori of $\mathbb{C}^{*r}$ translated by points of finite order[1].

1.4.3. *Essential for a given set of components tori*
By coordinate torus (corresponding to components $C_{i_1},\dots,C_{i_s}$) we shall mean a subtorus in $\mathbb{C}^{*r}$ given by

$$t_{i_1} = \cdots = t_{i_s} = 1. \tag{1.4.3.1}$$

The inclusion $I_{i_1,\dots,i_s} : \mathbb{C}^2 - \bigcup_{i=1,\dots,r} C_i \to \mathbb{C}^2 - \bigcup_{i \neq i_1,\dots,i_s} C_i$ induces a surjective map $\tilde{I}_{i_1,\dots,i_s} : \pi_1(\mathbb{C}^2 - \bigcup_{i=1,\dots,r} C_i) \to \pi_1(\mathbb{C}^2 - \bigcup_{i \neq i_1,\dots,i_s} C_i)$ with restriction $\tilde{I}'_{i_1,\dots,i_s}$: $\pi'_1(\mathbb{C}^2 - \bigcup_{i=1,\dots,r} C_i) \to \pi'_1(\mathbb{C}^2 - \bigcup_{i \neq i_1,\dots,i_s} C_i)$ which is also surjective. Indeed if $K = \operatorname{Ker} \pi_1(\mathbb{C}^2 - \bigcup_{i=1,\dots,r} C_i) \to H_1(\mathbb{C}^2 - \bigcup_{i \neq i_1,\dots,i_s} C_i)$ then the map $K \to \pi'_1(\mathbb{C}^2 - \bigcup_{i \neq i_1,\dots,i_s} C_i)$ is surjective. Since K' is a normal closure of $\pi'_1(\mathbb{C}^2 - \bigcup_{i=1,\dots,r} C_i)$ and loops trivial in $\pi_1(\mathbb{C}^2 - \bigcup_{i \neq i_1,\dots,i_s} C_i)$ (e.g. loops which consist of paths from the base point to a point in vicinity of C_i, $(i \neq i_1,\dots,i_s)$ and loops bounding small disk transversal to C_i) the surjectivity of $I'_{i_1,\dots,i_s}$ follows.

The latter gives rise to a surjective map of $\mathbb{C}[H_1(\mathbb{C}^2 - \bigcup_{i=1,\dots,r} C_i)]$-modules: $\pi'_1/\pi''_1(\mathbb{C}^2 - \bigcup_{i=1,\dots,r} C_i) \to \pi'_1/\pi''_1(\mathbb{C}^2 - \bigcup_{i \neq i_1,\dots,i_s} C_i)$ which induces an injection of corresponding characteristic varieties:

$$V_k(\mathbb{C}^2 - \bigcup_{i \neq i_1,\dots,i_s} C_i) \to V_k(\mathbb{C}^2 - \bigcup_{i=1,\dots,r} C_i) \tag{1.4.3.2}$$

(cf. Lemma 1.2.1). A component of $V_k(\mathbb{C}^2 - \bigcup_{i=1,\dots,r} C_i)$ which is an image of a component for some $i_1,\dots,i_s$ in (1.4.3.2) is called obtained via a pull back. A component of $V_k(\mathbb{C}^2 - \bigcup_{i=1,\dots,r} C_i)$ is called essential if it is not a pull back of component of a characteristic variety of a curve composed of irreducible components of C.

LEMMA 1.4.3. *Let V be a connected component of the characteristic variety V_1 of C having positive dimension and belonging to the coordinate torus $t_{i_1} = \cdots = t_{i_s} = 1$. Then it is obtained via a pull back of a component of characteristic variety for the union of components of $\bigcup C_i$ $(i \neq i_1,\dots,i_s)$.*

---

[1] Though the paper by Arapura considers only the case of the first characteristic variety (i.e. in his terminology characters ρ such that $\dim H^1(\rho) \geq 1$), D. Arapura communicated to the author that the statement is true for all V_k. Moreover it follows from his argument that the dimension of V_k, *containing the identity of the group of characters*, is $k+1$. Indeed by (1.4.2.1) for any local system in such irreducible component of positive dimension of $V_k(X)$ there exist L' on an appropriate curve C and the map $f : X \to C$ such that $L = f^* L'$. Moreover it follows from Proposition 1.7 in (Arapura) that for all but finitely many L one has $H_1(X,L) = H_1(C,L')$. But $\pi_1(C)$ is free and if $k+1$ is the number of its generators and L' is not trivial then $\dim H_1(C,L') = k$ (cf. 1.2.2.1).

PROOF. According to Arapura Theorem (cf. (1.4.2) in (Arapura)) component V defines a map $f:\mathbb{C}^2 - C \to C$ for some quasiprojective curve C such that for some local system $E \in \operatorname{Char}\pi_1(\mathbb{C}^2 - C)$ one has: $V = E \otimes f^*(\operatorname{Char} C)$ where $\operatorname{Char} C = \operatorname{Hom}(\pi_1(C), \mathbb{C}^*)$. We claim that f factors as follows:

$$\begin{array}{ccc} \mathbb{C}^2 - C & \xrightarrow{I_{i_1,\dots,i_s}} & \mathbb{C}^2 - \bigcup_{i \neq i_1,\dots,i_s} C_i \\ & \searrow f & \downarrow \tilde{f} \\ & & C \end{array} \tag{1.4.3.3}$$

The lemma is a consequence of existence of $\tilde{f}$. Indeed for almost all local systems L on C we have $H^1(E \otimes f^*(L)) = H^1(f_*(E) \otimes L)$ (cf. proof of Prop. 1.7 in (Arapura)). Moreover $H^1(f_*(E) \otimes L) = H^1(\tilde{f}_* \circ (I_{i_1,\dots,i_s})_*(E) \otimes L) = H^1((I_{i_1,\dots,i_s})_* E \otimes \tilde{f}^* L)$ and the latter has the same dimension for almost all L again by the same argument from the proof of Prop. 1.7 in (Arapura).

To show the existence of $\tilde{f}$, let $D = \bar{C} - C$ where $\bar{C}$ is a non singular compactification of C. Since for $j = i_1, \dots, i_s$ we have $t_j = 1$ on a translate (i.e. a coset) of $f^*(\operatorname{Hom}(H_1(C), \mathbb{C}^*))$ and hence $t_j = 1$ on the latter subgroup of $\operatorname{Char}(\pi_1(\mathbb{C}^2 - C))$ we have for any $\chi \in \operatorname{Char}(H_1(C, \mathbb{Z}))$ and $j = i_1, \dots, i_s$ the following: $\chi(f_*(\gamma_j)) = f^*(\chi)(\gamma_j) = t_j(f^*(\chi)) = 1$. Equivalently $f_*(\gamma_j) = 0$. Thus $f_*(\gamma) = 0$ in $H_1(C, \mathbb{Z})$ for $\gamma \in H_1(\mathbb{C}^2 - C, \mathbb{Z})$ if and only if γ belongs to the subgroup generated by $\gamma_{i_1}, \dots, \gamma_{i_s}$. Let us consider the pencil of curves on $\mathbb{P}^2$ formed by the fibres of f. f extends to the map from the complement to the base locus of this pencil to $\bar{C}$ [2]. Preimage of D in this extension is a union of components of C and we want to show that none of these components is C_i with $i = i_1, \dots, i_s$. But none of the components $C_i, i = i_1, \dots, i_s$ is taken by this extension into D since otherwise $f_*(\gamma_i) \neq 0$ for the corresponding γ_i. Hence domain of this extension of f contains all points of $C_{i_1} \cup \dots \cup C_{i_s}$ not belonging to the remaining components of C. □

Note that it can occur that $H^1(\mathbb{C}^2 - \bigcup_{i \neq i_1,\dots,i_s} C_i, L) \neq H^1(\mathbb{C}^2 - C, I^*_{i_1,\dots,i_s}(L))$ for isolated points of characteristic varieties, as is shown by examples in (Cohen and Suciu).

1.5. ADJOINTS FOR COMPLETE INTERSECTIONS

1.5.1. Let $F \subset \mathbb{P}^n$ be a surface which is a complete intersection given by the equations:

$$\bar{F}_1 = \cdots = \bar{F}_{n-2} = 0 \tag{1.5.1.1}$$

of degrees $d_1, \dots, d_{n-2}$ respectively. Let (cf. p. 242 in (Hartshorne))

$$\Omega_F = \mathcal{E}xt^{n-2}(\mathcal{O}_F, \Omega^n_{\mathbb{P}^n}))$$

[2] Incidentally, since the resolution $\tilde{X}$ of the base locus of this pencil is simply-connected one has $\bar{C} = \mathbb{P}^1$.

be the dualizing sheaf of F. From the latter and the Koszul resolution

$$\begin{aligned} 0 \to \mathcal{O}_{\mathbb{P}^n}(-d_1 - \cdots - d_{n-2}) \to \cdots \to \\ \to \mathcal{O}_{\mathbb{P}^n}(-d_1) \oplus \cdots \oplus \mathcal{O}_{\mathbb{P}^n}(-d_{n-2}) \to \mathcal{O}_{\mathbb{P}^n} \to \mathcal{O}_F \to 0 \end{aligned} \tag{1.5.1.2}$$

it follows that one can identify Ω_F with $\mathcal{O}_F(d_1 + \cdots + d_{n-2} - n - 1)$.

Let $f : \tilde{F} \to F$ be a resolution of singularities of F and $\tau : f_*(\Omega_{\tilde{F}}) \to \Omega_F$ the trace map (Blass and Lipman). It identifies sections of $f_*(\Omega_{\tilde{F}})$ over an open set with those meromorphic differentials on non singular part of the open set in F that when pulled back on a resolution $\tilde{F}$ admits a holomorphic extension over the exceptional set of f. The adjoint ideal $\mathcal{A}'$ is the annihilator of the cokernel of τ:

$$\mathcal{A}' = \mathcal{H}om_{\mathcal{O}_F}(\Omega_F, f_*(\Omega_{\tilde{F}})) = f_*(\Omega_{\tilde{F}})(-d_1 - \cdots - d_{n-2} + n + 1) \tag{1.5.1.3}$$

We define the sheaf of adjoint ideals on $\mathbb{P}^n$ as $\mathcal{A} = \pi^{-1}(\mathcal{A}')$ (also denoted as $AdjF$) where π is the most right map in (1.5.1.2). The degeneration of Leray spectral sequence for f (due to the Grauert-Riemenschneider vanishing theorem (Grauert and Riemenschneider, 1970) yields

$$\begin{aligned} H^i(\tilde{F}, \Omega_{\tilde{F}}) &= H^i(F, f_*(\Omega_{\tilde{F}})) = \\ &= H^i(F, \mathcal{A}'(d_1 + \cdots + d_{n-2} - n - 1) = \\ &= H^i(\mathbb{P}^n, \mathcal{A}(d_1 + \cdots + d_{n-2} - n - 1)) \end{aligned} \tag{1.5.1.4}$$

In particular the irregularity of $\tilde{F}$ i.e. $\dim H^1(\tilde{F}, \mathcal{O})$ can be found as the difference between the actual dimension $H^0(\mathbb{P}^n, \mathcal{A}(d_1 + \cdots + d_{n-2} - n - 1))$ and the "expected" dimension (i.e. $\chi(\mathcal{A}(d_1 + \cdots + d_{n-2} - n - 1)))$ of the adjoints (since $H^i(\mathcal{A}(d_1 + \cdots + d_{n-2} - n - 1)) = 0$ for $i \geq 2$).

1.5.2. *Local description of adjoint ideals*
Let

$$F_1(w_1, \ldots, w_n) = 0, \ldots, F_{n-2}(w_1, \ldots, w_n) = 0 \tag{1.5.2.1}$$

be a germ of a complete intersection of hypersurfaces in $\mathbb{C}^n$ having an isolated singularity at the origin O. For any two pairs $1 \leq i, j \leq n, i \neq j$ and $1 \leq k, l \leq n, k \neq l$ we have up to sign:

$$\frac{dw_i \wedge dw_j}{\frac{\partial(F_1, \ldots, F_{n-2})}{\partial(w_1, \ldots, \hat{w}_i, \ldots, \hat{w}_j, \ldots, w_n)}} = \frac{dw_k \wedge dw_l}{\frac{\partial(F_1, \ldots, F_{n-2})}{\partial(w_1, \ldots, \hat{w}_k, \ldots, \hat{w}_l, \ldots, w_n)}} \tag{1.5.2.2}$$

Indeed the Cramer's rule for the solutions of the system of equations:

$$\frac{\partial F_k}{\partial w_1} dw_1 \wedge dw_i + \cdots + \frac{\partial F_k}{\partial w_n} dw_n \wedge dw_i = 0 \quad (k = 1, \ldots, n-2)$$

when one views $dw_k \wedge dw_i (k = 1, \ldots, \hat{i}, ., n-1)$ as unknowns yields that up to sign:

$$dw_k \wedge dw_i = \frac{\frac{\partial(F_1,\ldots,F_{n-2})}{\partial(w_1,\ldots,\hat{w}_k,\ldots,\hat{w}_i,\ldots,w_n)} dw_n \wedge dw_i}{\frac{\partial(F_1,\ldots,F_{n-2})}{\partial(w_1,\ldots,\hat{w}_i,\ldots,w_{n-1})}} \tag{1.5.2.3}$$

(1.5.2.2) follows from this for any two pairs $(i,j),(k,l), i \neq j, k \neq l$.

Since $F_1 = \cdots = F_{n-2} = 0$ is a complete intersection with isolated singularity one of the Jacobians $\frac{\partial(F_1,\ldots,F_{n-2})}{\partial(w_1,\ldots,\hat{w}_i,\ldots,\hat{w}_j,\ldots,w_n)}$ is non vanishing in a neighborhood of the singularity everywhere except for the singularity itself. In particular each side (1.5.2.2) defines a holomorphic 2-form outside of the origin for any $(i,j), i \neq j$ or $(k,l), k \neq l$. In fact this form is just the residue of the log-form $\frac{dz_1 \wedge \cdots \wedge dz_n}{F_1 \cdots F_{n-2}}$ at non singular points (i.e. outside of the origin) of (1.5.2.1).

The adjoint ideal $\mathcal{A}_O$ in the local ring $\mathcal{O}_O$ of the origin of a germ of complete intersection (1.5.2.1), according to the description of the trace map 1.5.1 can be made explicit as follows. Let $f: \tilde{\mathbb{C}}^n \to \mathbb{C}^n$ be an embedded resolution of (1.5.1.1). Then $\mathcal{A}_O$ consists of $\phi \in \mathcal{O}_O$ such that $f^*(\phi \cdot \frac{dw_i \wedge dw_j}{\frac{\partial(F_1,\ldots,F_{n-2})}{\partial(w_1,\ldots,\hat{w}_i,\ldots,\hat{w}_j,\ldots,w_n)}})$ admits a holomorphic extension from $f^{-1}(\mathbb{C}^n - O)$ to $\tilde{\mathbb{C}}^n$.

Similarly, the elements of $H^0(\mathcal{A}(d_1 + \cdots + d_{n-2} - n - 1) \subset H^0(\Omega_{\mathbb{P}^n}(d_1 + \cdots + d_{n-2}))$ can be viewed as meromorphic forms with log singularities near non singular points of (1.5.1.1) having as residue a 2-form on a non singular locus of F and admitting a holomorphic extension on $\tilde{F}$.

2. Ideals and polytopes of quasiadjunction

2.1. IDEALS OF QUASIADJUNCTION

Let f be a germ of a *reduced* algebraic curve having a singularity with r irreducible branches at the origin of $\mathbb{C}^2$ near which it is given by local equation $f = f_1(x,y) \cdots f_r(x,y) = 0$. Let $\mathcal{O}$ be the local ring of the origin and A be an ideal in $\mathcal{O}$.

DEFINITION 2.1.1. *An ideal A is called an ideal of quasiadjunction of f with parameters $(j_1,\ldots,j_r|m_1,\ldots,m_r)$ (j_i, m_i are integers) if $A = \{\phi \in \mathcal{O} | z_1^{j_1} \cdots z_r^{j_r} \phi \in AdjV_{(m_1,f_1),\ldots,(m_r,f_r)}\}$ where $V_{(m_1,f_1),\ldots,(m_r,f_r)}$ is a germ at the origin of the complete intersection in $\mathbb{C}^{r+2}$ given by the equations:*

$$z_1^{m_1} = f_1(x,y), \ldots, z_r^{m_r} = f_r(x,y). \tag{2.1.1}$$

An ideal of quasiadjunction is an ideal in $\mathcal{O}$ which is an ideal of quasiadjunction for some system of parameters.

2.2. BASIC IDEAL

Let $A(f_1,\dots,f_r) \subset \mathcal{O}$ be the ideal generated by

$$\begin{aligned} &\frac{(f_i)_x}{f_i} f_1 f_2 \cdots f_r, \quad \frac{(f_i)_y}{f_i} f_1 f_2 \cdots f_r, (i=1,\dots,r), \\ &\frac{Jac(\frac{(f_i,f_j)}{(x,y)})}{f_i f_j} f_1 \cdots f_r, (i,j=1,\dots,r, i \neq j) \end{aligned} \tag{2.2.1}$$

(we shall call it the basic ideal).

Equating all polynomials (2.2.1) to zero yields a system of equations having $(0,0)$ as the only solution. Therefore $\mathcal{O}/A(f_1,\dots,f_r)$ is an Artinian algebra.

Moreover for any set of parameters $(i_1,\dots,i_r|m_1,\dots,m_r)$ the corresponding ideal of quasiadjuction contains $A(f_1,\dots,f_r)$. Indeed, if $F_i = z_i^{m_i} - f_i(x,y)$, then up to sign $\frac{\partial(F_1,\dots,F_r)}{\partial(z_1,\dots,\hat{z}_i,\dots,z_r,x)} = z_1^{m_1-1} \cdots \hat{z_i^{m_i-1}} \cdots z_r^{m_r-1} \cdot (f_i)_x$ and hence

$$\begin{aligned} \frac{(f_i)_x f_1 \cdots \hat{f_i} \cdots f_r dz_i \wedge dy}{\frac{\partial(F_1,\dots,F_r)}{\partial(z_1,\dots,\hat{z}_i,\dots,z_r,x)}} &= \frac{(f_i)_x f_1 \cdots \hat{f_i} \cdots f_r dz_i \wedge dx}{z_1^{m_1-1} \cdots \hat{z_i^{m_i-1}} \cdots z_r^{m_r-1} (f_i)_x} = \\ &= z_1 \cdots \hat{z}_i \cdots z_r dz_i \wedge dy \end{aligned} \tag{2.2.2}$$

which is holomorphic on $\mathbb{C}^{r+2}$. Similarly, one sees that the 2-forms corresponding to other generators of $A(f_1,\dots,f_r)$ coincide on $F_1 = \cdots = F_r = 0$ with the forms admitting a holomorphic extension to $\mathbb{C}^{r+2}$.

In particular, there are only finitely many ideals of quasiadjunction.

2.3. IDEALS OF QUASIADJUNCTION AND POLYTOPES

Let $\mathcal{U} = \{(x_1,\dots,x_r) \in \mathbb{R}^r | 0 \le x_i < 1\}$ be the unit cube with coordinates corresponding to the components of a curve C. Sometimes we shall denote this cube as $\mathcal{U}(C)$. If C' is formed by components of C then we shall view $\mathcal{U}(C')$ as the face of $\mathcal{U}(C)$ given by $x_j = 0$ where j runs through indices corresponding to components of C *not* belonging to C'.

By a *convex polytope* mean a subset of $\mathbb{R}^n$ [3] which is the convex hull of a finite set of points *with some faces possibly deleted.* By a *polytope* we mean a finite union of convex polytopes. Class of polytopes in this sense is closed under finite unions and intersections. A complement to a polytope within an ambient polytope is a polytope. By *face of maximal dimension* of a polytope we mean the intersection of the polytope's boundary with the hyperplane for which this intersection has the dimension equal to the dimension of the boundary. A *face* of

[3] Only subsets of $\mathcal{U}$ will occur below.

a polytope is an intersection of faces of maximal dimension. This is again a closed polytope.

Each $(x_1,\dots,x_r) \in \mathcal{U}$ defines the character $\chi(x_1,\dots,x_r)$ of $\pi_1(\mathbb{C}^2 - C)$ such that $\chi(x_1,\dots,x_r)(\gamma_i) = \exp(2\pi\sqrt{-1}x_i)$. We shall call it the *exponential map*. We also put $\overline{(x_1,\dots,x_r)} = (1-x_1,\dots,1-x_r)$ and call this conjugation since $\chi(1-x_1,\dots,1-x_r) = \overline{\chi(x_1,\dots,x_r)}$. This map is an involution of the interior of $\mathcal{U}$: $\mathcal{U}^\circ = \{(x_1,\dots,x_r) \in \mathcal{U} | x_i \neq 0, i = 1,\dots,r\}$.

Different arrays $(i_1,\dots,i_r|m_1,\dots,m_r)$ may define the same ideals of quasiadjunction. The next proposition describes when this is the case.

PROPOSITION 2.3.1. *Let A be an ideal of quasiadjunction. Then there exists a polytope $\bar{\Delta}(A)$, which is a open subset in $\mathcal{U}$, with the following property: for $(m_1,\dots,m_r) \in \mathbb{Z}^r$ and $(j_1,\dots,j_r) \in \mathbb{Z}^r, 0 \leq j_i < m_i, i = 1,\dots,r$ a holomorphic function $z_1^{j_1}\cdots z_r^{j_r}\phi$ belongs to the adjoint ideal of the germ of an abelian branched cover of the type $(m_1,\dots,m_r)$ of a neighborhood of the origin in $\mathbb{C}^2$ for any $\phi \in A$ if and only if $(\frac{j_1+1}{m_1},\dots,\frac{j_r+1}{m_r}) \in \bar{\Delta}(A)$.*

PROOF. Let $\rho : Y_f \to \mathbb{C}^2$ be an embedded resolution for the singularity given by $f = f_1(x,y)\cdots f_r(x,y) = 0$ at the origin. The complete intersection $V_{(m_1,f_1),\dots,(m_r,f_r)}$ (cf. (2.1.1)) provides a model with an isolated singularity of a branched abelian cover of a neighborhood of the origin in $\mathbb{C}^2$ with $f=0$ as its branching locus. Let $\pi_{(m_1,f_1),\dots,(m_r,f_r)} : V_{m_1,\dots,m_r} \to \mathbb{C}^2$ be the canonical projection. If $\bar{V}_{(m_1,f_1),\dots,(m_r,f_r)}$ is the normalization of $Y_f \times_{\mathbb{C}^2} V_{(m_1,f_1),\dots,(m_r,f_r)}$ then the projection $\rho_{(m_1,f_1),\dots,(m_r,f_r)}$: $\bar{V}_{(m_1,f_1),\dots,(m_r,f_r)} \to V_{(m_1,f_1),\dots,(m_r,f_r)}$ on the second factor is a resolution of the singularity at the origin in the category of V-manifolds. We have the diagram:

$$\begin{array}{ccc} \bar{V}_{(m_1,f_1),\dots,(m_r,f_r)} & \xrightarrow{\bar{\pi}_{(m_1,f_1),\dots,(m_r,f_r)}} & Y_f \\ \downarrow \rho_{(m_1,f_1),\dots,(m_r,f_r)} & & \downarrow \rho \\ V_{(m_1,f_1),\dots,(m_r,f_r)} & \xrightarrow{\pi_{(m_1,f_1),\dots,(m_r,f_r)}} & \mathbb{C}^2 \end{array} \tag{2.3.1}$$

Let $E = \bigcup_k E_k$ be the exceptional locus of ρ. Let $\bigcup E_{k,l}$ be the exceptional locus of $\rho_{(m_1,f_1),\dots,(m_r,f_r)}$, where $E_{k,l}$ is a cover of E_k.

Let $a_{k,i} = \text{mult}_{E_k}\rho^* f_i(x,y), (i = 1,\dots,r)$, $c_k = \text{mult}_{E_k}\rho^*(dx \wedge dy)$. For $\phi \in \mathcal{O}$ we put $f_k(\phi) = \text{mult}_{E_k}\rho^*(\phi)$. Finally let $g_{k,i} = g.c.d(m_i, a_{k,i}), (i = 1,\dots,r)$ and $s_k = g.c.d.(\frac{m_1}{g_{k,1}},\dots,\frac{m_r}{g_{k,r}}), (i = 1,\dots,r)$. The form $\frac{dx\wedge dy}{z_1^{m_1-1}\cdots z_r^{m_r-1}}$ is a non vanishing form on

$$V_{(m_1,f_1),\dots,(m_r,f_r)} - \text{Sing}\, V_{(m_1,f_1),\dots,(m_r,f_r)}$$

(cf. (1.5.2.2)). If A is an ideal of quasi-adjunction with parameters $(\bar{j}_1,\dots,\bar{j}_r| m_1,\dots,m_r)$ then the condition $\phi \in A$ is equivalent to the existence of a holomorphic extension of $\rho^*_{(m_1,f_1),\dots,(m_r,f_r)}(\frac{z_1^{j_1}\cdots z_r^{j_r}\phi\, dx\wedge dy}{z_1^{m_1-1}\cdots z_r^{m_r-1}})$ over the exceptional locus $\bigcup E_{k,l}$ in a neighborhood of each point of $\bigcup E_{k,l}$ not belonging to $E_{k,l} \cap E_{\bar{k},\bar{l}}$ for any

$(k,l),(\bar{k},\bar{l})$. This, in turn, is equivalent to:

$$\sum_{i=1}^{r}(\bar{j}_i - m_i + 1)\,\mathrm{mult}_{E_{k,l}}\rho^*_{(m_1,f_1),\ldots,(m_r,f_r)}(z_j) + \mathrm{mult}_{E_{k,l}}\rho^*_{(m_1,f_1),\ldots,(m_r,f_r)}\phi + \\ + \mathrm{mult}_{E_{k,l}}\rho^*_{(m_1,f_1),\ldots,(m_r,f_r)}(dx\wedge dy)) \geq 0 \tag{2.3.2}$$

for any pair of indices (k,l). On the other hand, we have the following equalities:

$$\begin{aligned} \mathrm{mult}_{E_{k,l}}\rho^*_{(m_1,f_1),\ldots,(m_r,f_r)}(z_i) &= \frac{m_1\cdots\hat{m}_i\cdots m_r\cdot a_{k,i}}{g_{k,1}\cdots g_{k,r}\cdot s_k}, \\ \mathrm{mult}_{E_{k,l}}\rho^*_{(m_1,f_1),\ldots,(m_r,f_r)}(\phi) &= \frac{f_k(\phi)\cdot m_1\cdots m_r}{g_{k,1}\cdots g_{k,r}\cdot s_k}, \\ \mathrm{mult}_{E_{k,l}}\rho^*_{(m_1,f_1),\ldots,(m_r,f_r)}(dx\wedge dy) &= \frac{c_k\cdot m_1\cdots m_r}{g_{k,1}\cdots g_{k,r}\cdot s_k} + \frac{m_1\cdots m_r}{g_{k,1}\cdots g_{k,r}\cdot s_k} - 1 \end{aligned} \tag{2.3.3}$$

To see (2.3.3), we can select local coordinates (u,v) on Y_f near a point belonging to a single component E_k in which the latter is given by the equation $u=0$. Then $\rho^*(f_i(x,y)) = u^{a_{k,i}}\cdot\varepsilon_i(u,v)$ where $\varepsilon_i(u,v)(i=1,\ldots,r)$ are units in the corresponding local ring. The fiber product $Y_f \times_{\mathbb{C}^2} V_{(m_1,f_1),\ldots,(m_r,f_r)}$ is a subvariety in $\mathbb{C}^{r+2}\times_{\mathbb{C}^2} Y_f$ given by the equations:

$$z_1^{m_1} = u^{a_{k,1}}\varepsilon_1(u,v),\ldots,z_r^{m_r} = u^{a_{k,r}}\varepsilon_r(u,v) \tag{2.3.4}$$

Each branch of (2.3.4) has the following local parameterization:

$$u = t^{\frac{m_1\cdots m_r}{g_{k,1}\cdots g_{k,r}\cdot s_k}},\quad z_i = t^{\frac{m_1\cdots \hat{m}_i\cdots m_r\cdot a_{k,i}}{g_{k,1}\cdots g_{k,r}\cdot s_k}},\ i=1,\ldots,r \tag{2.3.5}$$

(exponents are chosen so that their greatest common divisor will be equal to 1 and so that they will satisfy (2.3.4)). This yields the first equality in (2.3.3). We have

$$\mathrm{mult}_{E_{k,l}}\rho^*_{(m_1,f_1),\ldots,(m_r,f_r)}(\phi) = \mathrm{mult}_{E_{k,l}}\bar{\pi}_{(m_1,f_1),\ldots,(m_r,f_r)}u^{f_k(\phi)}.$$

Hence the second equality in (2.3.3) follows from (2.3.5).

Finally, since the map $\bar{\pi}_{(m_1,f_1),\ldots,(m_r,f_r)}$ is given locally by $(t,v)\to(t^{\frac{m_1\cdots m_r}{g_{k,1}\cdots g_{k,r}}},v)$, we have:

$$\begin{aligned} &\rho^*_{(m_1,f_1),\ldots,(m_r,f_r)}(dx\wedge dy|V_{(m_1,f_1),\ldots,(m_r,f_r)}) = \\ &= \rho^*_{(m_1,f_1),\ldots,(m_r,f_r)}\circ\pi^*_{(m_1,f_1),\ldots,(m_r,f_r)}(dx\wedge dy|\mathbb{C}^2) = \\ &= \bar{\pi}^*_{(m_1,f_1),\ldots,(m_r,f_r)}(u^{c_k}\,du\wedge dv) = t^{\frac{c_k\cdot m_1\cdots m_r}{g_{k,1}\cdots g_{k,1}\cdot s_k}+\frac{m_1\cdots m_r}{g_{k,1}\cdots g_{k,r}\cdot s_k}-1}\,dt\wedge dv \end{aligned}$$

which implies the last equality in (2.3.3).

Now it follows from (2.3.2) and (2.3.5) that $\phi\in A(\bar{j}_1,\ldots,\bar{j}_r|m_1,\ldots,m_r)$ if and only if for any k the multiplicity $f_k(\phi)$ satisfies:

$$\sum_{i=1}^{r}(\bar{j}_i - m_i + 1)\frac{m_1\cdots\hat{m}_i\cdots m_r\cdot a_{k,i}}{g_{k,1}\cdots g_{k,r}s_k} + \frac{m_1\cdots m_r\cdot f_k(\phi)}{g_{k,1}\cdots g_{k,r}\cdot s_k} + \\ + \frac{c_k\cdot m_1\cdots m_r}{g_{k,1}\cdots g_{k,r}\cdot s_k} + \frac{m_1\cdots m_r}{g_{k,1}\cdots g_{k,r}\cdot s_k} - 1 \geq 0 \tag{2.3.6}$$

For given k let $f_k(A)$ be the minimal *integer* solution, with $f_k(\phi)$ considered as unknown, for this inequality and ϕ_k be such that $f_k(\phi_k) = f_k(A)$. In other words $\phi \in A$ if and only if $f_k(\phi) \geq f_k(A)$. We have $f_k(A) = [(\Sigma_{i=1}^k (a_{k,i} - (\bar{j}_i + 1)a_{k,i}/m_i) - c_k] = \{\Sigma_{i=1}^k (a_{k,i} - (\bar{j}_i + 1)a_{k,i}/m_i) - c_k - 1\}$ where $\{r\}$ (resp. $[r]$) denotes the smallest integer that is strictly greater than (resp. the integer part of) r. We shall call $f_k(A)$ the multiplicity of A along E_k. This is the minimum of multiplicities along E_k of pull backs on Y_f of elements of A.

The same calculation shows that $z_1^{j_1} \cdots z_r^{j_r}\phi$, where ϕ belongs to an ideal of quasiadjunction A, is in the adjoint ideal of $z_1^{m_1} = f_1(x,y), \ldots, z_r^{m_r} = f_r(x,y)$ if and only if:

$$\begin{aligned}\sum_{i=1}^{r}(j_i - m_i + 1)\frac{m_1 \cdots m_r a_k}{g_{k,1} \cdots g_{k,r} \cdot m_i \cdot s_k} + \frac{m_1 \cdots m_r f_k(A)}{g_{k,1} \cdots g_{k,r} \cdot s_k} + \\ + \frac{c_k \cdot m_1 \cdots m_r}{g_{k,1} \cdots g_{k,r} \cdot s_k} + \frac{m_1 \cdots m_r}{g_{k,1} \cdots g_{k,r} s_k} - 1 \geq 0\end{aligned} \tag{2.3.7}$$

or equivalently:

$$\sum_{i=1}^{r} \frac{j_i + 1}{m_i} a_{k,i} > \sum_{i=1}^{r} a_{k,i} - f_k(A) - c_k - 1. \tag{2.3.8}$$

Indeed, if (2.3.7) holds, then, since $f_k(\phi) \geq f_k(A)$ for any $\phi \in A$, one can replace $f_k(A)$ in (2.3.7) by $f_k(\phi)$. This converts (2.3.7) into a necessary and sufficient condition for $z_1^{j_1} \cdots z_r^{j_r}\phi$ to belong to the adjoint ideal of (2.1.1) (cf. derivation of (2.3.6)). Vice versa, for ϕ_k satisfying $f_k(\phi_k) = f_k(A)$, the condition that $z_1^{j_1} \cdots z_r^{j_r}\phi_k$ is in the adjoint ideal of (2.1.1), is nothing else but (2.3.7). □

The polytope $\bar{\Delta}(A)$ satisfying the conditions of the proposition is the set of solutions of the inequalities:

$$\sum_{i=1}^{r} x_i a_{k,i} > \sum_{i=1}^{r} a_{k,i} - f_k(A) - c_k - 1, \quad \text{for all } k. \tag{2.3.9}$$

2.3.1. *Remarks*

1. If A_1 and A_2 are ideals of quasiadjunction and $A_1 \subset A_2$ then $\bar{\Delta}(A_2) \subset \bar{\Delta}(A_1)$
2.The polytope corresponding the basic ideal $A(f_1, \ldots, f_r)$ (cf. 2.2) is the whole unit cube $\mathcal{U}$.
3. $\mathcal{O}$ is considered as improper "ideal" of quasiadjunction since $A(m_1 - 1, \ldots, m_r - 1|m_1, \ldots, m_r) = \mathcal{O}$.

2.4. LOCAL POLYTOPES OF QUASIADJUNCTION AND FACES OF QUASIADJUNCTION

DEFINITION 2.4.1. *We say that two points in the unit cube $\mathcal{U}$ are equivalent if the collections of polytopes $\bar{\Delta}(A)$ containing each of the points coincide. A* *(local) polytope of quasiadjunction Δ is an equivalence class of points with this equivalence relation.*

DEFINITION 2.4.2. *A face of quasiadjunction is an intersection of a face (cf. (2.3)) of a local polytope of quasiadjunction and a (different) polytope of quasiadjunction. In particular, each face of quasiadjunction belongs to a unique polytope of quasiadjunction.*

For each face let us consider the system of equation defining the affine subspace of $\mathbb{Q}^r$ spanned by this face. One can normalize the system so that all coefficients of variables and the free term are *integers* and the g.c.d. of non zero minors of maximal order is equal to 1.

DEFINITION 2.4.3. *The order of a face of quasiadjunction is the g.c.d of minors of maximal order in the matrix of coefficients in the normalized system of linear equations defining this face.*

This is the order of the torsion of the quotient of $\mathbb{Z}^r$ by the subgroup generated by the vectors having as coordinates the coefficients of variables in the equations of the face. In particular this integer is independent of the chosen normalized system of equations and depends only on the face of quasiadjunction.

The (local) ideal of quasiadjunction corresponding to a face of quasiadjunction is the ideal of quasiadjunction corresponding to the polytope of quasiadjunction containing this face. The ideal corresponding to a face of quasiadjunction has the form $A(j_1,\ldots,j_r|m_1,\ldots,m_r)$ where $(\frac{j_1+1}{m_1},\ldots,\frac{j_r+1}{m_r})$ belongs to this face.

2.5. EXAMPLES

1. In the case of the branch with one component the ideals of quasiadjunction correspond to the constants of quasiadjunction, cf. (Libgober, 1983). Recall that for a germ ϕ the rational number κ_ϕ is characterized by the property $min\{i|z^i\phi \in Adj(z^n = f(x,y))\} = [\kappa_\phi \cdot n]$. The ideal of quasiadjunction corresponding to κ consists of ϕ such that $\kappa_\phi > \kappa$. For example for the cusp x^2+y^3 the only non zero constant of quasiadjunction is $1/6$, there are two polytopes of quasiadjunction i.e. $\Delta' = \{x \in [0,1] | 1 > x > 1/6\}$ and $\Delta'' = \{x \in [0,1] | 1/6 \geq x \geq 0\}$. $x = 1/6$ is the face of quasiadjunction and the corresponding ideal of quasiadjunction is the maximal ideal. For an arbitrary unibranched singularity we have $\bar{\Delta}(A_\kappa) = \{x | x > \kappa\}$. The order of the face $x = \kappa$ is equal to the order of the root of unity $\exp(2\pi i\kappa)$.

2. Let us consider a tacnode, locally given by the equation: $y(y-x^2)$ (i.e. $f(x,y) = y, g(x,y) = y - x^2$). Then the basic ideal is the maximal ideal and hence there is only one ideal of quasiadjunction. If $\mathcal{M}$ is the maximal ideal, then $\bar{\Delta}(\mathcal{M})$ is the whole unit square. To determine the polytope $\bar{\Delta}(\mathcal{O})$, note that after two blow-ups we obtain an embedded resolution which in one of the charts looks like: $x = uv, y = u^2v$ where $u = 0$ and $v = 0$ are the exceptional curves. Hence for the

component $u=0$ we obtain $a=b=2, f(\phi=1)=0, c=2$ and the corresponding polytope is $x+y>1/2$. The face of quasiadjunction is $x+y=1/2$ and the corresponding ideal is the maximal one.

3. For the ordinary singularity of multiplicity m: $(\alpha_1 x+\beta_1 y)\cdots(\alpha_m x+\beta_m y)=0$ the basic ideal is $\mathcal{M}^{m-2}$ where $\mathcal{M}\subset\mathcal{O}$ is the maximal ideal of the local ring at the origin. Since the resolution can be obtained by a single blow up, we have $a_{1,i}=1, c_1=1, i=1,\ldots,m$ i.e. the polytope $\bar{\Delta}(A)$ of an ideal of quasiadjunction A is:

$$x_1+\cdots+x_m>m-2-f_1(A) \tag{2.5.1}$$

Since $f_1(\phi)\geq f_1(A)$ is equivalent to $\phi\in\mathcal{M}^{f_1(A)}$ i.e. the latter is the ideal corresponding to the polytope (2.5.1). The faces of quasiadjunction are $x_1+\cdots+x_m=m-2-f_1(A)$ $(f_1(A)=0,\ldots,m-3)$ and the corresponding ideal of quasiadjunction is $\mathcal{M}^{f_1(A)+1}$.

Additional examples are discussed in (Libgober, 2001).

2.6. GLOBAL POLYTOPES AND SHEAVES OF QUASIADJUNCTION

Let $\mathbb{R}^r$, as in 2.3, be vector space coordinates of which are in one to one correspondence with the components of the curve $C=\bigcup_{i=1}^{i=r}C_i$. For a singular point p of C, let C_p be the collection of components of C passing through p. Each polytope of quasiadjunction $\Delta_p\subset\mathcal{U}(C_p)$ of p defines the polytope in $\mathcal{U}=\{(x_1,\ldots,x_r)|0\leq x_i\leq 1\}\subset\mathbb{R}^r$ consisting of points $\{(x_1,\ldots,x_r)|(x_1,\ldots,x_r)\in\mathcal{U},(x_{i_1},\ldots,x_{i_{r(p)}})\in\Delta_p\}$ where $(i_1,\ldots,i_{r(p)})$ are the coordinates corresponding to the components of C_p (i.e. passing through the singularity p). We shall use the notation $\Delta_p(\mathcal{U})$ for this polytope in $\mathcal{U}$.

2.6.1. *Definition*

A type of a point in $\mathcal{U}$ is the collection of polytopes $\Delta_p(\mathcal{U})\subset\mathcal{U}$ to which this point belongs, where p runs through all $p\in\operatorname{Sing}C$.

We call two points in $\mathcal{U}$ equivalent if they have the same type. *A (global) polytope of quasiadjunction* is an equivalence class of this equivalence relation. Global polytopes of quasiadjunction form a partition which is a refinement of every partitions of $\mathcal{U}$ defined by polytopes $\Delta_p(\mathcal{U})$ corresponding to local polytopes of quasiadjunction of singularities of C.

2.6.2. *Definition*

We shall call a face δ of quasiadjunction *contributing* if it belongs to a hyperplane $d_1x_1+\cdots+d_rx_r=l$ where $d_1,\ldots,d_r$ are the degrees of the components of C corresponding to respective coordinates $x_1,\ldots,x_r$. This hyperplane is called *contributing*, the integer $l=l(\delta)$ is called the *level* of both the contributing hyperplane and contributing face. The *order* of a global face of quasiadjunction is defined as in local case (cf. 2.4.2). *A polytope of quasiadjunction is called contributing* if it contains a contributing face.

A point $(x_1,\ldots,x_r) \in \delta$ is called *interior* C'-point for a curve C' formed by components of C if $x_i \neq 1$ if and only if i corresponds to a component of C' and $(x_1,\ldots,x_r)$ is in the interior of δ.

Remarks 2.6.2.1. In the case when $r=1$, e.g. for an irreducible curve, an order of a global face of quasiadjunction is the order of a root of the Alexander polynomial. Indeed, for a constant of quasiadjunction κ, $\exp(2\pi i\kappa)$ is a root of Alexander polynomial (Loeser and Vaquie).

2.6.2.2. The collection of orders of faces of quasiadjunction for reducible curves a priori cannot be determined just by the local types of all singular points. However it is *combinatorial* invariant of the curve in the sense that it depends only on local information about singularities and specification of components which contain specified singular points, e.g. it is independent of the geometry of the set of singular points in $\mathbb{P}^2$. In the case of arrangements of lines this is combinatorial invariant in the common sense of the word.

2.6.3. *Definition*

The sheaf of ideals $\mathcal{A}(\delta) \subset \mathcal{O}_{\mathbb{P}^2}$ such that $\mathrm{Supp}(\mathcal{O}_{\mathbb{P}^2}/\mathcal{A}) \subset \mathrm{Sing}\, C$ is called *the sheaf of ideals of quasiadjunction* corresponding to the face of quasiadjunction δ if the stalk $\mathcal{A}_p$ at each singular point $p \in C$ with local ring $\mathcal{O}_p$ is the ideal A of quasiadjunction corresponding to the face $\Delta_p = \Delta \cap H_p$ with $H_p \subset \mathbb{R}^r$ being given by $x_{i_j} = 0$ where i_j are the coordinates corresponding to the components of C not passing through p.

2.6.4. *Examples*

1. (Libgober, 1983) For an irreducible curve of degree d with nodes and the ordinary cusps as the only singularities the global polytope of quasiadjunction coincide with the local one of the cusp. The only face of quasiadjunction is $x = 1/6$. The contributing hyperplane is given by $dx = d/6$ and its level is $d/6$. The sheaf of quasiadjunction corresponding to this face of quasiadjunction is the ideal sheaf having stalks different from the local ring only at the points of $\mathbb{P}^2$ where the curve has cusps and the stalks at those points are the maximal ideals of the corresponding local rings.

2. Let us consider C which is an arrangement of lines. For a point P let m_P denotes the multiplicity. We consider only points with $m_P > 2$. Each global face of quasiadjunciton is a solution of a system of equations:

$$L_P: \quad x_{i_1} + \cdots + x_{i_m} = s_P \tag{2.6.1}$$

where $s_P = 1,\ldots,m_P - 2$ (cf. example 3 in 2.5). The indices of variables x correspond to the lines of the arrangement and x_i appears in L_P if and only if it correspond to a line passing through P. Each system (2.6.1) corresponding to a face of quasiadjunciton singles out a collection of vertices of the arrangement. This face is contributing if the equation

$$x_1 + \cdots + x_r = k, \quad (k \in NN) \tag{2.6.2}$$

is a linear combination of equations (2.6.1). The level of such contributing face is k. Its order is the g.c.d of minors of maximal order in system (2.6.1).

3. The first Betti number of an abelian cover

In this section we shall prove a formula for the irregularity of an abelian covers of $\mathbb{P}^2$ branched over C in terms of the polytopes of quasiadjunction introduced in the last section. More precisely we shall calculate the multiplicity of a character of the Galois group of the cover acting on the space $H^{1,0}(\tilde{V}_{m_1,\ldots,m_r})$ of holomorphic 1-forms. We translate this into an information about characteristic varieties of the fundamental group and consider several examples of characteristic varieties for the fundamental groups of the complements to arrangements of lines.

3.1. STATEMENT OF THE THEOREM

Let $C = \bigcup_{i=1}^{i=r} C_i$ be a reduced curve $f(u,x,y) = f_1(u,x,y) \cdots f_r(u,x,y)$ with r irreducible components and the degrees of components equal to $d_1, \ldots, d_r$ and $d = d_1 + \cdots + d_r$ be the total degree of $f(u,x,y) = 0$. Let L_∞ be the line $u = 0$ at infinity which, as above, we shall assume *transversal* to C (cf.1.1).

a) The irregularity of a desingularization $\tilde{V}_{m_1,\ldots,m_r}$ of an abelian cover of $\mathbb{P}^2$ branched over $C \cup L_\infty$ and corresponding to the surjection $\pi_1(\mathbb{P}^2 - C \cup L_\infty) \to \mathbb{Z}/m_1\mathbb{Z} \oplus \cdots \oplus \mathbb{Z}/m_r\mathbb{Z}$ is equal to

$$\sum_{C'} \left(\sum_{\delta(C')} N(\delta(C')) \cdot \dim H^1(\mathcal{A}_{\delta(C')}(d - l(\delta(C)) - 3)) \right) \tag{3.1.1}$$

where the summations are over all curves C' formed by the components of C and the contributing faces of quasiadjunction $\delta(C')$ respectively. Here $l(\delta(C'))$ is the level of the contributing face of $\delta(C')$ and $N(\delta(C'))$ is the number of interior C'-points $(\frac{i_1+1}{m_1}, \ldots, \frac{i_r+1}{m_r})$ in the contributing face of $\delta(C')$.

b) Let χ_j be the character of $\mathbb{Z}_{m_1} \oplus \cdots \oplus \mathbb{Z}_{m_r}$ taking on $(a_1, \ldots, a_j, \ldots, a_r)$ value $\exp(2\pi\sqrt{-1}\frac{a_j}{m_j})$. For a character χ of $\mathbb{Z}_{m_1} \oplus \cdots \oplus \mathbb{Z}_{m_r}$ let

$$H^{1,0}_\chi(\tilde{V}_{m_1,\ldots,m_r}) = \{x \in H^{1,0}(\tilde{V}_{m_1,\ldots,m_r}), g \in \mathbb{Z}_{m_1} \oplus \cdots \oplus \mathbb{Z}_{m_r} | g \cdot x = \chi(g) \cdot x\}.$$

If $(\frac{i_1+1}{m_1}, \ldots, \frac{i_r+1}{m_r})$ is an interior C-point (cf. 2.3.2) belonging to the contributing face δ then

$$\dim H^1_{\chi_1^{i_1} \cdots \chi_r^{i_r}}(\tilde{V}_{m_1,\ldots,m_r}) = \dim H^1(\mathcal{A}_\delta(d - 3 - l(\delta))) \tag{3.1.2}$$

c) Let $t_i = \exp(2\pi\sqrt{-1}x_i)$. For each contributing face δ belonging to $\mathcal{U}^\circ$ and its image $\bar{\delta}$ under the conjugation map (cf. 2.3), let $L_s(x_1,\ldots,x_r) = \beta_s$ be the system of equations defining it where $L_s(x_1,\ldots,x_r)$ is a linear form with integer coefficients such that g.c.d. of the minors of maximal order in the matrix of coefficients is equal to 1. Then the corresponding essential component of the characteristic variety of $\pi_1(\mathbb{P}^2 - C \cup L_\infty)$, which either has a positive dimension or is a torsion point, is the intersection of cosets given by the equations:

$$\exp(2\pi\sqrt{-1}L_s) = \exp(2\pi\sqrt{-1}\beta_s) \tag{3.1.3}$$

written in terms of t_i's. Vice versa, any essential component can be obtained in such way.

Note that c) implies that the essential components of the characteristic varieties are Zariski's closures of the images of the contributing faces under the exponential map. Indeed, since g.c.d. of minors of coefficients in L_s is 1 the intersection of subgroups $\exp(2\pi\sqrt{-1}L_s) = 1$ is connected and the closure of the image of the face of quasiadjunction is Zariski dense in the translation of this connected component given by (3.1.3).

3.2. PROOF OF THE THEOREM

We shall start with the case when $m_i \geq d_i$ for $i = 1,\ldots,r$. Let $\mathcal{A} \subset \mathcal{O}_{\mathbb{P}^{r+2}}$ be the sheaf of adjoint ideals of the complete intersection $V_{m_1,\ldots,m_r} \subset \mathbb{P}^{r+2}$ given by the equations (cf. (1.3.1.1)):

$$z_1^{m_1} = u^{m_1-d_1} f_1(u,x,y),\ldots,z_r^{m_r} = u^{m_r-d_r} f_r(u,x,y) \tag{3.2.1}$$

$V_{m_1,\ldots,m_r}$ provides a model of an abelian branched cover of $\mathbb{P}^2$ branched over $f_1 \cdots f_r = 0$ and the line at infinity. $V_{m_1,\ldots,m_r}$ has isolated singularities at the points of (3.2.1) which are above the singularities of C in $\mathbb{P}^2 - L_\infty$. The action of the Galois group of the cover is induced from the action of the product of groups of roots of unity $\mu_{m_1} \times \cdots \times \mu_{m_r}$ on the $\mathbb{P}^{r+2}$ via multiplication of corresponding z-coordinates.

Let H be the set of common zeros of $z_1,\ldots,z_r \in H^0(\mathbb{P}^{r+2},\mathcal{O}(1))$ and $\mathcal{A}_{i_1,\ldots,i_r}$ be the subsheaf of $\mathcal{O}_{\mathbb{P}^{r+2}}$ germs of section product of which with $z_1^{i_1} \cdots z_r^{i_r}$ belongs to $\mathcal{A}$. The action of $\mu_{m_1} \times \cdots \times \mu_{m_r}$ on $\mathbb{P}^{r+2}$ induces the action on $\mathcal{A}_{i_1,\ldots,i_r}$.

Let $\mathcal{J}_H$ be the ideal sheaf of the plane $H \subset \mathbb{P}^2$. We have the following $\mu_{m_1} \times \cdots \times \mu_{m_r}$-equivariant sequence:

$$\begin{aligned} 0 &\to \mathcal{A}_{i_1,\ldots,i_r}((m_1 - i_1) + \cdots + (m_r - i_r) - r - 3) \otimes \mathcal{J}_H \to \\ &\to \mathcal{A}_{i_1,\ldots,i_r}((m_1 - i_1) + \cdots + (m_r - i_r) - r - 3) \to \\ &\to \mathcal{A}_{i_1,\ldots,i_r}((m_1 - i_1) + \cdots + (m_r - i_r) - r - 3)|_H \to 0 \end{aligned} \tag{3.2.2}$$

Let

$$F(i_1,\dots,i_r) = \dim H^1(\mathcal{A}_{i_1,\dots,i_r}((m_1-i_1)+\dots+(m_r-i_r)-r-3));$$
$$F_\chi(i_1,\dots,i_r) = \dim\{x \in H^1(\mathcal{A}_{i_1,\dots,i_r}((m_1-i_1)+\dots+(m_r-i_r)+ -r-3)) \mid g\cdot x = \chi(g)x, \forall g \in \mu_{m_1}\times\cdots\mu_{m_r}\}. \tag{3.2.3}$$

In particular $F(0,\dots,0)$ is the irregularity of a nonsingular model of $V_{m_1,\dots,m_r}$.

Step 1. Degree of the curves in the linear system $H^0(\mathcal{A}_{i_1,\dots,i_r}((m_1-i_1)+\dots+(m_r-i_r)-r-3)|_H)$. Let us calculate the multiplicity of the line $L_\infty : z_1 = \dots = z_r = u = 0$ as the fixed component of the curves in the linear system cut on H by the hypersurfaces in the linear system $H^0(\mathcal{A}_{i_1,\dots,i_r}((m_1-i_1)+\dots+(m_r-i_r)-r-3))$. This multiplicity is the smallest k such that u^k belongs to the latter system of hypersurfaces. In appropriate coordinates $(z_1,\dots,z_r,u,v)$ at a point P of this line outside of $L_\infty \cap C$ (i.e. we have $f_1(P)\cdots f_r(P) \neq 0$) the local equation of $V_{m_1,\dots,m_r}$ is $z_1^{m_1} = u^{m_1-d_1},\dots,z_r^{m_r} = u^{m_r-d_r}$. Let

$$l = l.c.m.\left(\frac{m_1}{m_1-d_1}(m_1-d_1)\cdots(m_r-d_r),\dots,\right.$$
$$\left.\frac{m_r}{m_r-d_r}(m_1-d_1)\cdots(m_r-d_r),(m_1-d_1)\cdots(m_r-d_r)\right)$$

Then each branch of the normalization of $V_{m_1,\dots,m_r}$ has the parameterization (t,v) such that:

$$z_1 = t^{\frac{l(m_1-d_1)}{m_1(m_1-d_1)\cdots(m_r-d_1)}},\dots,z_r = t^{\frac{l(m_r-d_r)}{m_r(m_1-d_1)\cdots(m_r-d_r)}}, u = t^{\frac{l}{(m_1-d_1)\cdots(m_r-d_r)}}.$$

Therefore the pull back of the form $\frac{z_1^{i_1}\cdots z_r^{i_r}u^k du\wedge dv}{z_1^{m_1-1}\cdots z_r^{m_r-1}}|_{V_{m_1,\dots,m_r}}$ to the (t,v) chart is regular if and only if

$$k > \sum_{j=1}^{r}(m_j-d_j-(i_j+1)) + \frac{d_j(i_j+1)}{m_j} - 1. \tag{3.2.4}$$

The smallest k which satisfies this inequality, i.e. the multiplicity of the line $u=0$ as the component of a generic curve from $H^0(\mathcal{A}_{i_1,\dots,i_r}((m_1-i_1)+\dots+(m_r-i_r)-r-3)|_H)$, is equal to

$$\sum_j(m_j-d_j-(i_j+1)) + \left[\sum_j \frac{d_j(i_j+1)}{m_j}\right]. \tag{3.2.5}$$

As a consequence of this, the degree of the moving curves in the linear system $H^0(\mathcal{A}_{i_1,\dots,i_r}((m_1-i_1)+\dots+(m_r-i_r)-r-3)|H)$ is $\sum_j d_j - 3 - [\sum_j \frac{d_j(i_j+1)}{m_j}]$ and therefore the moving curves belong to the linear system $H^0(\mathcal{A}_\Delta((\sum_j d_j) - 3 - [\frac{d_j(i_j+1)}{m_j}])$ where Δ is the polytope of quasiadjunction containing $(\frac{i_1+1}{m_1},\dots,\frac{i_r+1}{m_r})$.

In fact the moving curves form a complete system since the cone over any curve in $H^0(\mathcal{A}_\Delta(\sum_j d_j - 3 - [\sum_j \frac{d_j(i_j+1)}{m_j}])$ belongs to $H^0(\mathcal{A}_{i_1,\dots,i_r}(\sum_j(m_j - i_j) - r - 3))$.

Step 2. A recurrence relation for $F(i_1,\dots,i_r)$ and $F_\chi(i_1,\dots,i_r)$.
Let $s(i_1,\dots,i_r) = \dim H^1(\mathcal{A}_\Delta(\sum_j d_j - 3 - [\sum_j \frac{d_j(i_j+1)}{m_j}])$ where Δ is the polytope of quasiadjunction of C containing $(\frac{i_1+1}{m_1},\dots,\frac{i_r+1}{m_r})$ and $\varepsilon_\chi(i_1,\dots,i_r) = 1$ (resp. 0) if $\chi = \chi_1^{i_1-m_1+1}\cdots\chi_r^{i_r-m_r+1}$ (resp. otherwise). We claim the following recurrence:

$$
\begin{aligned}
F(i_1,\dots,i_r) &= s(i_1,\dots,i_r) + \\
&\quad + \sum_{l=1}^{r}(-1)^{l+1} \sum_{i_{j_1}<\dots<i_{j_l}} F(\dots,i_{j_1}+1,\dots,i_{j_l}+1,\dots); \\
F_\chi(i_1,\dots,i_r) &= \varepsilon_\chi(i_1,\dots,i_r)s(i_1,\dots,i_r) + \\
&\quad + \sum_{l=1}^{r}(-1)^{l+1} \sum_{i_{j_1}<\dots<i_{j_l}} F_{\chi(\chi_{j_1}\cdots\chi_{j_l})^{-1}}(\dots,i_{j_1}+1,\dots,i_{j_l}+1,\dots)
\end{aligned}
\tag{3.2.6}
$$

Equivalently the first of equalities (3.2.6) can be written as

$$
s(i_1,\dots,i_r) = \sum_{l=0}^{l=r}(-1)^l \sum_{i_{j_1}<\dots<i_{j_l}} F(\dots,i_{j_1}+1,\dots,i_{j_l}+1,\dots)
$$

and similarly for the second. This identity will be derived from the following. For h such that $1 \le h \le r$ let

$$
\begin{aligned}
&F(i_1,\dots,i_r|q_1,\dots,q_h) = \\
&= \dim H^1(\mathcal{A}_{i_1,\dots,i_r}((m_1-i_1)+\dots+(m_r-i_r)-r-3)|_{H_{q_1}\cap\dots H_{q_h}})
\end{aligned}
$$

where H_s is the hyperplane $z_s = 0$ in $\mathbb{P}^{r+2}$ while for $h=0$ we let $F(i_1,\dots,i_r|\emptyset) = F(i_1,\dots,i_r)$. In particular $s(i_1,\dots,i_r) = F(i_1,\dots,i_r|1,\dots,r)$. Similarly one defines $F_\chi(i_1,\dots,i_r|q_1,\dots,q_h)$. We shall prove by induction over h:

$$
\begin{aligned}
&F(i_1,\dots,i_r|q_1,\dots,q_h) = \sum_{l=0}^{h}(-1)^l \sum_{\substack{i_{j_1}<\dots<i_{j_l} \\ i_{j_1},\dots,i_{j_l}\subset(q_1,\dots,q_h)}} F(\dots,i_{j_1},\dots,i_{j_l},\dots) \\
&F_\chi(i_1,\dots,i_r|q_1,\dots,q_h) = \\
&= \sum_{l=0}^{h}(-1)^l \sum_{\substack{i_{j_1}<\dots<i_{j_l} \\ (i_{j_1},\dots,i_{j_l})\subset(q_1,\dots,q_h)}} F_{\chi\cdot\chi_{j_1}^{-1}\cdots\chi_{j_l}^{-1}}(\dots,i_{j_1}+1,\dots,i_{j_l}+1,\dots)
\end{aligned}
\tag{3.2.7}
$$

The identity (3.2.6) is a special case of (3.2.7) when $q_i = i$. For any $(i_1, \dots, i_r \mid q_1, \dots, q_h)$, $(h \ge 0)$ from the exact sequence (in which the left map is the multiplication by $z_{q_{h+1}}$):

$$\begin{aligned}0 \to \mathcal{A}_{\dots,i_{q_{h+1}}+1,\dots}(\dots + (m_{q_{h+1}} - i_{q_{h+1}} - 1) + \dots - r - 3)|_{H_{q_1}\cap\dots\cap H_{q_h}} \to \\ \to \mathcal{A}_{i_1,\dots,i_r}((m_1 - i_1) + \dots + (m_r - i_r) - r - 3)|_{H_{q_1}\cap\dots\cap H_{q_h}} \to \\ \to \mathcal{A}_{i_1,\dots,i_r}((m_1 - i_1) + \dots + (m_r - i_r) - r - 3)|_{H_{q_1,\dots,q_h,q_{h+1}}} \to 0\end{aligned} \tag{3.2.8}$$

we obtain

$$\begin{aligned}F(i_1,\dots,i_r|q_1,\dots,q_h,q_{h+1}) &= F(i_1,\dots,i_r|q_1,\dots,q_h) + \\ &\quad - F(i_1,\dots,i_{q_{h+1}}+1,\dots,i_r|q_1,\dots,q_h) \\ F_\chi(i_1,\dots,i_r|q_1,\dots,q_h,q_{h+1}) &= F_\chi(i_1,\dots,i_r|q_1,\dots,q_h) + \\ &\quad - F_{\chi\cdot\chi_{q_{h+1}}^{-1}}(i_1,\dots,i_{q_{h+1}+1},\dots,i_r|q_1,\dots,q_h).\end{aligned} \tag{3.2.9}$$

Indeed, the following map is surjective

$$\begin{aligned}H^0(\mathcal{A}_{i_1,\dots,i_r}((m_1 - i_1) + \dots + (m_r - i_r)) - r - 3)|_{H_{q_1,\dots,q_h}} \to \\ \to H^0(\mathcal{A}_{i_1,\dots,i_r}((m_1 - i_1) + \dots + (m_r - i_r)) - r - 3)|_{H_{q_1,\dots,q_h,q_{h+1}}},\end{aligned} \tag{3.2.10}$$

because the cone in $H_{q_1} \cap \dots \cap H_{q_h}$ over the hypersurface in $H^0(\mathcal{A}_{i_1,\dots,i_r}((m_1 - i_1) + \dots + (m_r - i_r) - r - 3)|_{H_{q_1,\dots,q_h,q_{h+1}}})$ belongs to $H^0(\mathcal{A}_{i_1,\dots,i_r}((m_1 - i_1) + \dots + (m_r - i_r) - r - 3)|_{H_{q_1,\dots,q_h}})$. Moreover for $q_i = i$, $i = 1,\dots,r$, we have $F_\chi(i_1, \dots, i_r \,|\, 1, \dots, r) = \varepsilon_\chi(i_1,\dots,i_r)s(i_1,\dots,i_r)$ since to $\phi(x,y) \in H^0(\mathcal{A}_{i_1,\dots,i_r})$ corresponds the form $\psi = z_1^{m_1-i_1-1}\dots z_r^{m_r-i_r-1}\pi^*\phi$ holomorphic on $\tilde{V}_{m_1,\dots,m_r}$ and satisfying: $g^*(\psi) = \chi_1^{i_1-m_1+1}\dots\chi_r^{i_r-m_r+1}\psi$. This shows (3.2.7) for $h = 1$ and that validity of (3.2.7) for the array $(q_1,\dots,q_{h+1})$ provided it is valid for all $(q_1,\dots,q_h)$.

Step 3. An explicit formula for $F(i_1,\dots,i_r)$. Let $C(j_1,\dots,j_s) = C_{j_1} \cup \dots \cup C_{j_s}$ be a curve formed by a union of the components of C and let

$$\begin{aligned}F_{C(j_1,\dots,j_s)}(i_1,\dots,i_s) = \dim H^1(\mathcal{A}(C(j_1,\dots,j_s))_{i_1,\dots,i_s}((m_{j_1} - i_1) + \dots + \\ + (m_{j_s} - i_s) + s - 3).\end{aligned}$$

Note that

$$i_j = m_j - 1 (j \neq j_1,\dots,j_s) \Rightarrow F(i_1,\dots,i_r) = F_{C(j_1,\dots,j_s)}(i_1,\dots,i_s) \tag{3.2.11}$$

since the local conditions defining both sheaves coincide, indeed:

$$\frac{z_1^{i_1}\dots z_{j_k}^{m_{j_k}-1}\dots z_r^{i_r}\,dx\wedge dy}{z_1^{m_1-1}\dots z_r^{m_r-1}} = \frac{z_1^{i_1}\dots z_{i_k}^{\hat{m_{i_k}}-1}\dots z_r^{i_r}\,dx\wedge dy}{z_1^{m_1-1}\dots z_{i_{j_k}}^{\hat{m_{j_k}}-1}\dots z_r^{m_r-1}},$$

as well as the degrees of the curves in the corresponding linear systems. Moreover

$$F_{C(j_1,\dots,j_s)}(0,\dots,0)$$

is the irregularity of the cover of $\mathbb{P}^2$ branched over $C(j_1,\ldots,j_s)$ and having the ramification index m_i over the component $C_i (i = j_1,\ldots,j_s)$.
We solve the recurrence relation (3.2.6) subject to the "initial condition" (3.2.11). It is convenient to view each relation (3.2.6) as the one connecting the values of the function defined at the vertices of the integer lattice in the parallelepiped $0 \le x_i \le m_i, (i = 1,\ldots,r)$. Each equation connects the values of this function at the vertices of a parallelepiped with sides equal to 1. It is clear that the sum of all equations (3.2.6) yields:

$$F(0,\ldots,0) = \sum_{0 \le i_s < m_s - 1} s(i_1,\ldots,i_r) + \sum_{l=1}^{r-1} \sum_{j_1 < \cdots < j_l} F_{C(j_1,\ldots,j_l)}(0,\ldots,0)$$

$$F_\chi(0,\ldots,0) = \sum_{0 \le i_s < m_s - 1} \varepsilon_\chi(i_1,\ldots,i_r) s(i_1,\ldots,i_r) + \qquad (3.2.12)$$

$$+ \sum_{l=1}^{r-1} \sum_{j_1 < \cdots < j_l} F_{C(j_1,\ldots,j_l)_\chi}(0,\ldots,0)$$

Remark. Alternative derivation of (3.2.12).
Sheaves $\mathcal{A}_\Delta(d_1 + \cdots + d_r - r - 2 - [\sum_j \frac{d_j(i_j+1)}{m_j}])$ admit the following interpretation also yielding (3.2.12). Let us consider the global version of the diagram (2.3.1):

$$\begin{array}{ccc} \bar{V}_{m_1,\ldots,m_r} & \xrightarrow{\bar{\pi}} & Y_C \\ \downarrow \bar{\rho} & & \downarrow \rho \\ V_{m_1,\ldots,m_r} & \xrightarrow{\pi} & \mathbb{P}^2 \end{array} \qquad (3.2.13)$$

Here $\rho : Y_C \to \mathbb{P}^2$ is an embedded resolution of singularities of C which are worse than nodes, $\bar{V}_{m_1,\ldots,m_r}$ is the normalization of $V_{m_1,\ldots,m_r} \times_{\mathbb{P}^2} Y_C$ and $\bar{\pi}, \bar{\rho}$ are the obvious projections. Let

$$\bar{\pi}_*(\mathcal{O}_{\bar{V}_{m_1,\ldots,m_r}}) = \oplus \mathcal{L}^{-1}_{\chi_1^{i_1} \cdots \chi_r^{i_r}} \qquad (3.2.14)$$

be the decomposition by the characters of the Galois group acting on $\bar{\pi}_*(\mathcal{O}_{\bar{V}_{m_1,\ldots,m_r}})$. Then we have:

$$\mathcal{A}_\Delta\Big(\sum_j d_j - 3 - \Big[\sum_j \frac{d_j(i_j+1)}{m_j}\Big]\Big) =$$
$$= \rho_*(\bar{\pi}_*(\Omega_{\bar{V}_{m_1,\ldots,m_r}}) \otimes \mathcal{L}_{\chi(m_1-(i_1+1),\ldots,m_r-(i_r+1))}) \qquad (3.2.15)$$

where Δ is the polytope of quasiadjunction containing $(\frac{i_1+1}{m_1},\ldots,\frac{i_r+1}{m_r})$. Indeed it follows from (2.3.8) that a germ ϕ of a holomorphic function belongs to the sheaf in the left side of (3.2.15) if and only if the order of ϕ along an exceptional curve $E_k \subset Y_C$ satisfies: $ord_{E_k}\phi \ge \sum a_{k,j}(\frac{m_j-(i_j+1)}{m_j} - c_k)$ and the sheaf on the left is a subsheaf of $\mathcal{O}_{\mathbb{P}^2}([\sum_j d_j(\frac{m_j-(i_j+1)}{m_j})])$ with the quotient having a zero-dimensional support. One readily sees that the sheaf on the right has the same local description. This identity also implies (3.2.12) as follows from Serre's duality and (3.2.14).

Step 4. A vanishing result.
If Δ is a polytope of quasiadjunction, $\Xi_k = \{(x_1,\ldots,x_r) \in \mathcal{U} | k \le d_1x_1 + \cdots + d_rx_r < k+1\}$ and k is such that $\Delta \cap \Xi_k \neq \emptyset$ then

$$H^1(\mathcal{A}_\Delta(d-r-2-k)) = 0 \qquad (3.2.16)$$

unless Δ is a contributing polytope of quasiadjunction and $\Delta \cap \Xi_k$ is a face of quasiadjunction.

If Δ isn't contributing (cf. 2.6.2), then the intersection $\Delta \cap \Xi_k$ has a positive volume. If $X(n)$ is the number of points $(\frac{i}{n},\ldots,\frac{i}{n})$ in the latter, it follows from (3.2.12) that we have $b_1(C,n) \ge \dim H^1(\mathcal{A}_\Delta(d-r-2-k)) \cdot X(n)$. We have $X(n) > C \cdot n^r$ for some non zero constant C. Therefore we get contradiction with Corollary (1.3.3) unless $\dim H^1(\mathcal{A}_\Delta(d-r-2-k)) = 0$.

Step 5. End of the proof. Step 4 and the formula (3.2.12) give a) and b) of the theorem in the case $m_i \ge d_i$ for $i = 1,\ldots,i_r$.

If χ is a character of $\mathbb{Z}_{m_1} \oplus \cdots \oplus \mathbb{Z}_{m_r}$ acting on $H^0(\Omega^1_{\tilde{V}_{m_1,\ldots,m_r}}) = H^{1,0}(\tilde{V}_{m_1,\ldots,m_r})$ then $\bar{\chi}$ is a character with eigenspace of the same dimension for the action of $\mathbb{Z}_{m_1} \oplus \cdots \oplus \mathbb{Z}_{m_r}$ on $H^{0,1}(\tilde{V}_{m_1,\ldots,m_r})$. Hence part b) and Sakuma formula (cf. 1.3.2) imply that a points $(\ldots \frac{i_j+1}{m_j} \ldots)$ belongs to a contributing face of Δ or its conjugate if and only if $(\ldots, \exp(2\pi\sqrt{-1}\frac{i_j+1}{m_j}),\ldots)$ belongs to i-th characteristic variety with $i = \dim H^1(\mathcal{A}_\Delta(d_1 + \cdots + d_r - r - 3 - k(\Delta))$. Since a characteristic variety is a translated by a point of finite order subtorus (cf. 1.4.2) this implies c). Now the remaining cases of the formula a) follows from Sakuma's result (1.3.2.2).

3.3. EXAMPLES

In 2.6.4 we did describe systems of equations for faces of quasiadjunction in the case of arrangements of lines. To determine if a set of solutions of the system corresponding to a face δ actually corresponds to a component of characteristic variety one should

a) calculate the superabundance (3.1.2) of the corresponding linear system and

b) decide the "amount of translation" i.e. to normalize the system of equations so that the g.c.d. of minors of the left hand sides of (2.6.1) will be equal to one.

In any event, if superabundance is not zero, then clearly the component of characteristic variety will be a connected component of the subgroup given by the equations: $\exp(L_P) = 1$ with P running through all vertices singled out by the face of quasiadjunction.

Example 1. Let us calculate the irregularity of the abelian cover of $\mathbb{P}^2$ branched over the arrangement $L : uv(u-v)w = 0$ and corresponding to the homomorphism

$H_1(\mathbb{P}^2 - L) = \mathbb{Z}^3 \to (\mathbb{Z}/n\mathbb{Z})^3$. The only nontrivial ideal of quasiadjunction is the maximal ideal of the local ring with corresponding polytope of quasiadjunction: $x+y+z>1$. Hence the irregularity of the abelian cover is $Card\{(i,j)|0<i<n, 0<j<n, \frac{i}{n}+\frac{j}{n}+\frac{k}{n}=1\} \cdot \dim H^1(\mathcal{J}(3-3-1))$ where $\mathcal{J} = \operatorname{Ker} \mathcal{O} \to \mathcal{O}_P$ where $P: u=v=0$. $\mathcal{J}$ has the following Koszul resolution:

$$0 \to \mathcal{O}(-2) \to \mathcal{O}(-1) \oplus \mathcal{O}(-1) \to \mathcal{J} \to 0$$

which yields $H^1(\mathcal{J}(-1)) = H^2(\mathcal{O}(-3)) = \mathbb{C}$. Now the counting points on $x+y+z=1$ yields $\frac{n^2-3n+2}{2}$ as the irregularity of the abelian cover.

Example 2. Let us consider the arrangement formed by the sides of an equilateral triangle (x_1,x_2,x_3) and its medians (x_4,x_5,x_6) arranged so that the vertices are the intersection points of $(x_1,x_2,x_4),(x_2,x_3,x_5)$ and x_3,x_1,x_6 respectively (Ceva arrangement cf. (Barthel, Hirzebruch and Hofer)). It has 6 lines, 4 triple and 3 double points. The polytopes of quasiadjunction are the connected components of the partition of $\mathcal{U} = \{(x_1,\ldots,x_6)|0 \le x_i \le 1, i=1,\ldots,6\}$ by the hyperplanes:

$$x_1+x_2+x_4=1, x_2+x_3+x_5=1, x_3+x_1+x_6=1, x_4+x_5+x_6=1 \qquad (3.3.1)$$

The only face of a polytope of quasiadjunction which belongs to a hyperplane $H_k: x_1+x_2+x_3+x_4+x_5+x_6=k, k\in\mathbb{Z}$ is formed by set of solutions of the system of *all* 4 equations (3.3.1). This face belongs to H_2 and is the only contributing face. Hence the irregularity is equal to $N\cdot\dim H^1(\mathcal{J}(6-3-2))$ where N is the number of solutions (3.3.1) of the form $x_i=\frac{j}{n}$. To calculate $\dim H^1(\mathcal{J}(6-3-2))$ notice that 4 triple points form a complete intersection of two quadrics. This yields $H^1(\mathcal{J}(1)) = H^2(\mathcal{O}(-3)) = \mathbb{C}$.

By (3.3.1), the only essential torus is a component of subgroup:

$$t_1t_2t_4=1, \quad t_2t_3t_5=1, \quad t_1t_3t_6=1, \quad t_4t_5t_6=1. \qquad (3.3.2)$$

This subgroup has two connected components:

$$(u,v,u^{-1}v^{-1},u^{-1}v^{-1},u,v),\ (-u,-v,-u^{-1}v^{-1},u^{-1}v^{-1},u,v),\ u,v\in\mathbb{C}^*. \qquad (3.3.3)$$

The second component is a translation of the first by $(1,1,1,-1,-1,-1)$, a point of order 2. Since (3.3.1) admits an integral solution image under the exponential map of the contributing face does contains trivial character and hence the subgroup in (3.3.3) is the essential torus.

There are also 4 nonessential tori corresponding to each of triple points:

$$\begin{matrix} t_1t_2t_3=1, t_i=1, i\neq 1,2,3, & t_5t_2t_3=1, t_i=1, i\neq 5,2,3, \\ t_4t_6t_3=1, t_i=1, i\neq 4,6,3, & t_4t_5t_6=1, t_i=1, i\neq 4,5,6. \end{matrix} \qquad (3.3.4)$$

Let us consider the abelian cover of $\mathbb{C}^2$ corresponding to the homomorphism $H_1(\mathbb{C}^2 - C) \to (\mathbb{Z}/n\mathbb{Z})^6/\mathbb{Z}/n\mathbb{Z}$ (embedding of the quotiented subgroup is diagonal). Then each of five tori contributes the same number into irregularity equal to $\frac{n^2-3n+2}{2}$, i.e. the irregularity of the abelian cover of $\mathbb{P}^2(\mathbb{C})$ is $5\frac{n^2-3n+2}{2}$ (e.g., for $n = 5$ the irregularity is 30, cf. (Ishida)).

Example 3. Let us calculate the characteristic varieties of the union of 9 lines which are dual to nine inflection points on a non singular cubic curve $C \subset \mathbb{P}^2(\mathbb{C})$. This arrangement in $\mathbb{P}^{2*}(\mathbb{C})$ has 12 triple points corresponding to 12 lines determined by the pairs of the inflection points of C. One can view inflection points of C as the points of $\mathbb{F}_3^2$ ($\mathbb{F}_3$ is the field with 3 elements) i.e. as the points of the affine part in a projective plane $\mathbb{P}^2(\mathbb{F}_3)$. The triple points of this arrangement then can be viewed as lines in $\mathbb{P}^2(\mathbb{F}_3)$ different from the line at infinity (i.e. the complement to the chosen affine plane). In dual picture one identifies triple points of this arrangement with points of the dual plane $\mathbb{P}^{2*}(\mathbb{F}_3)$ different from a fixed point P corresponding to the line at infinity. The lines of this arrangement in $\mathbb{P}^{2*}(\mathbb{C})$ are identified with the lines in $\mathbb{P}^{2*}(\mathbb{F}_3)$ not passing through the fixed point P.

Each essential component corresponds to a collection of vertices $\mathcal{S}$ (cf. (2.6.4), example 2). The structure of the system of equations (2.6.1) shows that $|\mathcal{S}|/k = r/m = 3$. Hence one has either:

a) $|\mathcal{S}| = 3, k = 1$ or
b) $|\mathcal{S}| = 6, k = 2$ or
c) $|\mathcal{S}| = 9, k = 3$ or
d) $|\mathcal{S}| = 12, k = 4$.

Cases a) and b) will not define non empty tori since in this case $r^2 > 9|\mathcal{S}|$ (cf. corollary 4.1).

In the case c) each collection $\mathcal{S}$ is determined by one of 4 choices of a line ℓ through P and consists of 9 points in $\mathbb{P}^{2*}(\mathbb{F}_3)$ in the complement to the chosen line. In this case the corresponding homogeneous system has rank 7 i.e. a 2-dimensional space of solutions. Moreover, $\dim H^1(\mathbb{P}^{2*}(\mathbb{C}), I(9-3-3)) = 1$ since the points on $\mathbb{P}^{2*}(\mathbb{C})$ corresponding to 9 points in $\mathbb{P}^{2*}(\mathbb{F}_3)$ in the complement to a line $p \subset \mathbb{P}^{2*}(\mathbb{F}_3)$ form a complete intersection of two cubics. These cubics formed by the unions of triples of lines in $\mathbb{P}^{2*}(\mathbb{C})$ corresponding to triple of lines in $\mathbb{P}^{2*}(\mathbb{F}_3)$ passing through a point of ℓ. Indeed, for a given $P_1, P_2 \in \ell$ and a point Q on $\mathbb{P}^{2*}(\mathbb{F}_3)$ outside of ℓ, there are exactly 2 lines in $\mathbb{P}^{2*}(\mathbb{F}_3)$ intersecting at this point and passing respectively through P_1 and P_2. The same incidence relation is valid on $\mathbb{P}^{2*}(\mathbb{C})$.

In the case d) the homogeneous system has rank 9, i.e. the corresponding system does not define a torus.

Non essential tori correspond to subarrangements with number of lines divisible by $m = 3$ (cf. 2.6.3). There are 12 triples of lines corresponding to each of triple points each defining a 2-torus. A collection of 6 lines should have 4 triple points but the arrangement of this example does not contain such subarrangements.

Therefore we have 16 2-dimensional tori. In the abelian cover of $\mathbb{C}^2$ which sends each generator of $H_1(\mathbb{C}^2 - C)$ to a generator of $\mathbb{Z}/n\mathbb{Z}$ contributing tori are the essential torus of this arrangement and subtori corresponding to subarrangements formed by triple of lines defined by the triple points. Each torus contributes $(n-1)(n-2)/2$ to the Betti number i.e. the total Betti number of this cover is $16 \times (n-1)(n-2)/2$. These tori can be explicitly described as follows. Defining equations of non essential tori are products of 3 generators t_i's corresponding to a triple of points in $\mathbb{F}_3^2$ belonging to a line with the rest of t_i is 1. Each of essential tori is given by 9 equations $t_i t_j t_k = 1$ where (i,j,k) are the triples of points $\mathbb{F}_3^2$ (which interpreted as the lines of the arrangement) which belong to lines not passing through a fixed point at infinity.

If the point at infinity is $(1,-1,0)$, then the lines not passing through it are: $x+z=0, x-z=0, x=0, y=0, x-y=0, x-y+z=0, x-y-z=0, y+z=0, y-z=0$ i.e. the corresponding torus satisfies:

$$\begin{aligned} &t_{20}t_{21}t_{22}=1, t_{10}t_{11}t_{12}=1, t_{00}t_{01}t_{02}=1, t_{00}t_{10}t_{20}=1, t_{00}t_{11}t_{22}=1, \\ &t_{01}t_{12}t_{20}=1, t_{02}t_{10}t_{21}=1, t_{02}t_{12}t_{22}=1, t_{01}t_{11}t_{21}=1 \end{aligned} \tag{3.3.5}$$

where the points of the complement to $z=0$ (i.e. the lines in $\mathbb{P}^{2*}(\mathbb{C})$) are labeled as:

$$(0,0),(0,1),(0,2),(1,0),(1,1),(1,2),(2,0),(2,1),(2,2).$$

The corresponding torus can be parameterized as

$$\begin{aligned} &t_{00}=t, t_{01}=s, t_{02}=t^{-1}s^{-1}, t_{10}=s, t_{11}=t^{-1}s^{-1}, \\ &t_{12}=t, t_{20}=t^{-1}s^{-1}, t_{21}=t, t_{22}=s. \end{aligned}$$

The equations for other essential tori, corresponding to choices of the point at infinity as respectively: $(1,1,0),(1,0,0),(0,1,0)$ can be obtained from (3.3.5) by applying linear transformation to the indices which takes $(1,-1,0)$ to respective point. For $(x,y) \to (x,-y)$ which takes $(1,-1,0)$ to $(1,1,0)$ we obtain:

$$\begin{aligned} t_{20}t_{22}t_{21} &= t_{10}t_{12}t_{11} = t_{00}t_{02}t_{01} = t_{00}t_{10}t_{20} = t_{00}t_{12}t_{21} = \\ &= t_{02}t_{11}t_{20} = t_{01}t_{10}t_{22} = t_{01}t_{11}t_{21} = t_{02}t_{12}t_{21} = 1. \end{aligned}$$

For $(x,y) \to (x,x+y)$ which takes $(1,-1,0)$ to $(1,0,0)$ we obtain:

$$\begin{aligned} t_{22}t_{20}t_{21} &= t_{11}t_{12}t_{10} = t_{00}t_{01}t_{02} = t_{00}t_{11}t_{22} = t_{00}t_{12}t_{21} = \\ &= t_{01}t_{10}t_{20} = t_{02}t_{11}t_{20} = t_{02}t_{10}t_{21} = t_{01}t_{12}t_{20} = 1. \end{aligned}$$

For $(x,y) \to (x+y,y)$ which takes $(1,-1,0)$ to $(0,1,0)$ we have:

$$\begin{aligned} t_{20}t_{01}t_{12} &= t_{10}t_{21}t_{02} = t_{00}t_{11}t_{22} = t_{00}t_{10}t_{20} = t_{00}t_{21}t_{12} = \\ &= t_{11}t_{01}t_{20} = t_{22}t_{10}t_{01} = t_{22}t_{02}t_{12} = t_{11}t_{21}t_{01}. \end{aligned}$$

Example 4. Let us consider the curve of degree 4 which has one ordinary point of multiplicity 4. Faces of the polytopes of quasiadjunction are $H_1 : x_1 + x_2 + x_3 +$

$x_4 = 1$ (resp. $H_2 : x_1 + x_2 + x_3 + x_4 = 2$). The number of points $(\frac{i_1}{n}, \frac{i_2}{n}, \frac{i_3}{n}, \frac{i_4}{n})$ on H_1 (resp. H_2) is $(n-1)(n-2)(n-3)/6$ (resp. $(n-1)(2n^2-4n+3)/3$). The ideal corresponding to the polytope of quasiadjunction with the face H_1 (resp. H_2) is $\mathcal{M}^2$ (resp. $\mathcal{M}$ the maximal ideal of the local ring) and the level of the supporting face H_1 (resp. H_2) is 1 (resp. 2). Moreover $\dim H^1(\mathbb{P}^2, \mathcal{I}_{\mathcal{M}^{3-l}}(4-3-l))$ is 2 (resp. 1) for $l = 1$ (resp. for $l = 2$). Hence the irregularity of the cover corresponding to homomorphism $H_1(\mathbb{P}^2 - \bigcup_{i=1,2,3,4} L_i) \to \mathbb{Z}/n\mathbb{Z}$ is equal to

$$2\frac{(n-1)(n-2)(n-3)}{6} + \frac{(n-1)(2n^2-4n+3)}{3} + 4\frac{(n-1)(n-2)}{2} = \qquad (3.3.6)$$
$$= (n-1)(n^2-n-1).$$

This implies that the characteristic variety in this case is just

$$t_1t_2t_3t_4 = 1.$$

The latter contains $(n-1)^3 - (n-1)(n-2) = (n-1)(n^2-3n+3)$ points with coordinates in μ_n and the Betti number of the branched cover from Sakuma's formula is equal to $2(n-1)(n^2-3n+3)+4(n-1)(n-2) = 2(n-1)(n^2-n-1)$.

Example 5. Let us consider the arrangement formed by 12 lines which compose 4 degenerate fibers in a Hesse pencil of cubics formed by a non singular cubic curve and its Hessian. For example one can take the following pencil:

$$x^3 + y^3 + z^3 - 3\lambda xyz = 0.$$

This arrangement has 9 points of multiplicity 4 (inflection points of non singular cubic). In ${\mathbb{C}^*}^{12}$ there are 10 tori of dimension 3 which are defined by 9 quadruples of lines corresponding to 9 quadruple points and one 3-torus corresponding to the whole configuration. Contribution into the first Betti number an abelian cover also comes from 94 tori of dimension 2: 2-tori corresponding to triples of lines forming each of 9 quadruple points (total 36 2-tori), 2-tori corresponding to configurations of 9 lines formed by triples of 4 special fibers of the pencil (total 4 2-tori) and 54 2-tori corresponding to configurations of 6 lines passing through 4 inflection points no three of which belong to a line (since the choice of 4 points must be made among points of affine space over $\mathbb{F}^3$ the ordered collection can be made in $9 \times 8 \times 6 \times 3$ way and $54 = 9 \cdot 8 \cdot 6 \cdot 3/24$). In particular, the irregularity of the cover with the Galois group $(\mathbb{Z}/3\mathbb{Z})^2$ is equal to 154, cf. (Ishida). Indeed the contribution of each 2-torus into the first Betti number is 2 and in the case of 3-tori the contribution is 6, since the 3-torus contains 6 points with coordinates $i/3$. Since the depth of 3-tori is 2 the first Betti number is equal to $6 \times 10 \times 2 + 94 \times 2$.

4. The structure of characteristic varieties of algebraic curves

In this section we describe sufficient conditions for the vanishing of cohomology of linear systems which appear in description of characteristic varieties given in section 3. This, therefore, yields conditions for absence of essential components. In the cyclic case one obtains triviality of Alexander polynomial.

4.1. ABSENCE OF CHARACTERISTIC VARIETIES FOR CURVES WITH SMALL NUMBER OF SINGULARITIES

THEOREM 4.1.1. *Let C be a plane curve as above. Suppose that $\rho : Y \to \mathbb{P}^2$ is obtained by a sequence of blow ups such that the proper preimage $\tilde{C}$ of C in Y has only normal intersection with the exceptional set and satisfies $\tilde{C}^2 > 0$. Then C has no essential characteristic subvarieties.*

COROLLARY 4.1.2. *1. Let C be an irreducible curve which has ordinary cusps and nodes as the only singularities. If the number of cusps is less than $d^2/6$ then the Alexander polynomial of C is equal to 1.*

2. Let $\mathcal{H}$ be an arrangement consisting of d lines and which has N points of multiplicity m. Let $l(\delta)$ be the level of a face of quasiadjunction for the complement to $\mathcal{H}$. If $d^2 > m^2N$ then the superabundance is zero for the system of curves of degree $d-3-l(\delta)$ which local equations belong to the ideal of quasiadjunction corresponding to δ at the points of multiplicity m.

Remark. One can compare corollary 1 with results of (Nori). The latter yields that the fundamental group of the complement to a curve of degree d with δ nodes and κ cusps is abelian if $d^2 > 6\kappa + 2\delta$ while a weaker inequality $d^2 > 6\kappa$ yields the triviality of the Alexander polynomial. For example, for the branching curve of a generic projection of a smooth surface of degree N in $\mathbb{P}^3$ one has $d^2 > 6\kappa$ for $N > 4$ but $d^2 < 6\kappa + 2\delta$ for $N > 2$. The fundamental groups of these curves are non abelian for $N > 2$ and the Alexander polynomial for $N = 3, 4$ is equal to $t^2 - t + 1$, cf. (Libgober, 1983).

PROOF OF THE THEOREM. We should show that for any contributing face of the quasi-adjunction δ we have $\dim H^1(\mathbb{P}^2, \mathcal{A}_\delta(\sum d_i - 3 - l(\delta))) = 0$. If $\rho : Y \to \mathbb{P}^2$ is a blow up of $\mathbb{P}^2$, satisfying conditions of the theorem, then we have:

$$\mathcal{A}_\delta(\sum_i d_i - 3 - l(\delta)) = \rho_*(\omega_Y \otimes \mathcal{O}_Y(\gamma\tilde{C}) \otimes \mathcal{O}_Y(\sum \varepsilon_k E_k)) \qquad (4.1.1)$$

for some rational $\gamma > 0$ and $0 \le \varepsilon < 1$. More precisely, $\gamma = \frac{1}{\Sigma_i d_i} \cdot \Sigma_i d_i (1 - \frac{j_i+1}{m_i})$ for some choice of $(\ldots, (j_i+1)/m_i, \ldots)$ belonging to the face δ (with ε_k a priori depending on this choice). Indeed, from (2.3.6) and the discussion after, the multiplicity $f_k(\phi)$ along an exceptional curve E_k of the pull back on Y of a germ in the ideal of quasiadjunction with parameters $(j_1, \ldots, j_r | m_1, \ldots, m_r)$ such that $(\frac{j_1+1}{m_1}, \ldots, \frac{j_r+1}{m_r}) \in \delta$ satisfies:

$$f_k(\phi) \ge \left[\sum_i a_{k,i}\left(1 - \frac{j_i+1}{m_i}\right) - c_k\right]. \qquad (4.1.1.1)$$

Hence

$$A_\delta = \rho_*\Big(\otimes_k \mathcal{O}_Y\Big(\Big(c_k - \Big[\sum_i a_{k,i} - a_{k,i}\Big(\frac{j_i+1}{m_i}\Big)\Big]\Big)E_k\Big)\Big). \tag{4.1.1.2}$$

We have: $l(\delta) = \sum_i d_i \frac{j_i+1}{m_i}$ and $\mathcal{O}_{\mathbb{P}^2}(C_i) = \mathcal{O}_{\mathbb{P}^2}(d_i)$. Therefore

$$\begin{aligned} A_\delta\Big(\sum_i d_i - 3 - l(\delta)\Big) &= \mathcal{O}_{\mathbb{P}^2}\Big(\sum_i C_i\Big(1 - \frac{j_i+1}{m_i}\Big)\Big) \otimes \mathcal{O}_{\mathbb{P}^2}(-3) \otimes \\ &\otimes \rho_*\Big(\otimes_k \mathcal{O}_Y\Big(\Big(c_k - \Big[\sum_i a_{k,i} - a_{k,i}\Big(\frac{j_i+1}{m_i}\Big)\Big]\Big)E_k\Big)\Big). \end{aligned} \tag{4.1.1.3}$$

Since $\omega_Y = \otimes_k \mathcal{O}_Y(c_k E_k) \otimes \rho^*(\mathcal{O}_{\mathbb{P}^2}(-3))$ and

$$\otimes_k \mathcal{O}_Y\Big(a_{k,i}\Big(1 - \frac{j_i+1}{m_i}\Big)E_k\Big) \otimes \mathcal{O}(\tilde{C})^{\frac{d_i}{\sum d_i}} = \rho^*\Big(\mathcal{O}_{\mathbb{P}^2}\Big(C_i\Big(1 - \frac{j_i+1}{m_i}\Big)\Big)\Big) \tag{4.1.1.4}$$

(because $\mathcal{O}_{\mathbb{P}^2}(C_i) = \mathcal{O}_{\mathbb{P}^2}(1)^{d_i} = \mathcal{O}_{\mathbb{P}^2}(C)^{\frac{d_i}{\sum_i d_i}}$), we see that (4.1.1.3) yields (4.1.1) with $\varepsilon_k = \{\sum_i a_{k,i}(1 - \frac{j_i+1}{m_i})\}$ where $\{x\} = x - [x]$ is the fractional part.

The Kawamata-Viehweg vanishing theorem (cf. for example (Kollar)) implies that the cohomology of the sheaf $\omega_Y \otimes \mathcal{O}_Y(\gamma\tilde{C}) \otimes \mathcal{O}_Y(\sum \varepsilon_k E_k)$ is trivial in positive dimensions if $\tilde{C}$ is big and nef. But this follows from the assumptions of the theorem. Finally, the exact sequence $0 \to E_2^{1,0} \to H^1(Y, \mathcal{F})$ of lower degree terms in the Leray spectral sequence $H^p(\mathbb{P}^2, R^q\rho_*\mathcal{F}) \Rightarrow H^{p+q}(Y, \mathcal{F})$ for the sheaf $\mathcal{F} = \omega_Y \otimes \mathcal{O}_Y(\gamma\tilde{C}) \otimes \mathcal{O}_Y(\sum \varepsilon_k E_k)$ yields $H^1(\mathbb{P}^2, \mathcal{A}_\delta(\sum_i d_i - 3 - l(\delta))) = 0$. □

PROOF OF THE COROLLARY. For each blow up at an ordinary point of multiplicity m of a curve C we have $\tilde{C}^2 = C^2 - m^2$ where $\tilde{C}$ is the proper preimage of C. Hence $\mathcal{O}(\tilde{\mathcal{H}})$ is big if $d^2 > m^2 N$. The case of ordinary cusps is similar. □

4.1.2. *Dimensions of components of characteristic varieties*

This theorem imposes restrictions on the dimensions of the contributing faces and hence on the dimensions of characteristic varieties. For example, let us consider an arrangement of lines with at most triple points as singularities. Then each contributing face is the intersection of hyperplanes defining the only local polytope of an ordinary triple point. These hyperplanes are given by the equations of the form $x_i + x_j + x_k = 1$ where (i, j, k) are the indices corresponding to the lines through the triple point. The matrix of this system therefore has the property that in each row only 3 non zero entries are equal to 1, any two rows have at most one non zero entry in the same column and the number of rows is at least $d^2/9$ (since by the corollary only in this case one can get a contributing face with $H^1 \neq 0$). In particular, the number of non zero entries in the matrix is at least $d^2/3$. The rank of this system is at least $d/3$. Indeed, the matrix contains a column with at least $d/3$

1's. On the other hand, if a column s contains k non zero entries in rows $v_1,\dots,v_k$, then these rows are linearly independent since the left hand side of a relation $\sum\lambda_i v_i = 0$ has λ_i as the entry of the column different from the s-th and in which v_i has a non zero entry. In particular in the arrangement of d lines, the dimension of characteristic variety is at most $2d/3$. For the components containing a trivial character, the dimension is at most 2, since by (Arapura) such component induces the map onto $\mathbb{P}^1$ minus three point such that pull backs of rank one local systems from the latter form the component (cf. footnote in section 1 and (Libgober and Yuzvinsky)).

PROBLEM. Let m be the maximal number of components which meet at a singular point of C. Is it true that the dimension of each component of the characteristic variety of C is at most $m-1$?

4.2. TRANSLATIONS OF THE TORI FORMING CHARACTERISTIC VARIETIES AND THE DEGREES OF IRREDUCIBLE COMPONENTS OF C

THEOREM 4.2.1. *Let a_δ be g.c.d. of non zero minors of maximal order in a system of equations with integer coefficients whose solution set contains a face of quasi-adjunction δ with the dimension of the solution set equal to* $\dim\delta$. *Let $d_i = \deg C_i$.*
a) Each irreducible essential component of Char C *belongs to a coset of a subtorus of* $\mathrm{Hom}(H_1(\mathbb{C}^2 - C),\mathbb{C}^*)$ *of order dividing a_δ. In particular each irreducible essential component belongs to a coset of order equal to the order of the face (cf. 2.4.2 and 2.6.2).*
b) Each essential component lies in a subtorus of codimension 1 in $\mathrm{Hom}(H_1(\mathbb{C}^2 - C),\mathbb{C}^*)$ *translated by a point of the order dividing $\rho = g.c.d.(d_1,\dots,d_r)$.*

PROOF. Any component of a characteristic variety is a Zariski closure of the image under the exponential map of a contributing face δ with the equation $d_1x_1 + \cdots + d_rx_r = l(\delta)$. This yields b) (it also follows from Cor. 3.3 from (Libgober, 1992)). By Theorem 3.1, c), irreducible essential component is a connected component of the subgroup of $\mathrm{Hom}(H_1(\mathbb{C}^2 - C),\mathbb{C}^*)$ belongs to a subgroup

$$\chi_1 = \cdots = \chi_r = 1 \tag{4.2.1.1}$$

where χ_i is the character of $\mathrm{Hom}(H_1(\mathbb{C}^2 - C),\mathbb{C}^*)$ having form $\exp(L_i)$ where L_i is the form with integer coefficients such that $L_i = l_i, l_i \in \mathbb{Z}$ are the equations defining the face of quasiadjunction. The order of the group of cosets of the group (4.2.1.1) by its connected component of identity is the order of the torsion of its group of characters, $\mathrm{Char}(\mathrm{Hom}(H_1(\mathbb{C}^2 - C),\mathbb{C}^*))/(\chi_1,\dots,\chi_r)$. This yields a). □

4.2.2. *Linear systems corresponding to different faces of quasiadjunction*
Another byproduct of results in section 3 is equality of superabundances of linear systems of curves defined by rather different local conditions.

PROPOSITION 4.2.2. *a) Let* δ *and* δ' *be two faces of global polytopes quasiadjunction such that the Zariski closures of* $\exp(\delta)$ *and* $\exp(\delta')$ *coincide. Then if* δ *is a contributing face then* δ' *is also contributing and* $H^1(\mathcal{A}_\delta(d-3-l(\delta)) = H^1(\mathcal{A}_{\delta'}(d-3-l(\delta'))$.
b) Let $\alpha \in \mathbb{Q}$ *is such that* $\alpha \cdot g.c.d(d_1,\ldots,d_r)$ *is the level of a face of quasiadjunction* δ *and* $\sigma \in \mathrm{Gal}(\mathbb{Q}(\exp(2\pi i\alpha))/\mathbb{Q})$ *such that* $\sigma(\exp(2\pi i\alpha)) = \exp(2\pi i\beta)$ *with* $0 < \beta < 1$. *Then* β *is equal to* $\frac{l(\delta')}{g.c.d(d_1,\ldots,d_r)}$ *for some face of quasiadjunction* δ' *and* $H^1(\mathcal{A}_\delta(\sum_i d_i - 3 - l(\delta))) = H^1(\mathcal{A}_{\delta'}(d-3-l(\delta')))$.

PROOF. a) Since Zariski closures of $\exp(\delta)$ and $\exp(\delta')$ are the same the corresponding to δ and δ' components of the characteristic variety are the same. If the depth of this component of characteristic variety is i, then the dimension of each of the cohomology group in the statement equals to i and the result follows.
b) Since i-th characteristic variety is defined over $\mathbb{Z}$, the Galois group $\mathrm{Gal}(\bar{\mathbb{Q}}/\mathbb{Q})$ acts on the set of its irreducible components. The irreducible component corresponding to δ is a translation by $\exp(2\pi\sqrt{-1}\alpha)$ of a subgroup of $H^1(\mathbb{C}^2 - C, \mathbb{C}^*)$ defined over $\mathbb{Z}$, therefore σ takes the component corresponding to δ into translation of the same subgroup by $\exp(2\pi\sqrt{-1}\beta)$. This translation is a Zariski closure of $\exp(2\pi\sqrt{-1}\delta')$ for some face δ'. It does satisfy the conclusions of b). □

Remarks.
1. For irreducible curves the order of each face of quasiadjunction is a root of a local Alexander polynomial. So the divisibility theorem from (Libgober, 1982) is a special case of 4.2.1.

2. One of the consequences of 4.2.1 is a non trivial restriction on an abstract group which is necessary to satisfy in order that the group can be realized as the fundamental group of an arrangement. For example $t_1 \cdots t_r = -1$ cannot be a component of characteristic variety of arrangement of r lines since it is cannot belong to an intersection of subgroups of $\mathbb{C}^{*r}$.

3. As an illustration to 4.2.2b), let us consider an irreducible curve of degree d with singularities locally isomorphic to singularity $x^2 = y^5$. If the linear system consisting of curves of degree $d - 3 - d/10$ with local equations belonging to ideals of quasiadjunction of all singular points corresponding to the constant of quasiadjunction $1/10$ is superabundant, then the linear system of curves of degree $d - 3 - 3d/10$ with local equations in the ideals of quasiadjunction corresponding to $3/10$ is also superabundant and the superabundances are equal.

4. An example of faces of quasiadjunction with the same Zariski closure of the images of the exponential map as in a) of the proposition is given by $x_1 + \cdots + x_m = i, x_1 + \cdots + x_m = j, 0 < i, j \leq m-2$ which are the faces of quasiadjunction for the complement to m lines through a point (cf. 2.5 example 3).

5. Resonance conditions for rank one local systems on complements to line arrangements

5.1. COMPLEXES ASSOCIATED WITH ARRANGEMENTS

Let $\mathcal{L} = \bigcup_{i=1,\ldots,r} L_i$ be an arrangement of lines in $\mathbb{C}^2$. We shall assume for convenience (cf. (1.2.3)) that the line at infinity is transversal to all lines in $\mathcal{L}$. Let $l_i(x,y) = 0$ be the equation of L_i and $\eta_i = \frac{1}{2\pi i}\frac{dl_i}{l_i}$. Let $A^i(i = 0,1,2)$ be the subspace generated by the forms $\eta_{j_1} \wedge \cdots \wedge \eta_{j_i}$ in the space $H^0(\Omega^i(*\mathcal{L}))$ of meromorphic forms with poles along $\mathcal{L}$. Let

$$\omega = \sum \eta_i \cdot s_i, s_i \in \mathbb{C} \tag{5.1.1}$$

The exterior product with ω defines the complex:

$$A^{\bullet}_{\omega} : 0 \to A^0 \to A^1 \to A^2 \to 0 \tag{5.1.2}$$

If $s_i = 0, (i = 1,\ldots,r)$, then the cohomology groups of $A^{\bullet}$ are isomorphic to the cohomology groups of $\mathbb{C}^2 - \mathcal{L}$ (Brieskorn). On the other hand, the collection $s_* = (s_1,\ldots,s_r)$ defines the map $\pi_1(\mathbb{C}^2 - \mathcal{L}) \to \mathbb{C}^*$ which sends γ_i (cf. (1.1)) to $\exp(2\pi\sqrt{-1}s_i)$ and hence the local system which we shall denote $\mathcal{A}_{s_*}$. A theorem from (Esnault, Schechtman and Viehweg, p. 558), in the case of line arrangements, asserts that

$$H^i(\mathcal{A}_{s_*}) = H^i(A^{\bullet}_{\omega}) \tag{5.1.3}$$

provided the following non resonance condition is satisfied. For any point singular point P of $\mathcal{L}$ of multiplicity $m > 2$, if the lines through P are $l_{i_1},\ldots,l_{i_m}$, then

$$s_{i_1} + \cdots + s_{i_m} \neq n \in \mathrm{NN} - 0. \tag{5.1.4}$$

THEOREM 5.1.1. *The isomorphism (5.1.3) takes place, provided*

$$(\exp(2\pi i s_1),\ldots,\exp(2\pi i s_r)) \tag{5.1.5}$$

does not belong to the characteristic variety Char_1 *of* $\mathbb{C}^2 - \mathcal{L}$.

5.2. REMARKS

1. It is easy to construct examples of local systems for which (5.1.4) is violated but for which (1.5.3) takes place. Indeed, the image under the exponential map onto the torus $\mathbb{C}^{*r}$ of those $(s_1,\ldots,s_r)$ which violate (5.1.4) is a union of codimension 1 tori in $\mathbb{C}^{*r}$. On the other hand, the characteristic varieties typically have rather small dimension relative to r (cf. (3.3) and (4.1)).
2. For arrangements of arbitrary dimension in $\mathbb{C}^n$ (with l_i denoting the equations of hyperplanes of the arrangement, rather than lines) we have

$$H^1(A^{\bullet}(\mathcal{L}),\omega) = 0, \tag{5.2.1}$$

provided condition (5.1.5) of the theorem is met. Indeed, for a generic plane $H \subset \mathbb{C}^n$ the map $A^i(\mathcal{L}) \to A^i(\mathcal{L} \cap H)$ induced by inclusion is isomorphism for $i = 0, 1$ and injective for $i = 2$. The latter follows from the Lefschetz Theorem since $A^i(\mathcal{L}) = H^i(\mathbb{C}^n - \mathcal{L})$ by (Brieskorn). This yields the isomorphism of cohomology of the complexes $A^\bullet$ for $\mathcal{L}$ and $\mathcal{L} \cap H$ and hence (5.2.1). Therefore the theorem from (5.1) (and also from (5.4)) holds for arbitrary arrangements.

5.3. PROOF

We will derive this theorem from the following:
a)$H^0(A^\bullet_\omega) = H^1(A^\bullet_\omega) = 0$
b)The euler characteristics of both $A^\bullet_\omega$ and $\mathcal{A}_{s_*}$ are equal to $e(\mathbb{C}^2 - \mathcal{L})$.
To show b) note that the euler characteristic of $A^\bullet$ is $e(\mathbb{C}^2 - \mathcal{L})$ by (Brieskorn) and for $H^i(\mathcal{A}_{s_*})$ this can be seen by looking at the cochain complex of $\mathcal{A}_{s_*}$, i.e. (here χ_{s_*} is the character of the fundamental group defining the local system $\mathcal{A}$)

$$C^i((\mathbb{C}^2 - \mathcal{L})^\sim \otimes_{\chi_{s_*}} \mathbb{C}) = \mathbb{C}^{b_i} \otimes \mathbb{C}[H_1(\mathbb{C}^2 - \mathcal{L})] \otimes_{\chi_{s_*}} \mathbb{C} \tag{5.3.1}$$

since the multiplicity of a representation in the regular representation of an abelian group is 1.

On the other hand a) follows since $\dim H^1(A^\bullet, \omega) \leq \dim H^1(\mathcal{A}_{s_*})$ (Libgober and Yuzvinsky) and the latter group is trivial if (5.1.5) is satisfied.

Now Theorem 5.1 is a consequence of a),b) and (1.4.1.1).

5.4. COMPLEXES (5.1.2) WITH NON VANISHING COHOMOLOGY

Complexes $A^\bullet$ for a fixed arrangement $\mathcal{L}$ are parameterized by the space $H^0(\mathbb{P}^2, \Omega^1(\log(\mathcal{L} \cup L_\infty)) = H^1(\mathbb{C}^2 - \mathcal{L}, \mathbb{C})$. Let

$$\mathcal{V}_i = \{\omega \in H^0(\mathbb{P}^2, \Omega^1(\log(\mathcal{L} \cup L_\infty)))| H^1(A^\bullet, \omega) \geq i\} \tag{5.4.1}$$

THEOREM 5.4.1. *$\mathcal{V}_i$ is a union of linear space of dimension $i+1$. There is the one to one correspondence between these linear spaces and irreducible components of the characteristic variety V_i containing the trivial character which have a positive dimension. For such an irreducible component of $\mathcal{V}_i$ the map which assigns to a form ω the point $(\ldots, \exp(2\pi i Res_{L_i} \omega), \ldots)$ in $H^1(\pi_1(\mathbb{C}^2 - \mathcal{L}), \mathbb{C}^*)$, by identification (1.4.1), is the universal cover of the corresponding component of V_i.*

PROOF. If $\dim H^1(A^\bullet, \omega) > 0$, i.e. there exist linearly independent with ω form η such that $\eta \wedge \omega = 0$, then for any form ω' in the space spanned by ω and η one has $\dim H^1(A^\bullet, \omega') > 0$. Let $\mathcal{V}$ be an irreducible component of $\mathcal{V}_i$ in (5.4.1) containing ω and having dimension $k \geq 2$. For the local system $L_{\omega'}$ corresponding to each $\omega' \in \mathcal{V}$ we have $\dim H^1(L_{\omega'}) > 0$. Indeed we can assume that ω' is generic since this only decrease $H^1(L_{\omega'})$, cf. (Libgober and Yuzvinsky). On the other hand for generic ω', according to (Esnault, Schechtman and Viehweg), we have $\dim H^1(A^\bullet,\omega') = \dim H^1(L_{\omega'})$. Therefore L_ω belongs to an irreducible component, say V, of the characteristic variety of $\mathbb{C}^2 - \mathcal{L}$. Since the exponential map is a local homemorphism this component has the dimension equal to at least k. In fact the dimension of this component is exactly k. Assume to the contrary that this dimension is $l > k$ and let $f : \mathbb{C}^2 - \mathcal{L} \to \mathbb{P}^1 - \bigcup_{i=1}^{i=k+1} p_i$ be the map on a curve of general type (cf. (1.4.2) and (Arapura, Prop. 1.7)) corresponding to the component V. Then the pull back of form $H^0(\mathbb{P}^1,\Omega^1(\log(\cup p_i)))$ gives l-dimensional space of forms on $\mathbb{C}^2 - \mathcal{L}$ for which the wedge with ω is zero (note that the map f^* is injective on H^1) and we have a contradiction. Let $t_1^{a_{1,j}} \cdots t_r^{a_{r,j}} = 1 (j = 1,\ldots,s)$ be the equations defining V (cf. 4.2). Then ω belongs to the union of affine subspaces of $H^0(\mathbb{P}^2,\Omega^1(\log(\mathcal{L} \cup L_\infty))$ given by $Q_j = \sum a_{i,j}x_i = n_j, n_j \in \mathbb{Z}, j = 1,\ldots,s$. Since $\dim H^1(A^\bullet,\lambda\omega) = \dim H^1(A^\bullet,\omega), \lambda \in \mathbb{C}^*$ we see that $n_j = 0$ for any j. Hence $\mathcal{V}$ is a linear space of dimension k (and $i = \dim H^1(A^\bullet,\omega) = \dim H^1(L_\omega) = k-1$). $\square$

5.5. COMBINATORIAL CALCULATION OF CHARACTERISTIC VARIETIES

A consequence of the Theorem 5.4 is that the irreducible components of the characteristic varieties containing the identity element of $H^1(\mathbb{C}^2 - \mathcal{L},\mathbb{C}^*)$ are determined by the cohomology of the complex (5.1.2). $H^1(A^\bullet,\omega)$ is the quotient of $\{\eta \in A^1 | \eta \wedge \omega = 0\}$ by the subspace spanned by ω and can be calculated as follows.

It is easy to see that a 2-form is cohomologous to zero iff its integrals over all 2-cycles belonging to small balls about the multiple points of the arrangement are zeros. The group of such 2-cycles near a point which is the intersection of the lines $l_{i_1},\ldots,l_{i_m}$ are generated by $\gamma_{i_j} \times (\gamma_{i_1} + \cdots + \gamma_{i_m})$. If $(\sum A_i\eta_i) \wedge \omega$ is cohomologous to zero in $\Omega^2(\mathbb{C}^2 - \mathcal{L})$ then vanishing of $\int A_j\eta_j \wedge s_i\eta_i$ over those 2-cycles yields:

$$A_j(\sum s_i) - (\sum A_j)s_j = 0 \tag{5.5.1}$$

Therefore we obtain

$$A_j = C_\upsilon s_j, \qquad \text{if } \sum_{\upsilon \in l_j} s_j \neq 0, \tag{5.5.2}$$

$$\sum A_j = 0, \qquad \text{if } \sum_{\upsilon \in l_j} s_j = 0, \tag{5.5.3}$$

for vertices υ of the arrangements. If we are looking for essential components of the characteristic variety (which we always can assume) then $s_i \neq 0$ and condition (5.5.2) can be replaced by

$$\frac{A_j}{s_j} = \frac{A_{j'}}{s_{j'}}, \qquad \text{if } \sum_{\upsilon \in l_j} s_j \neq 0. \tag{5.5.4}$$

Now for each subset of the set of vertices such that the system of equation (5.5.3) and supplementing it by equations (5.5.4) for vertices outside of selected subset has a solution non proportional to $(s_1, \ldots, s_r)$ we obtain a component $\mathcal{V}$ and hence corresponding component of the characteristic variety. We leave as an exercise to the reader to work out calculations of the characteristic varieties for the examples from section 3 using this method.

References

Arapura, D. (1997) Geometry of cohomology support loci for local systems, I, *J. Alg. Geom.* **6**, 563–597.

Buchsbaum, D., and Eisenbud, D. (1977) What annihilates a module?, *J. Algebra* **47**, 231–243.

Barthel, G., Hirzebruch, F., and Hofer, T. (1987) *Geradenkonfiguarationen und Algebraishe flächen*, Vieweg Publishing, Wiesbaden.

Blass, P., and Lipman, J. (1979) Remarks on adjoints and arithmetic genus of algebraic varieties, *Amer. J. of Math.* **101**, 331–336.

Brieskorn, E. (1973) Sur les groups des tresses, in *Sem. Bourbaki*, Lect. Notes in Math. 317, Springer, Berlin, pp. 21–44.

Cartan, H., and Eilenberg, S. (1956) *Homological Algebra*, Princeton University Press, Princeton.

Cohen, D., and Suciu, A. (1995) On Milnor fibrations of arrangements, *J. London Math. Soc.* **51**, 105–119.

Cohen, D., and Suciu, A. (1999) Characteristic varieties of arrangements, *Math. Proc. Cambridge Phil. Soc.* **127**, 33–53.

Deligne, P. (1971) Theorie de Hodge II, *Publ. Math. I.H.E.S.* **40**, 5–58.

Esnault, H., Schechtman, V., and Viehweg, E. (1992) Cohomology of local systems on the complement to hyperplanes, *Inv. math.* **109**, 557–561.

Falk, M. (1997) Arrangements and cohomology, *Ann. Comb.*, **1**, 135–157.

Grauert, H., and Riemenschneider, O. (1970) *Inv. Math.* **11**, 263–292.

Hartshorne, R. (1977) *Algebraic Geometry*, Springer Verlag, Berlin.

Hillman, J. (1981) *Alexander Ideals of Links*, Lecture Notes in Math. 895, Springer Verlag, Berlin.

Hironaka, E. (1997) Multi-polynomial invariants of plane algebraic curves, in *Singularities and Complex Geometry*, Studies in Advanced Mathematics 5, AMS and International Press, 67–74.

Ishida, M. (1983) The irregularity of Hirzebruch's examples of surfaces of general type with $c_1^2 = 3c_2$, *Math. Ann.* **262**, 407–420.

van Kampen, E.R. (1933) On the fundamental group of an algebraic curve, *Amer. J. of Math.* **55**, 255–260.

Kollar, J. (1995) *Shafarevich maps and automorphic forms*, Princeton University Press.

Libgober, A. (1982) Alexander polynomials of plane algebraic curves and cyclic multiple planes, *Duke Math. J.* **49**, 833–851.

Libgober, A. (1983) Alexander invariants of plane algebraic curves, *Proc. Symp. Pure Math.* **40**, AMS. Providence, RI, 135–143.

Libgober, A. (1992) On homology of finite abelian coverings, *Topology and Applications* **43**, 157-166.

Libgober, A. (1994) Groups which cannot be realized as fundamental groups of the complements to hypersurfaces in $\mathbb{C}^N$, in C. Bajaj (ed.), *Algebraic Geometry and Application*, Springer Verlag, Berlin, pp. 203–207.

Libgober, A. (1996) Position of singularities of hypersurfaces and the topology of their complements, *J. Math. Sciences* **82**, 3194–3210.

Libgober, A. (1999) Abelian Covers of projective plane, in *Singularity theory (Liverpool, 1996)*, xxi, London Math. Soc. Lecture Note Ser. 263, Cambridge Univ. Press, Cambridge, pp. 281–289.

Libgober, A. (2001) Hodge decomposition of Alexander invariants, Preprint.

Libgober, A., and Yuzvinsky, S. (2000) Cohomology of the Brieskorn-Orlik-Solomon algebra and local systems, in *Arrangements—Tokyo 1998*, Adv. Stud. Pure Math. 27, Kinokuniya, Tokyo, pp. 169–184.

Loeser, F., and Vaquie, M. (1990) Le polynome d'Alexander d'une courbe plane projective, *Topology* **29**, 163–173.

Nori, M. (1983) Zariski's conjecture and related problems, *Ann. Sci. Ecole Norm. Sup.* **16**, 305–344.

Sabbah, C. (1990) Module d'Alexander et $\mathcal{D}$-modules, *Duke Math. Journal* **60**, 729–814.

Sakuma, M. (1995) Homology of abelian coverings of links and spatial graphs, *Canadian Journal of Mathematics* **17**, 201–224.

Serre, J.P. (1975) *Algebre locale. Multiplicites*, Lecture Notes in Mathemematics 11, Springer Verlag, Berlin.

Steenrod, N. (1945) Homology with local coefficients, *Ann. Math.* **44**, 610–627.

Sumners, D., and Woods, J. (1977) The monodromy of reducible plane curves, *Inv. Math.* **40**, 107–141.

Zariski, O. (1971) *Algebraic surfaces*, Chapter 8, Springer Verlag, Berlin.

Zuo, K. (1989) Kummer Oberlagerungen algebraischer Flächen, *Bonner Mathematische Schriften*, Bonn.

COMMUNICATION NETWORKS AND HILBERT MODULAR FORMS

R. LIVNÉ
The Hebrew University of Jerusalem, Israel

Abstract. Ramanujan graphs, defined and constructed by Lubotzky, Phillips and Sarnak, allow the design of efficient communication networks. In joint work with B. Jordan we gave a higher-dimensional generalization. Here we explain how one could use this generalization to construct efficient communication networks which allow for a number of verification protocols and for the distribution of information along several channels. The efficiency of our network hinges on the Ramanujan-Petersson conjecture for certain Hilbert modular forms. We obtain this conjecture in sufficient generality to apply to some particularly appealing constructions, which were not accessible before.

Key words: Ramanujan local systems, cubical complexes, quaternion algebras, spectrum of the Laplacian.

Mathematics Subject Classification (2000): 68R05, 11R80, 11R52.

Introduction

The concept of a Ramanujan graph was introduced and studied by (Lubotzky, Phillips and Sarnak), shortly (LPS). These are r-regular graphs for which the nontrivial eigenvalues $\lambda \neq \pm k$ of the adjacency matrix satisfy the bounds $|\lambda| \leq 2\sqrt{r-1}$. In many aspects these bounds are optimal and natural. For example, the adjacency matrix is the combinatorial analog of the Laplace operator, and the bounds parallel the (conjectured) Selberg bounds for the Laplacian on Riemann surfaces. The main result of (LPS) was an explicit construction of $p+1$-regular such graphs, $p \equiv 1 \pmod 4$ a prime, through the arithmetic of quaternion algebras over the rational numbers. From the point of view of Communication Network Theory, the arithmetic examples are particularly interesting: all Ramanujan graphs are super-expanders; but in addition the examples have many other useful properties, for example very good expansion constants and large girth. Thus they can be used to design efficient communication networks.

The Ramanujan property for the (LPS) examples hinges on the truth of the Ramanujan-Petersson conjecture for an appropriate space of modular forms of weight 2 over $\mathbb{Q}$. Let f be a weight 2 holomorphic cuspidal Hecke eigenform on

C. Ciliberto et al. (eds.),
***Applications of Algebraic Geometry to Coding Theory, Physics and Computation,* 255–270.**

a congruence subgroup. The conjecture is that for any prime p not dividing the level the Hecke eigenvalue a_p satisfies $|a_p| \leq 2\sqrt{p}$. Eichler and Shimura reduced the conjecture to Weil's results on the absolute value of Frobenius eigenvalues for curves over a finite fields. They achieved this for a fixed form and all but a finite, unspecified set of primes p. The proof requires a deep study of the reduction modulo p of modular curves and Hecke correspondences. Igusa subsequently showed that the method applied for all primes not dividing the level of the form, see (Deligne, 1969) for a more modern exposition and a generalization to forms over $\mathbb{Q}$ of any weight. Igusa's part is essential for the applications to communication networks because there one first chooses the prime p and only then an increasing sequence of levels prime to it. This gives an infinite family of Ramanujan graphs of *fixed* regularity $p+1$.

In a joint work with B. Jordan (2000) we gave a higher dimensional generalization of this theory to $(r_1,\ldots,r_g)$-regular cubical complexes. These are cell complexes locally isomorphic to a product of r_i-regular trees. Here there are partial Laplacians, one for each tree factor. The parallel bounds for their eigenvalues define the notion of being Ramanujan. We then constructed infinite towers of explicit arithmetic examples whenever each r_i is of the form q_i+1 for a prime power q_i. We used quaternion algebras over totally real fields and hence our generalization required the Ramanujan-Petersson conjecture for Hilbert modular forms of multi-weight $(2,\ldots,2)$. In place of the work of Eichler, Shimura and Igusa we invoked analogous results of (Carayol) on the reduction modulo p of Shimura curves over totally real number fields, which as before allow to use Weil's results.

Carayol's method imposes on the form a technical assumption which forced us to exclude certain natural and particularly appealing examples. Building on ideas of Langlands, (Brylinski and Labesse) obtained results which are free from such assumptions, by reducing the conjecture to Deligne's bounds for Frobenius eigenvalues on the intersection homology of Hilbert-Blumenthal varieties. However in their proof they used in an essential way the Satake-Baily-Borel compactification which was available only over $\mathbb{Q}$. This forced them to exclude for any fixed form an unspecified finite set of primes, which makes their results inapplicable to us as was explained above. One of our aims here is to show that a modification of these arguments proves the following

THEOREM 0.1. *Let F be a totally real field of degree d over $\mathbb{Q}$. Let f be a holomorphic Hilbert cuspidal Hecke eigenform over F of multi-weight $(k_1,\ldots,k_d)$, where the k_i's are integers ≥ 2 and all of the same parity. Let v be a prime of F of degree $d_v = [F_v : \mathbb{Q}_p]$ whose residual characteristic p is prime to the discriminant* $\mathrm{Disc}\,F$ *of F and to the level $N = N(f)$ of f. Let $\lambda_v(f)$ denote the eigenvalue of the Hecke operator T_v belonging to f. Then the Ramanujan-Petersson conjecture holds for f and v, namely we have $|\lambda_v(f)| \leq 2p^{(k-1)d_v/2}$, where $k = \max_i k_i$.*

For the theorem to be meaningful we must specify the normalizations made in defining T_v and λ_v. However Hilbert modular forms are best approached through representation theory, and we have in fact opted to define T_v and λ_v only representation-theoretically. This spares us the tedious task of making "classical" definitions and then comparing them to the representation-theoretic ones. The version of the theorem we will actually prove is therefore Theorem 2.4 below.

Here is an outline of the article. In Section 1 we discuss the significance of the higher dimensional theory of (Jordan and Livné, 2000) to communication networks, especially in the two-dimensional case. In Section 2 we prove the above theorem. This enables us to handle in Section 3 a particularly pretty example which was left unsettled in (Jordan and Livné, 2000). The result is a mixture of graph theory, automorphic forms, algebraic geometry, and the arithmetic of quaternions over number fields. In our attempt to make this work accessible to a mixed audience and yet not too lengthy we have undoubtedly failed to give the right level of detail to any particular reader. We can only ask for indulgence in this matter.

The debt this work owes to my long term collaboration with B. Jordan, in particular to the paper (2000), should be evident. I started thinking about this problem at the instigation of P. Sarnak. Different approaches to the problem were subsequently suggested by D. Blasius and C.-L. Chai (independently). P. Deligne suggested to use compactly supported cohomology. Conversations with them and with J. Bernstein, G. Faltings, G. Harder, and N. Katz were encouraging and helpful. Y. Glasner made useful comments on a previous version on the manuscript. It is a pleasure to thank them all.

1. Cubical complexes and communication networks

A basic problem in communication networks is to design explicit *super-expanders*. For a given, arbitrarily large set of nodes one wants to connect each node to a fixed number r of "neighbor" nodes, so that information can spread fast over the resulting network, i.e. N. A commonly used quantitative measure of efficiency is the expansion constant — the largest real constant $c > 0$ so that each subset A of the nodes of N having $|A| \leq |N|/2$ nodes has at least $c|A|$ (new) neighbors. One seeks sequences of networks N_k where $|N_k| \to \infty$ with k, while c is independent of k and is as large as possible. Even though one can define the meaning of a *random* network on n vertices and prove that random networks are good expanders, it is quite hard to get explicit examples. The problem whether a random network is Ramanujan is open, although it is known to be "almost Ramanujan" as $r \to \infty$. More precisely, $\lambda \leq 2\sqrt{r-1} + \log(r-1) + \text{Const}$ with probability $\to 1$ with N, see Friedman.

In this work we shall explain how to design any number $g \geq 1$ of efficient communication networks on the same arbitrarily large set N_k of nodes. Thinking of the connections of each network as having a different color, each node will have a fixed number r_i of color i neighbors. Each color will be a Ramanujan network, and hence a super-expander. But in addition, any set of h edges of different colors starting at the same vertex can be completed to an h-cube. For example, given a red edge e_r and a blue edge e_b emanating from the same vertex, there will be a unique pair of red and blue edges e'_r, e'_b completing them to a square. In other words, the origins $o(.)$ and ends $t(.)$ of the edges will satisfy $o(e'_b) = t(e_r)$, $o(e'_r) = (e_b)$, and $t(e'_b) = t(e'_r)$. We say that the network satisfy the cube property (or the square property for 2 colors). Like the (LPS) examples, the ones we give have good expansion properties and large girth.

In practical applications the cube property permits a variety of potential applications

1. To send information along (say) the red network, and verify, using a hash function, a compatibility condition via the blue channel.
2. To send divide the information to g parts and send each part individually. One could arrange so that only the combination of the various elements would lead to a meaningful whole.

It seems a difficult problem to construct by random methods systems of $(k_1, \ldots, k_g)$-regular networks satisfying the cube condition or to parametrize the space of such systems. The key point is to realize that the cube condition makes natural the introduction of higher dimensional cubical complexes into the situation. Let $r_i \geq 3$ be a sequence of length g of integers, let $\mathbf{T}_{r_i}$ be a regular tree of regularity r_i, and set $\mathbf{T} = \prod_i \mathbf{T}_{r_i}$. Then an $(r_1, \ldots, r_g)$-regular cubical complex in the sense of (Jordan and Livné, 2000) is a complex X in which each connected component is isomorphic to a quotient $\Gamma \backslash \mathbf{T}$ for a discrete, torsion free subgroup of $\prod_i \mathrm{Aut}\,\mathbf{T}_{r_i}$. For simplicity we also require the *parity condition*: the ith component of any element of Γ moves each vertex of $\mathbf{T}_{r_i}$ to an even distance. The 1-skeleton of X_I is a collection of g graphs, or communication networks, on the same set of vertices. The parity assumption makes these graphs bipartite. The vertices have a "multiparity", an element of $\{0,1\}^g$; the ith graph is r_i-regular, and its 2^{g-1} connected components consist of the vertices in which all parities except for the ith have been fixed. Its edges come from the ith tree factor of $\mathbf{T}$. Most importantly, the g graphs satisfy the cube property.

Theorem 3.1 of loc.cit. provides us with the following general class of arithmetic examples:

THEOREM 1.1. *Let B be a totally definite quaternion algebra over a totally real field F, and let $S = \{v_1, \ldots, v_g\}$ be a nonempty set of g distinct finite primes of of F so that B is split at each v_i. Let q_i be the cardinality of the residue field of F_{v_i} and set $r_i = q_i + 1$. let Γ be an S- arithmetic torsion-free congruence subgroup of the algebraic group over $\mathbb{Q}$ of the norm 1 elements in B. Then Γ acts on $\mathbf{T} = \prod_i \mathbf{T}_{r_i}$ and the resulting quotient $\Gamma \backslash \mathbf{T}$ is an irreducible $(r_1, \ldots, r_g)$-regular complex with parities. Moreover if an appropriate space of Hilbert cusp forms* *(of weight* $(2,\dots,2)$*) satisfies the Ramanujan-Petersson conjecture, then* $\Gamma\backslash\mathbf{T}$ *is Ramanujan.*

The assumptions on B mean that $B\otimes F_v$ is isomorphic to the Hamilton quaternion algebra $\mathbb{H}$ for any infinite prime v of F, and that $B\otimes F_{v_i}$ is isomorphic to $GL(2,F_{v_i})$. Let $\mathcal{O}$ be the order of elements of F integral away from $v_1,\dots,v_g$ and let $\mathcal{M}$ be a maximal $\mathcal{O}$-order of B. Then Γ is taken to be a sufficiently small congruence subgroup of the group of norm 1 elements of $\mathcal{M}$.

Our complexes, defined as quotients of infinite complexes by infinite groups, also admit a finite description, which is more elementary and very convenient for explicit calculation. This finite description takes a particularly simple shape if certain assumptions are made, see (Jordan and Livné, 2000) for the details: let us assume there is an ideal $N_0\neq 0$ of $\mathcal{O}_F$, prime to the v_j's (we allow $N_0=\mathcal{O}_F$), such that the following holds:

CONDITIONS 1.2. *1. Every ideal of F has a totally positive generator that is congruent to* 1 *modulo* N_0.
2. *The class number of B is* 1.
3. *For a maximal order* $\mathcal{M}$ *of B, the map* $\mathcal{M}^\times\to(\mathcal{M}/N_0\mathcal{M})^\times$ *is onto, with kernel contained in the center* $\mathcal{O}_F^\times$ *of* $\mathcal{M}^\times$.

Fix a totally positive generator π_j which is $\equiv 1 \pmod{N_0}$ for each prime ideal v_j. We then have the following:

PROPOSITION 1.3. 1. *For each* $1\le j\le g$ *there are exactly* r_j *(principal) ideals* $P_{j,i}$, $1\le i\le r_j$ *of* $\mathcal{M}$ *with norm* v_j. *We can moreover choose generators* $\varpi_{j,i}\equiv 1 \pmod{N_0\mathcal{M}}$ *for* $P_{j,i}$ *whose norm is* π_j.
2. *For every permutation* σ *of* $\{1,\dots,g\}$ *and any sequence of indices* $i_1,\dots,i_g$, *with* $1\le i_j\le r_j$, *there is a (unique) sequence* $i'_1,\dots,i'_g$, *with* $1\le i'_j\le r_j$, *and a (unique) unit* $u\in\mathcal{O}_F^\times$, *satisfying* $u\equiv 1 \pmod{N_0}$, *so that we have the following arithmetic counterpart of the cube condition:*

$$\varpi_{\sigma(1),i_1}\cdots\varpi_{\sigma(g),i_g}=u\varpi_{1,i'_1}\cdots\varpi_{g,i'_g}\,. \tag{1.1}$$

Let N_1 be a prime ideal of $\mathcal{O}_F$ prime to the v_j's and to N_0, and set $N=N_0N_1$. Let A be the subgroup of $(\mathcal{O}_F/N_1)^\times$ generated by the images modulo N_1 of the π_j's. Let B be the subgroup of scalars in $(\mathcal{M}/N_1\mathcal{M})^\times$ generated by the images modulo $N_1\mathcal{M}$ of the π_j's and by those units of $\mathcal{O}_F$ which are $\equiv 1 \pmod{N_0}$. Set

$$H=\{g\in(\mathcal{M}/N_1\mathcal{M})^\times\mid \mathrm{Nm}(g)\in A\}/B,$$

that is isomorphic either to $SL_2(\mathcal{O}_F/N_1)$, $PSL_2(\mathcal{O}_F/N_1)$, or $PGL_2(\mathcal{O}_F/N_1)$. (The examples of Section 3 are of the PSL_2 type.) We now have the following

PROPOSITION 1.4. 1. *The vertices of* $X(N)$ *are the elements of* H.
2. *The (oriented) edges of direction* j *of* $X(N)$ *are the pairs* (v,i_j), *where* $v\in H=\mathrm{Ver}\,X(N)$ *and* $1\le i_j\le r_j$.

In other words, each of the graphs is the Cayley graph for *the same* group H for a different set of generators. This description is particularly convenient for explicit calculation. Let us emphasize that a finite description exists even when condition (1.2) does not hold (see (Jordan and Livné, 1997) for the form it takes in the prototypical one-dimensional case), but it is messier. By (Jordan and Livné, 2000, Theorem 3.1(4)) the resulting graphs $\mathrm{Gr}_i(X)$ are Ramanujan provided that appropriate spaces of cusp forms satisfy the Ramanujan-Petersson conjecture. The statement that fixing the parities gives connectedness follows from Theorem 3.1(1) there.

2. Hilbert modular forms

2.1. HILBERT-BLUMENTHAL SCHEMES

For the generalities on Hilbert-Blumenthal schemes which follow see (Deligne and Pappas; Rapoport). Let $\mathcal{H}_\pm = \mathbb{P}^1(\mathbb{C}) - \mathbb{P}^1(\mathbb{R})$ be the union of the upper and the lower half complex planes. Let Σ be the set of the d embeddings of F into a Galois closure F^{Gal}, itself considered as a subfield of $\mathbb{R}$. Denote by $\mathbb{G}$ the group scheme $\mathrm{Res}_{\mathcal{O}_F/Z}\mathrm{GL}_2$ over Z. Then $\mathbb{G}(\mathbb{R})$ is isomorphic to $\mathrm{GL}_2(\mathbb{R})^\Sigma$, and $\mathbb{G}(\mathbb{R})$ acts on $\mathcal{D} = (\mathcal{H}_\pm)^\Sigma$ componentwise through the resulting d Möbius transformations. Let U be a compact open subgroup of the finite adèles $\mathbb{G}(\mathbb{A}_\mathrm{f}) = \mathrm{GL}_2(\mathbb{A}_\mathrm{f} \otimes F)$. Then the Hilbert-Blumenthal complex space $\mathcal{X}_U^{\mathrm{an}} = \mathbb{G}(\mathbb{Q})\backslash(\mathbb{G}(\mathbb{A}_\mathrm{f})/U \times \mathcal{D})$ is nonsingular if U is sufficiently small. It has a natural structure of a quasi-projective variety which admits a canonical model over $\mathbb{Q}$ in the sense of Shimura (Deligne, 1971). In fact, if U contains the principal congruence subgroup of level N and $D = \mathrm{Disc}\,F$ then there is a moduli-theoretic interpretation of $\mathcal{X}_U^{\mathrm{an}}$ as a coarse moduli space, parametrizing principally polarized abelian d-folds with an $\mathcal{O}_F$-action and some level N structure. This yields a model $\mathcal{X}_U$ over $\mathbb{Z}[1/ND]$ so that $\mathcal{X}_U \times \mathbb{Q}$ is Shimura's canonical model. If U is sufficiently small then $\mathcal{X}_U$ is a fine moduli scheme, smooth and quasi-projective over $\mathbb{Z}[1/ND]$. The Hecke algebra $\mathbb{T}_U$ of distributions on $U\backslash\mathbb{G}(\mathbb{A}_\mathrm{f})/U$ acts on $\mathcal{X}_U \times \mathbb{Q}$ through correspondences. We shall suppose that U is a product U_pU^p of a part $U_p \subset \mathbb{G}(\mathbb{Q}_p)$ at p and a part $U^p \subset \mathbb{G}(\mathbb{A}_f^p)$ away from p. Then an element of $\mathbb{T}_U$ coming from $U_p\backslash\mathbb{G}(\mathbb{Q}_p)/U_p$ (which we will call a Hecke operator at p) acts on $\mathcal{X}_U \times \mathbb{Z}[1/NDp]$ through its standard moduli interpretation.

Now let p be a rational prime not dividing ND. Then $\mathcal{X}_U$ has good reduction modulo p. In (Langlands) the number of fixed points $T \times \mathrm{Frob}_p$ on $\mathcal{X}_U \times \overline{\mathbb{F}}_p$ was computed for any Hecke operator T away from p, i.e. T coming from $U^p\backslash\mathbb{G}(\mathbb{A}_f^p)/U^p$. In fact Langlands allows more general groups than GL_2 and he also allows algebraic local system on $\mathcal{X}_U$. This second generalization enables one to handle cusp forms of weight $(k_1,\ldots,k_d)$ rather than $(2,\ldots,2)$. Even though the trivial local system in the GL_2 case is all that we need here, there is no advantage in restricting to it, and it has a potential use for the Ramanujan local systems of (Jordan and Livné, 2000). On the other hand we decided to restrict to the GL_2 case so as to simplify our exposition by avoiding all mention of L-packets and endoscopy. The interested reader can see (Brylinski and Labesse; Langlands) that this restriction is unnecessary. Thus let ξ be an irreducible algebraic representation of $\mathbb{G}$ (all are defined over the Galois closure F^{Gal} of F). We will always assume that U is sufficiently small and that ξ is trivial on the (Zariski closure) of the units $\mathcal{O}_F^\times$ in the center Z_G of $\mathbb{G}$. Then ξ defines a local system

$$\mathbb{V}_\xi^{\mathrm{an}} = \mathbb{G}(\mathbb{Q})\backslash(\mathbb{G}(\mathbb{A}_f)/U \times \mathcal{D} \times \xi)$$

on $\mathcal{X}_U^{\mathrm{an}}$. The possible ξ's can be described explicitly: identify $\mathbb{G} \times F^{\mathrm{Gal}} \simeq GL_2^d$ and let Symm^k be the representation of GL_2 on the polynomials of degree $k \geq 0$ in two variables. Then ξ is a tensor product $\otimes_{i=1}^d \xi_i$ of irreducible representations ξ of the GL_2 factors, and $\xi_i \simeq \mathrm{Symm}^{k_i-2} \otimes \det^{\alpha_i}$, with $k_i \geq 2$. The condition of being trivial on the units of Z_G translates into $k_i - 2 + 2\alpha_i = C$ with C a constant. The k_i's must all have the same parity as C, and $\alpha_i = (C + 2 - k_i)/2$. Observe that the central character of ξ maps $t = (t_1, \ldots, t_d)$ to $\mathrm{Nm}(t)^C$, where $\mathrm{Nm}(t) = t_1 \ldots t_d$.

The ξ's above have a moduli theoretic interpretation and hence ℓ-adic analogs $\mathbb{V}_{\xi,\lambda}$, which are lisse sheaves of 2-dimensional F_λ^{Gal} vector spaces over $\mathcal{X}_U[1/\ell]$. Here λ is any prime of F^{Gal} lying above ℓ. For this let $f : \mathcal{A} \to \mathcal{X} = \mathcal{X}_U[1/\ell]$ be the universal abelian variety (which we may assume exists by adding auxiliary level structure). Then $R^1 f_* \mathbb{Q}_\ell$ is a lisse sheaf of $F \otimes \mathbb{Q}_\ell$-modules of rank 2. For every $\sigma \in \Sigma$ we have an $F \otimes \mathbb{Q}_\ell$-action on F_λ^{Gal}, and we set $\mathbb{V}_{\sigma,\lambda} = R^1 f_* \mathbb{Q}_\ell \otimes_{F \otimes \mathbb{Q}_\ell} F_\lambda^{\mathrm{Gal}}$, with the tensor product taken relative to this action. Then for ξ as before we set $\mathbb{V}_{\xi,\lambda} = \otimes_i (\mathrm{Symm}^{k_i} \mathbb{V}_{\sigma_i,\lambda} \otimes (\det \mathbb{V}_{\sigma_i,\lambda})^{\alpha_i})$ with $\sigma_i \in \Sigma$ the embedding corresponding to the index $1 \leq i \leq d$. The moduli description gives that $\mathbb{V}_{\xi,\lambda}$ is pure (in the sense of (Deligne, 1981)) of weight $\sum_{i=1}^d (k_i - 2 + 2\alpha_i) = dC$.

2.2. THE DUAL GROUP

Let p be a rational prime not dividing $\mathrm{Disc}\, F$ and fix some embedding of $\mathbb{R}$ into $\overline{\mathbb{Q}}_p$. Then $\mathbb{G}(\mathbb{Q}_p) \simeq \prod_{v|p} \mathbb{G}_v(\mathbb{Q}_p)$, where $\mathbb{G}_v = \mathrm{Res}_{F_v/\mathbb{Q}_p} GL_2$. The set Σ is a disjoint union $\sqcup_v \Sigma_v$, where Σ_v is the set of the $d_v = [F_v : \mathbb{Q}_p]$ embeddings of F_v in $\overline{\mathbb{Q}}_p$. The connected component ${}^L\mathbb{G}_v(\mathbb{Q}_p)^0$ of the Langlands dual to $\mathbb{G}_v(\mathbb{Q}_p)$ is then the quotient of $GL_2^{\Sigma_v}$ by the Σ_v-tuples $(\ldots, z_\sigma, \ldots)_{\sigma \in \Sigma_v}$ of scalar matrices whose product is the identity. Similarly ${}^L\mathbb{G}(\mathbb{Q}_p)^0$ is the quotient of GL_2^Σ by the Σ-tuples $(\ldots, z_\sigma, \ldots)_{\sigma \in \Sigma}$ of scalar matrices whose product is the identity. In particular, ${}^L\mathbb{G}(\mathbb{Q}_p)^0$ is a quotient of $\prod_{v|p} {}^L\mathbb{G}_v(\mathbb{Q}_p)^0$. The Galois group $\mathrm{Gal}_{\mathbb{Q}_p} = \mathrm{Gal}(\overline{\mathbb{Q}}_p/\mathbb{Q}_p)$ acts naturally on each Σ_v and hence on their union Σ, and it is clear that we get an action of $\mathrm{Gal}_{\mathbb{Q}_p}$ on ${}^L\mathbb{G}_v(\mathbb{Q}_p)$ and on ${}^L\mathbb{G}(\mathbb{Q}_p)$. Then ${}^L\mathbb{G}_v(\mathbb{Q}_p)$ is the semidirect product ${}^L\mathbb{G}_v(\mathbb{Q}_p)^0 \rtimes \mathrm{Gal}_{\mathbb{Q}_p}$ and likewise ${}^L\mathbb{G}(\mathbb{Q}_p) = {}^L\mathbb{G}(\mathbb{Q}_p)^0 \rtimes \mathrm{Gal}_{\mathbb{Q}_p}$. It is clear that ${}^L\mathbb{G}_v(\mathbb{Q}_p)^0$ acts naturally on $\mathbb{C}^{2^{d_v}}$ and that this extends to an embedding r_v of ${}^L\mathbb{G}_v(\mathbb{Q}_p)$ into $\mathrm{GL}_{2^{d_v}}(\mathbb{C})$. Likewise we get an injection

$$r: {}^L\mathbb{G}(\mathbb{Q}_p) \to \mathrm{GL}_{2^d}(\mathbb{C}).$$

Now let π_p be an irreducible admissible representation of $\mathbb{G}(\mathbb{Q}_p)$ unramified at p. Then π_p is a product $\prod_{v|p} \pi_v$ of unramified principal series representations $\pi_v \simeq \pi(\mu_{1,v}, \mu_{2,v})$ of $\mathbb{G}_v(\mathbb{Q}_p) = \mathrm{GL}_2(F_v)$, with $\mu_{i,v}$ unramified characters of $F_v^\times$, see (Jacquet and Langlands, Ch. I.3). We will always assume that each $\pi(\mu_{1,v}, \mu_{2,v})$ is infinite-dimensional. Let ϖ_v be a uniformizer at v and put $a_v = \mu_{1,v}(\varpi_v)$ and $b_v = \mu_{2,v}(\varpi_v)$. Assume that the central character of π_v maps each $t \in F_v^\times$ to t^C. Then $|a_v b_v| = p^{d_v C}$. We shall be interested in the size $|\lambda_v|$ of the Hecke eigenvalue $\lambda_v = a_v + b_v$ on π_v. Recall that π_p is *tempered* if and only if all the π_v, $v|p$ are. This means that $|a_v| = |b_v|$ for all $v|p$. If this holds — which is the assertion of the Ramanujan-Petersson conjecture if π_p is a local component of a cuspidal automorphic representation — then $|\lambda_v| \leq 2p^{d_v C/2}$.

Set ${}^0t_v = \begin{bmatrix} a_v & 0 \\ 0 & b_v \end{bmatrix}$. Then the Satake isomorphism associates to π_v the conjugacy class of the image in ${}^L\mathbb{G}_v(\mathbb{Q}_p)$ of the element

$$t(\pi_v) = ({}^0t'_v, \mathrm{Frob}_p) \in \mathrm{GL}_2(\mathbb{C})^{d_v} \rtimes \mathrm{Gal}_{\mathbb{Q}_p},$$

where ${}^0t'_v = ({}^0t_v, \mathrm{Id}_{2\times 2}, \ldots, \mathrm{Id}_{2\times 2})$, see e.g. (Breen and Labesse, Exp. VI.4. It follows that the Satake isomorphism associates to π_p the element

$$t(\pi_p) = ({}^0t_p, \mathrm{Frob}_p) \in {}^L\mathbb{G}(\mathbb{Q}_p)^0 \rtimes \mathrm{Gal}_{\mathbb{Q}_p},$$

where ${}^0t_p$ is the image of $(\ldots, {}^0t'_v, \ldots)$ in ${}^L\mathbb{G}(\mathbb{Q}_p)^0$.

We now recall the formula for the trace of $r(t(\pi_p))^m$ for any integer $m \geq 1$. Put $l_v = \gcd(d_v, m)$, and write $m = q_v d_v + r_v$ with integers $q_v \geq 0$ and $0 \leq r_v < d_v - 1$. We then have

$$t'(\pi_v)^m = ((\underbrace{{}^0t_v^{q_v+1}, \ldots, {}^0t_v^{q_v+1}}_{r_v \text{ times}}, \underbrace{{}^0t_v^{q_v}, \ldots, {}^0t_v^{q_v}}_{d_v - r_v \text{ times}}), \mathrm{Frob}_p^m).$$

A straightforward calculation (see (Langlands) or loc. cit.) then gives

$$\mathrm{Tr}\, r(t(\pi_p))^m = \prod_v \mathrm{Tr}\, r_v(t(\pi_v))^m = \prod_v (a_v^{m/l_v} + b^{m/l_v})^{l_v}. \tag{1}$$

2.3. THE NAIVE LEFSCHETZ NUMBER

Write $U = U^p U_p$ with p as before, so that in particular $U_p \subset \mathbb{G}(\mathbb{Q}_p)$ and $U^p \subset \mathbb{G}(\mathbb{A}_f^p)$. Let $m \geq 0$ be an integer and take $a \in \mathbb{G}(\mathbb{A}_f^p)$. Following (Langlands; Brylinski and Labesse) one defines functions f_ξ on $\mathbb{G}(\mathbb{R})$, $\phi_a = \phi_{U^p a U^p}$ on $\mathbb{G}(\mathbb{A}_f^p)$, h_p^m on $\mathbb{G}(\mathbb{Q}_p)$ and $f^G(m,a) = f_\xi \phi_a^p h_p^m$ on $\mathbb{G}(\mathbb{A}) = \mathbb{G}(\mathbb{R})\mathbb{G}(\mathbb{A}_f^p)\mathbb{G}(\mathbb{Q}_p)$ having the following properties:

1. $\mathrm{Tr}\,\pi_\infty(f_\xi)$ is the multiplicity of ξ in π_∞ for each representation π_∞ of $\mathbb{G}(\mathbb{R})$
2. ϕ_a is the characteristic function of $U^p a U^p$;
3. $\mathrm{Tr}\,\pi_p(h_p^m) = p^{md/2}\,\mathrm{Tr}\,r(t(\pi_p)^m)$ for each unramified infinite-dimensional representation π_p.

Let $T(a)$ be the Hecke operator $T(a)$ associated to $U^p a U^p$. Then $T(a) \times \mathrm{Frob}_p^m$ acts as a correspondence on $\mathcal{X}_U \times \overline{\mathbb{F}}_p$. If m is sufficiently large then the set of fixed points of this correspondence is finite, and its graph on $(\mathcal{X}_U \times \overline{\mathbb{F}}_p)^2$ is transversal to the diagonal. This enables to define its Lefschetz number by

$$\mathrm{Lef}(T(a) \times \mathrm{Frob}_p^m, \mathcal{X}_U \times \overline{\mathbb{F}}_p, \mathbb{V}_{\xi,\lambda}) = \sum_t \mathrm{Tr}(T(a) \times \mathrm{Frob}_p^m, (\mathbb{V}_{\xi,\lambda})_t),$$

the sum taken over the (finite) fixed point set of $T(a) \times \mathrm{Frob}_p^m$ on $S(\mathcal{X}_U \times \overline{\mathbb{F}}_p)$.

Recall that the central character of ξ is $\omega_\xi(t_1, \ldots, t_d) = (t_1 \ldots t_d)^{\mathrm{C}})$. Let us denote by $\mathrm{L}_2^{\mathrm{dis}}(\mathbb{G}(\mathbb{Q})(Z(\mathbb{A}_f) \cap U)\backslash\mathbb{G}(\mathbb{A}), \omega_\xi)$ the space of functions f on $\mathbb{G}(\mathbb{A})$ satisfying

1. $f(g_\mathbb{Q} z_U z_\infty g) = \omega_\xi(z_\infty)^{-1} f(g)$ for all $g_\mathbb{Q} \in \mathbb{G}(\mathbb{Q})$, $z_U \in Z(\mathbb{A}_f) \cap U$, $z_\infty \in Z(\mathbb{R})$ and $g \in \mathbb{G}(\mathbb{A})$.
2. $f|\det|^{\mathrm{C}/2}$ is square integrable on $Z(\mathbb{R})(Z(\mathbb{A}_f) \cap U)\mathbb{G}(\mathbb{Q})\backslash\mathbb{G}(\mathbb{A})$.

As the notation indicates, $\mathrm{L}_2^{\mathrm{dis}}(\mathbb{G}(\mathbb{Q})(Z(\mathbb{A}_f) \cap U)\backslash\mathbb{G}(\mathbb{A}), \omega_\xi)$ is a discrete sum of representations of $\mathbb{G}(\mathbb{A})$.

Using the moduli interpretation (see (Milne) and (Breen and Labesse, Exposé V)), the results of Langlands (loc.cit.) together with (Brylinski and Labesse, Section 3.3) give the following key

THEOREM 2.1. *We have*

$$\mathrm{Lef}(T(a) \times \mathrm{Frob}_p^m, \mathcal{X}_U \times \overline{\mathbb{F}}_p, \mathbb{V}_{\xi,\lambda}) = \sum_\Pi \mathrm{Tr}\,\Pi(f^G(m,a)),$$

where the sum is over all irreducible representations Π *of* $\mathbb{G}(\mathbb{A})$ *which occur in the discrete spectrum* $\mathrm{L}_2^{\mathrm{dis}}(\mathbb{G}(\mathbb{Q})(Z(\mathbb{A}_f) \cap U)\backslash\mathbb{G}(\mathbb{A}), \omega_\xi)$.

(In this formula we are implicity using the strong multiplicity one theorem for GL_2.)

2.4. THE TRUE LEFSCHETZ NUMBER

The problem now is to identify Langlands's "naive" Lefschetz number as the sum of the local terms in an actual Lefschetz trace formula. An important technical issue is the contribution of the boundary. A conjecture of Deligne asserts it is 0 for $T(a) \times \mathrm{Frob}_p^m$, with a given a, if p is sufficiently large. In his thesis Rapoport constructed smooth (toroidal) compactifications of $\mathcal{X}_U$ over $\mathbb{Z}[1/ND]$ in which the complement of $\mathcal{X}_U$ is a relative normal crossing divisor (Rapoport, Corollaire 5.3). For all but finitely many primes p there also exists the Baily-Borel compactification, whose boundary is the finite set of cusps, and the toroidal compactification is a blow-up of it. At such primes p, Deligne's conjecture has been proved by (Brylinski and Labesse, Theorem 2.3.3). According to C.-L. Chai (private communication) this holds for all p prime to ND. Alternatively we can use the results of (Pink), since Condition 7.2.1 there, which suffices for Deligne's conjecture, holds for these compactifications, provided we know also that the monodromy of our sheaves around the toroidal resolutions of the cusps are tame. This is manifest from the higher dimensional Mumford-Raynaud-Tate parametrization near the cusps (Rapoport, 5.1 and 4.11). The idea is that p-adically near a cusp, the ℓ-adic Tate module of a universal abelian variety admits a canonical exact sequence having an étale quotient and a multiplicative sub-object (which are moreover Cartier dual to one another). The ℓ-adic monodromy is therefore manifestly tame at p. Finally, we could appeal to (Fujiwara), where Deligne's conjecture is proved in general. Either way we get the following

COROLLARY 2.2. *For every a as above there exists an integer $m_0(a)$ such that for each integer $m \geq m_0(a)$ we have*

$$\mathrm{Lef}(T(a) \times \mathrm{Frob}_p^m, \mathcal{X}_U \times \overline{\mathbb{F}}_p, \mathbb{V}_{\xi,\lambda}) = \mathrm{Tr}(T(a) \times \mathrm{Frob}_p^m | H_c^*(\mathcal{X}_U \times \overline{\mathbb{F}}_p, \mathbb{V}_{\xi,\lambda})),$$

where as usual $\mathrm{Tr}(\cdot|H_c^*)$ *means the alternating sum* $\sum_{i=0}^{2d}(-1)^i \mathrm{Tr}(\cdot|H_c^i)$.

2.5. CUSPIDALITY AND COMPACT SUPPORT

Assume that $\pi = \otimes_p \pi_p$ is an irreducible *cuspidal* representation of $\mathbb{G}(\mathbb{A})$. In particular each π_p (and π_v) is infinite dimensional, and $\pi \otimes |\det|^{-C/2}$ is unitary. Then π contributes to $H^*(\mathcal{X}_U, \mathbb{V}_\xi)$ if and only if π has a U-invariant vector and if π_∞ is isomorphic to

$$\prod_{i=1}^{d} \pi(|\cdot|^{(1-k_i)/2}, |\cdot|^{(k_i-1)/2} \mathrm{sign}^{k_i}) \otimes \det{}^{C}.$$

We will then say that π is of *type* ξ. In this case π contributes to $H^i(\mathcal{X}_U, \mathbb{V}_\xi)$ if and only if $i = d$. Taking the inverse limit over U we set

$$U^d(\pi_f, \xi) = \mathrm{Hom}_{\mathbb{G}(\mathbb{A}_f)}(\pi_f, \varprojlim_U H^d(\mathcal{X}_U, \mathbb{V}_\xi)).$$

Then $U^d(\pi_f,\xi)$ is a 2^d-dimensional space (see (Brylinski and Labesse, Section 3.4)), and notice that by the strong multiplicity 1 theorem for GL_2 we do not have multiplicities in our case). The ℓ-adic analog likewise gives a 2^d-dimensional $Gal_{\mathbb{Q}} = Gal(\overline{\mathbb{Q}}/\mathbb{Q})$ representation $U^d_\lambda(\pi_f,\xi)$. We now need the following

PROPOSITION 2.3. *If π is a cusp form as above, then $U^d(\pi_f,\xi)$ comes from the compactly supported cohomology; in other words, the natural map of forgetting supports*

$$\mathrm{Hom}_{\mathbb{G}(\mathbb{A}_f)}(\pi_f, \varprojlim_U H^d_c(\mathcal{X}_U, \mathbb{V}_\xi)) \to U^d(\pi_f,\xi)$$

is an isomorphism.

PROOF. See (Borel, Cor. 5.5), and also the last comment in (Harder). Here is the general idea of the proof. For any cusp c of $\mathcal{X}_U$ let $\mathcal{X}_{U,c}$ be the boundary component corresponding to c in the Borel-Serre compactification of $\mathcal{X}_U$. The cuspidality of π implies that the (de-Rham) cohomology classes of the vector-valued differential forms $\{\omega\}$ associated to it restrict to 0 on each $\mathcal{X}_{U,c}$. This can be seen also by verifying that the periods $\int_\beta \omega$ of these differential forms around each (vector-valued Borel-Moore) d-dimensional cycle β is 0. In fact, we can move β towards the cusp; then on the one hand the period stays constant, but on the other hand the cuspidality of π makes the ω and with it the period decrease exponentially. This forces the period to be 0, and we conclude that our $\{\omega\}$'s are cohomologous to compactly supported forms. □

Hence we get as a corollary that $U^d(\pi_f,\xi)$, which initially was derived from $H^d(\mathcal{X}_U,\mathbb{V}_\xi)$, is in the image of the "forget support" map from $H^d_c(\mathcal{X}_U,\mathbb{V}_\xi)$. This image is the *parabolic* cohomology $\tilde{H}^d(\mathcal{X}_U,\mathbb{V}_\xi)$, and it exists in ℓ-adic cohomology. As the sheaf $\mathbb{V}_{\xi,\lambda}$ is pure of weight dC, a weight argument (see (Deligne, 1969)) gives that $\tilde{H}^d(\mathcal{X}_U,\mathbb{V}_\xi)$ is pure of weight $d(C+1)$. The same is therefore true for $U^d_\lambda(\pi_f,\xi)$.

2.6. THE RAMANUJAN-PETERSSON CONJECTURE

We partially recall our previous notation. With ξ as before (in particular, of central character $\omega_\xi = \mathrm{Nm}^C$), let $\pi = \prod_p \pi_p$ be an irreducible cuspidal representation of $\mathbb{G}(\mathbb{A})$ of infinity type ξ, so that π_∞ is the discrete series representation corresponding to ξ as above. Let p be a rational prime which is prime to the level of π and to $\mathrm{Disc}\,F$. For a prime v of F above p let d_v denote the degree of $F_v/\mathbb{Q}_p$, and let T_v be the Hecke operator at v, defined as the action of the double class $U_v \begin{bmatrix} \varpi_v & 0 \\ 0 & 1 \end{bmatrix} U_v$, with $U_v \simeq GL_2(\mathcal{O}_{F,v})$ having measure 1. Here ϖ_v is a uniformizer at v. The Satake parameters of π_v are given as before in terms of a_v and b_v (up to order). The eigenvalue of T_v on π (or on π_v) is $\lambda_v = a_v + b_v$. We now obtain the following version of the Ramanujan-Petersson Conjecture:

THEOREM 2.4. *The representations π_p and π_v for $v|p$ are tempered. More precisely, each of a_v, b_v is a Weil number of weight $d_v + C$, i.e. a_v and b_v are algebraic numbers such that under any embedding into $\mathbb{C}$ we have*

$$|a_v| = |b_v| = p^{d_v C/2}.$$

Consequently, $|\lambda_v| \leq 2p^{d_v C/2}$.

PROOF. Using our previous notation, there exists a finite linear combination $T = \sum c_i T(g_i)$ of Hecke operators away from p which acts as projection onto the U^p invariants of the π_f^p-isotypical part. By the strong multiplicity one theorem this is a projection onto the U^p invariants of the π_f isotypical part. Let m_0 be an integer such that Deligne's conjecture holds for $T(g_i) \times \mathrm{Frob}_p^m$ on $\mathcal{X}_U \times \overline{\mathbb{F}}_p$ for all i and for all $m \geq m_0$. From Theorem 2.1 and Corollary 2.2 applied to $T \times \mathrm{Frob}_p^m$, with T viewed as the above linear combination, and from the definition of $U_\lambda^d(\pi_f, \xi)$ we get

$$\mathrm{Tr}(\mathrm{Frob}_p^m | U_\lambda^d(\pi_f, \xi)) = \mathrm{Tr}\,\pi_p(h_p^m).$$

The purity of the parabolic cohomology and the property of h_p^m from Section 2.2.3 give the existence of $2^d = \dim U_\lambda^d(\pi_f, \xi)$ Weil numbers $\{ w_i \}_i$ of weight $d(C+1)$ so that for every large enough m we have

$$\sum_i w_i^m = p^{md}\,\mathrm{Tr}\,r(t(\pi_p)^m) = p^{md} \prod_{v|p} (a_v^{m/l_v} + b_v^{m/l_v})^{l_v},$$

by equation (1), where as before $l_v = \gcd(m, d_v)$. As in (Brylinski and Labesse, Theorem 3.4.6) this implies that each π_v is tempered, namely that $|a_v| = |b_v|$ for every embedding of a_v, b_v into $\mathbb{C}$. That the weight is as it should be then follows from our knowledge of the product $|a_v b_v|$, as was explained in Section 2.2.2. □

Theorem 0.1 follows when one makes the necessary normalizations. In particular one takes $C = \max_i k_i$.

3. Special cases

We now give two special cases when the conditions 1.2 are satisfied. The first one was accessible in (Jordan and Livné, 2000), while the second one was not.

(A) Take $F = \mathbb{Q}$ and let B be a quaternion algebra over $\mathbb{Q}$ of discriminant 2 (it is unique up to an isomorphism). A model for B is given by the usual rational quaternion algebra generated over $\mathbb{Q}$ by $\mathbf{i}, \mathbf{j}$ satisfying $\mathbf{i}^2 = \mathbf{j}^2 = -1$ and $\mathbf{ij} = -\mathbf{ji}$. It is very easy to see that Conditions 1.2 are satisfied with $N_0 = 2\mathbb{Z}$. With $g = 1$, take a prime p satisfying $p \equiv 1 \pmod 4$, one gets the Lubotzky-Phillips-Sarnak graphs (LPS). These are Cayley graphs on $\mathrm{PSL}(2, \mathbb{F}_N)$, for any prime $N \equiv 1 \pmod 4$

which is different from p and a square modulo p. The set of generators is the reductions modulo N of the $p+1$ norm p integral quaternions $\equiv 1 \pmod 2$.

The case $g=2$, of two primes $p \neq q$ satisfying $p \equiv q \equiv 1 \pmod 4$ appeared in a different context in (Mozes). The resulting two (blue and red) communication networks are the same Cayley graphs as before on $\mathrm{PSL}(2,\mathbb{F}_N)$, for any prime $N \equiv 1 \pmod 4$ which is different from p and q and is a square modulo both. The sets of generators are the reductions modulo N of the $p+1$ (respectively $q+1$) norm p (respectively norm q) integral quaternions $\equiv 1 \pmod 2$. The square property holds, and each node has $p+1$ blue neighbors and $q+1$ red ones.

(B) We shall now construct a blue and a red Ramanujan communication networks having the square property over the same set of vertices $\mathrm{PSL}(2,\mathbb{F}_N)$ in which each node has $p+1$ blue and $p+1$ red neighbors. Here p is any prime $\equiv 1,9 \pmod{20}$ and $N \neq p$ is a prime satisfying $N \equiv 1,9 \pmod{20}$ and which is a square modulo p. There are other conditions which N must satisfy. These cannot be stated as simply, but we will see that they are satisfied by a set of primes N of positive density which can be explicitly described.

Take $F = \mathbb{Q}(\sqrt{5}) \subset \mathbb{R}$. Let $\infty_1 = \mathrm{Id}$, ∞_2 be the real primes of F. The fundamental unit of F is $\tau = (1+\sqrt{5})/2$. We will need the following lemma from Class Field Theory:

LEMMA 3.1. *1. Let N be a rational prime such that $\left(\frac{-1}{N}\right) = \left(\frac{5}{N}\right) = 1$, in other words $N \equiv 1,9 \pmod{20}$. It is then possible to write $N = a^2 - 20b^2$ with a,b integers and $a>0$ odd. Then τ, considered modulo N via either choice of $\sqrt{5}$ mod N, is a square modulo N if and only if $a+2b \equiv 1 \pmod 4$, and this holds for a set of primes N of Dirichlet density $1/8$.*
2. Let $p \neq N$ be a rational prime satisfying with $\left(\frac{5}{p}\right) = \left(\frac{-1}{p}\right) = 1$ and use part 1. to write $p = a^2 - 20b^2$ for integers a, b. Choose a $\sqrt{5}$ modulo N. Then the set of primes N satisfying the criterion in part 1. and for which both $a \pm 2b\sqrt{5}$ are squares modulo N has Dirichlet density $1/32$. Such primes N satisfy in particular $\left(\frac{p}{N}\right) = 1$.

PROOF. If 5 is a square modulo N then N splits in F, and since the class number of F is 1 we get $\pm N = \nu\bar{\nu}$, with ν and $\bar{\nu}$ conjugate integers of F. Since -1 is the norm of τ and since τ reduces modulo 2 (which is inert in F) to a generator of $\mathbb{F}_4^\times$, with $\mathbb{F}_4 = \mathcal{O}_F/2\mathcal{O}_F$, we can assume that the sign is $+$ and that $\nu \equiv 1 \pmod{2\mathcal{O}_F}$. Writing $\nu = a + b'\sqrt{5}$, we get that a and b' are integers with a odd and $b' = 2b$ even. Multiplying by -1 we may also assume that $a>0$. A choice of a square root of 5 modulo N then determines the sign of b, or equivalently the prime $\nu = a + 2b\sqrt{5}$ above N. We then ask when is the ideal $\nu\mathcal{O}_F$ split in $L = F(\sqrt{\tau})$. Since $N \equiv 1 \pmod 4$, it splits in $\mathbb{Q}(\sqrt{-1})$, so that ν (and $\bar{\nu}$) split in $F(\sqrt{-1})$. Therefore ν splits in $F(\sqrt{\tau})$ if and only if N is completely split in $\mathbb{Q}(\sqrt{5},\sqrt{-1},\sqrt{\tau})$, which is Galois of degree 8 over $\mathbb{Q}$. By the Čebotarev density theorem, this happens for a set of primes N of Dirichlet density $1/8$ (namely, for half the primes $\equiv 1,9 \pmod{20}$). It remains to determine the condition for ν to split in $F(\sqrt{\tau})$.

The Idèle class group character χ of F which cuts L is of order 2 and unramified outside 2. Since $\infty_1(\tau) > 0$ and $\infty_2(\tau) < 0$, we see that χ is trivial on $F_{\infty_1}^{\times}$ but not on $F_{\infty_2}^{\times}$. For a prime v of F set $U_v = \mathcal{O}_{F,v}^{\times}$ and let $U_{v,n}$ be its subgroup of elements congruent to 1 modulo v^n for any $n \geq 1$. Then χ must be trivial on the product U_∞ of the connected components of $F_{\infty_1}^{\times}, F_{\infty_2}^{\times}$ and on $U = \prod_{\neq 2} U_v$. The units $\mathcal{O}_F^{\times} = \pm\tau^{\mathbb{Z}}$ surject onto $U_2/U_{2,1} \times \prod_i \pi_0(F_{\infty_i})$, and the kernel of this surjection is $\tau^{6\mathbb{Z}}$. Since the class number of F is 1, χ is determined by its values on

$$C = F^{\times} \backslash \mathbb{A}_F^{\times} / UU_\infty \simeq \tau^{6\mathbb{Z}} \backslash U_{2,1}/(U_{2,1})^2 .$$

First we calculate C. Through the 2-adic logarithm map, $U_{2,1} \simeq \mu(F) \oplus L$, where $L = 2\mathbb{Z}_2 + 4\mathcal{O}_{F,2}$ and $\mu(F) = \{\pm 1\}$ is the group of the roots of unity of F. Since $\tau^6 = 9 + 4\sqrt{5} \equiv 5 \pmod{8\mathcal{O}_F}$ we get that

$$C \simeq \mu(F) \oplus 2\mathcal{O}_F/(2\mathbb{Z} + 4\mathcal{O}_F) \simeq (\mathbb{Z}/2\mathbb{Z})^2 ,$$

with generators $\overline{-1}$ and $2\overline{\tau}$ for the respective two factors.

To determine χ we use the product formula $\prod_v \chi_v(x) = 1$ for any $x \in F^{\times}$, the product taken over all places of F. For $x = -1$ we get $\chi(\overline{-1}) = \chi_2(-1) = \chi_{\infty_2}(-1) = -1$, and since $\tau^3 = 2 + \sqrt{5} = 1 + 2\tau$, we also get

$$\chi(2\overline{\tau}) = \chi_2(1 + 2\tau) = \chi_2(\tau)^3 = -1 .$$

In particular, for $x, y \in \mathbb{Z}_2$ we have $\chi_2(1 + 2(x + y\tau)) = (-1)^{x+y}$. It follows the prime $\nu\mathcal{O}_F$ splits in $F(\sqrt{\tau})$ if and only if $\chi_\nu(\nu) = -1$. But

$$\chi_\nu(\nu) = \chi_2(\nu)\chi_{\infty_2}(\nu) = (-1)^{(a-1)/2+b} \cdot 1 = (-1)^{(a-1)/2+b} .$$

As claimed. (We write χ_ν etc. for the local component of χ at the place defined by the prime $\nu\mathcal{O}_F$.)

2. Write $p = \pi\overline{\pi}$, with (say) $\pi = a + 2\sqrt{b}$. The field $F(\sqrt{\pi}, \sqrt{\overline{\pi}})$ is Galois over $\mathbb{Q}$ and contains $\sqrt{p}$. As an extension of F it is ramified only over π and $\overline{\pi}$: this is because $(1 + \sqrt{\pi})/2$ is an algebraic integer, as its trace and norm to F are 1 and $(1 - a)/2 - b\sqrt{5}$. Hence it is linearly disjoint from $F(\sqrt{-1}, \sqrt{\tau})$ over F. The density statement follows by applying the Čebotarev density theorem to the following degree 32 extension of $\mathbb{Q}$:

$$\mathbb{Q}(\sqrt{-1}, \sqrt{5}, \sqrt{p}, \sqrt{\tau}, \sqrt{\pi}). \qquad \square$$

Now set $B_F = B \otimes F$, with B the rational quaternions as before. For the facts we need about B_F see (Vignéras, Chapter 5). In particular, B_F is ramified precisely at the two infinite primes of F.

The class number of B_F is 1, and all maximal orders in B_F are conjugate. To describe the sets of generators of our red and blue networks we need to explicitly compute in one maximal order $\mathcal{M}$. We take for $\mathcal{M}$ the $\mathcal{O}_F$-submodule of B_F generated by

$$\begin{aligned} e_1 &= (1+\tau^{-1}\mathbf{i}+\tau\mathbf{j})/2, & e_2 &= (\tau^{-1}\mathbf{i}+\mathbf{j}+\tau\mathbf{k})/2, \\ e_3 &= (\tau\mathbf{i}+\tau^{-1}\mathbf{j}+\mathbf{k})/2, & e_4 &= (\mathbf{i}+\tau\mathbf{j}+\tau^{-1}\mathbf{k})/2. \end{aligned} \tag{3.1}$$

We have $\mathcal{M}/2\mathcal{M} \simeq \mathrm{Mat}_{2\times 2}(\mathbb{F}_4)$, with $\mathbb{F}_4 = \mathcal{O}_F/2\mathcal{O}_F$ a field with 4 elements. The reduction map induces an exact sequence

$$1 \to (\mathcal{O}_F + 2\mathcal{M})^\times \to \mathcal{M}^\times \to \mathrm{PGL}(2,\mathbb{F}_4) \to 1\,.$$

Our choice of $\mathcal{M}$ implies that $\mathcal{O}_F + 2\mathcal{M}$ is the ($\mathcal{O}_F$-) sub-order of $\mathcal{M}$ given by

$$\{a+b\mathbf{i}+c\mathbf{j}+d\mathbf{k} \mid |a,b,c,d \in \mathcal{O}_F,\ b+\tau c+\tau^{-1}d \in 2\mathcal{O}_F\}\,.$$

Since τ reduces modulo 2 to a generator (of order 3) of $\mathbb{F}_4^\times$, we obtain the following

LEMMA 3.2. *Every element in $\mathcal{M}$ whose norm to $\mathcal{O}_F$ is odd, namely not in $2\mathcal{O}_F$, can be uniquely written as $u(a+b\mathbf{i}+c\mathbf{j}+d\mathbf{k})$, with $u \in \mathcal{M}^\times$ and $a,b,c,d \in \mathcal{O}_F$ satisfying*

1. $b+\tau c+\tau^{-1}d \in 2\mathcal{O}_F$;
2. $a \equiv 1 \pmod{2\mathcal{O}_F}$;
3. $1 \le a < \tau^3$.

When $u = 1$ we shall say that our element is in normal form.

As a corollary we see that Conditions 1.2 hold with $N_0 = 2\mathcal{O}_F$.

Now write $4p = a^2 - 20b^2$, with a, b integers (say, minimal positive), necessarily of the same parity. The elements $\pi = (a+2b\sqrt{5})/2$ and $\overline{\pi} = (a-2b\sqrt{5})/2$ are the primes above p in $\mathcal{O}_F$. Let $\mathcal{M}_{0,F} = \mathcal{O}_F[\mathbf{i},\mathbf{j}]$ be the order of $\mathcal{O}_F$-integral quaternions. We shall say that an element $x \in \mathcal{M}_{0,F}$ is in normal form if $x \equiv 1 \pmod 2$ and $\mathrm{Tr}\,x = \mathrm{Tr}_{B_F/F}\,x$ is totally positive (namely positive for both real embeddings) and satisfies $\tau^{-3} < \mathrm{Tr}\,x \le \tau^3$. Then Proposition 1.3 implies the following

LEMMA 3.3. 1. *There are precisely $p+1$ elements in normal form γ_l (respectively $\overline{\gamma}_l$), $1 \le l \le p+1$, in $\mathcal{M}_{0,F}$ whose norm is π (respectively $\overline{\pi}$).*
2. *For any indices $1 \le l,m \le p+1$ there exist unique indices $1 \le l',m' \le p+1$ and a unique unit $u \in \mathcal{M}_{0,F}{}^\times = \langle \pm\pi^{3\mathbb{Z}}p^{\mathbb{Z}}\rangle$ such that*

$$\gamma_l\overline{\gamma}_m = \overline{\gamma}_{m'}\gamma_{l'}u\,.$$

Now let N be a rational prime congruent to 1 or 9 modulo 20. Choosing square roots of -1 and of 5 in the prime field $\mathbb{F}_N$ allows us to map $\mathcal{M}$ homomorphicallly onto $\mathrm{Mat}_{2\times 2}(\mathbb{F}_N)$ by

$$\sqrt{5} \mapsto \begin{bmatrix} \sqrt{5} & 0 \\ 0 & \sqrt{5} \end{bmatrix}, \quad \mathbf{i} \mapsto \begin{bmatrix} \sqrt{-1} & 0 \\ 0 & -\sqrt{-1} \end{bmatrix}, \quad \text{and} \quad \mathbf{j} \mapsto \begin{bmatrix} 0 & -1 \\ 1 & 0 \end{bmatrix}.$$

Proposition 1.4 now gives us that we have constructed two Ramanujan communication networks as the two Cayley graphs on $\mathrm{PSL}(2,\mathbb{F}_N)$ having the reductions modulo N of the γ_l's (respectively the $\bar{\gamma}_m$'s) for generators.

References

Borel, A. (1983) Stable real cohomology of arithmetic groups II, in *Oeuvres, vol. III*, Springer-Verlag, pp. 650–684, and also in *Prog. in Math.* **14**, Birkhauser, Boston, 1990, pp. 21–55.

Breen, L., and Labesse, J.P. (eds., 1979), *Variétés de Shimura et Fonctions L*, 2nd edition, Publications Mathématiques de l'Université Paris VII.

Brylinski, J.-L., and Labesse, J.-P. (1984) Cohomologie d'intersection et fonctions L de certaines variétés de Shimura, *Ann. Scient. Éc. Norm. Sup.* **17**, 361–412.

Carayol, H. (1986) Sur les représentations ℓ-adiques associées aux formes modulaires de Hilbert, *Ann. scient. Ec. Norm. Sup.* **19**, 409–468.

Deligne, P. (1969) Formes modulaires et représentations ℓ-adiques, in *Sém. Bourbaki 355 (1968/9)*, Lecture Notes in Math., vol. 179, Springer-Verlag, Berlin, 1971, pp. 139–172.

Deligne, P. (1971) Travaux de Shimura, in *Sém. Bourbaki 389 (1970/71)*, pp. 139–172.

Deligne, P. (1981) La conjecture de Weil II, *Publ. Math. Inst. Hautes Études Sci.* **52**, 313–428.

Deligne, P., and Pappas, G. (1994) Singularités des espaces de modules de Hilbert en les caractéristiques divisant le discriminant, *Compositio Math.* **90**, no. 1, 59–79.

Friedman, J. (1991) On the second eigenvalue and random walks in random d-regular graphs, *Combinatorica* **11**, no. 4, 331–362.

Fujiwara, K. (1997) Rigid Geometry, Lefschetz-Verdier trace formula and Deligne's conjecture, *Invent. math.* **127**, no. 3, 489–533.

Harder, G. (1975) On the cohomology of discrete arithmetically defined groups, in *Discrete Subgroups of Lie Groups and Applications to Moduli, Internet. Colloq., Bombay, 1973*, Oxford Univ. Press, Bombay, pp. 129–160.

Jacquet, H., and Langlands, R.P. (1970) *Automorphic Forms on* GL(2), Lecture Notes in Math., vol. 114, Springer-Verlag, Berlin.

Jordan, B.W., and Livné, R. (1997) Ramanujan local systems on graphs, *Topology* **36**, 1007–1024.

Jordan, B.W., and Livné, R. (2000) The Ramanujan property for regular cubical complexes, *Duke Math. J.* **105**, 85–103.

Langlands, R. (1979) On the zeta function of some simple Shimura varieties, *Canadian J. math.* **31**, 1121–1216.

Lubotzky, A. (1994) *Discrete Groups, Expanding Graphs and Invariant Measures*, Birkhäuser, Berlin.

Lubotzky, A., Phillips, R., and Sarnak, P. (1988) Ramanujan graphs, *Combinatorica* **8**, 261–278.

Milne, J.S. (1979) Points on Shimura varieties mod p, in *Automorphic Forms, Representations and L-functions*, Proc. Symp. in Pure Math., vol 33, part 2, AMS, R.I., pp. 165–184.

Mozes, S. (1995) Actions of Cartan subgroups, *Israel J. of Math.* **90**, 253–294.

Pink, R. (1992) On the calculation of local terms in the Lefschetz-Verdier trace formula and its application to a conjecture of Deligne, *Ann. of Math.* (2) **135**, no. 3, 483–525.

Rapoport, M. (1978) Compactifications de l'espace des modules de Hilbert-Blumenthal, *Comp. Math.* **36**, 255–335.

Vignéras, M.-F. (1980) *Arithmétique des Algèbres de Quaternions*, Lecture Notes in Math., vol. 800, Springer-Verlag, Berlin.

COMPACT KÄHLER THREEFOLDS WITH SMALL PICARD NUMBERS

TH. PETERNELL
University of Bayreuth, Germany

Abstract. We review the current state of Mori theory on compact Kähler manifolds, in particular threefolds, and apply it to threefolds with Picard number ≤ 1.

Key words: Kähler manifold, Mori theory, nef line bundle.

Mathematics Subject Classification (2000): 32J17, 32J27.

1. Introduction

Mori theory explains the numerical behaviour of the canonical bundle $K_X = \det T_X^*$ of a projective manifold in terms of geometry: very roughly one can say that if K_X has some negativity, then one finds rational curves (and in case of enough negativity X will be covered by rational curves) whereas if K_X is "semipositive", then some multiple of K_X is generated by global sections. Therefore Mori theory—especially in dimension 3 where the theory is fully developped—gives important insight in the global structure of algebraic varieties. The analytic analogues of algebraic varieties are Kähler manifolds (or spaces) and it is very tempting to ask whether the theory works also here in this larger category. Even for algebraic geometry it is important to understand Kähler varieties; e.g. they occur naturally in deformation and moduli problems. But of course Kähler geometry has its own importance.

Many methods of algebraic geometry fail in the Kähler case: there are no ample line bundles, hence vanishing theorems vanish; there might only be a few curves and so on. This means that new methods are needed. Even central definitions have to be changed. For example on an algebraic variety a line bundle L is nef, if $L \cdot C \geq 0$ for all curves C. This does not make much sense on Kähler varieties since there are in general not so many curves. Instead nefness has to be defined in terms of curvature. The starting point of Mori theory in the algebraic context is to look at varieties X whose cannonical bundles K_X are not nef, i.e. there

C. Ciliberto et al. (eds.),
Applications of Algebraic Geometry to Coding Theory, Physics and Computation, 271–290.

is a curve C with $K_X \cdot C < 0$. Then as a main result there exists a contraction, i.e. a map contracting curves C' with $K_X \cdot C' < 0$ (possibly different from C, whose classes are extremal.

Now in the Kähler case, say for threefolds, it is completely unclear whether in case K_X is not nef (in some analytic sense!) there is any curve C with $K_X \cdot C < 0$. In fact, for an arbitrary line bundle this is false and it is clear that new deep methods are necessary to construct C. If however C exists, then one can even choose C rational and one can construct the contraction.

In dimension 3 one knows a lot on the global bimeromorphic structure of non-algebraic manifolds; important tools are the algebraic reduction, the Albanese map and Hodge theory. This knowledge can be used to construct C in almost all cases. There is however a mysterious case which is still untractable at the moment and which presents an unknown dark area in the landscape of Kähler threefolds. Namely, one says that a threefold X is *simple*, if there is no proper compact subvariety through a very general point of X. In particular X does not carry any meromorphic function and that there is no non-trivial meromorphic fibration from X. What can one say about the structure of such a simple threefold? Looking for examples, we have a lot of them in dimension 2: tori and K3 surfaces without meromorphic functions. In dimension 3 there are essentially only tori: Calabi-Yau threefolds are algebraic. Of course, one can modify a torus: one can divide by a finite group and also make bimeromorphic transformations. Varieties arising in such a way are called Kummer. Now the conjecture is that simple Kähler threefolds should be *Kummer*. Virtually nothing is known here, except for a result of Campana saying that a smooth simple threefold, which is not Kummer, has finite fundamental group. Now Mori theory would provide a positive answer to the conjecture: if we know the existence of contractions on (possibly slightly singular) Kähler threefolds and if we know abundance then any simple Kähler threefold would be Kummer.

The aim of this paper is to describe the present knowledge on analytic Mori theory on Kähler threefolds (sect. 2) and then to apply the theory to study threefolds with relatively simple structure, i.e with small Picard number namely 0 or 1 (sect. 4 and 5). We also use a dual version of the Picard number, defined by the dimension of the cone of curves (this is discussed in sect. 3). The paper presumes knowledge on algebraic Mori theory, basic references being (Kawamata, Matsuda and Matsuki), (Kollaŕ et al.) and (Miyaoka and Peternell).

2. The Status of Mori Theory on Kähler Threefolds

In all of this section X denotes a normal compact Kähler threefold which is $\mathbb{Q}$-factorial with at most terminal singularities. Recall that a normal complex space is Kähler, if the regular part of X admits a Kähler metric h such that every singular point of X has a small neighborhood U and an embedding of U into an open subset V of some $\mathbb{C}^N$ such that h is the restriction of a Kähler form on V.

We will say that X is minimal if K_X is nef. The relevant definition is

DEFINITION 2.1. *Let X be a normal compact Kähler space and L a line bundle on X.*
(1) Suppose X is smooth. Then L is nef, if $c_1(L)$ is in the closure of the Kähler cone of X, i.e. in the closure of the cone generated by the Kähler forms in $H^{1,1}(X)$.
(2) In general we take a desingularisation $\pi : \hat{X} \longrightarrow X$ and say that L is nef, if $\pi^*(L)$ is nef.

By (Paun, 1998), and (Peternell, 1998) in dimension 3, this definition does not depend on the choice of π. One can see that in case X is projective we get back the old definition: L is nef if and only if $L \cdot C \geq 0$ for all curves C (Demailly, Peternell and Schneider, 1994).

Here are the main questions in Mori theory for Kähler threefolds.

MAIN QUESTIONS 2.2. *1. Suppose K_X not nef. Then there exists a curve $C \subset X$ such that $K_X \cdot C < 0$ and there exists a map $\phi : X \to Y$ with connected fibers such that Y is a normal Kähler space, $-K_X$ is ϕ-ample and such that $b_2(X) = b_2(Y) + 1$.*
2. If X is minimal, i.e. K_X is nef, then some multiple mK_X is generated by global sections.

Since the minimal model program works once we know the existence of contractions (the reason being the local analytic nature of the proof of the existence of flips (Mori, 1988)), we have the following

CONSEQUENCE 2.3. *If K_X is not nef, then X is birational via a finite sequence of contractions and flips to another threefold X' (with the same properties) such that either X' is minimal or X' is uniruled.*

As already indicated in the introduction, a compact threefold X is said to be *simple*, if there is no proper positive dimensional subvariety through the very general point of X. And a variety X is Kummer, if it is bimeromorphic to T/G with T a torus and G a finite group.

COROLLARY 2.4. *Suppose that the Main Questions 2.2 have positive answers. Then every simple threefold is Kummer.*

PROOF. (Peternell, 1998). □

In the sequel we collect the known results towards 2.2, all in dimension 3.

THEOREM 2.5. *1. Suppose X is not at the same time simple and non-Kummer with $\kappa(X) = -\infty$. Suppose K_X is not nef. Then there exists a curve C such that $K_X \cdot C < 0$.*

2. *Suppose X smooth. Suppose there exists a curve C with $K_X \cdot C < 0$, then there exists a contraction $\phi : X \to Y$ to a normal compact complex space such that $-K_X$ is ϕ-ample and such that $b_2(X) = b_2(Y) + 1$.*

PROOF. (Campana and Peternell, 1997), (Peternell, 1998 and 2000). The proof of the first part uses structure theorems on threefolds which are not simple, the second part proceeds by deforming the curve C and by investigating non-splitting families of curves. □

Non-splitting families of curves are created by deforming a curve C with $K_X \cdot C < 0$. Here we need the smoothness of X in order to guarantee the deformability of C. This is still true in the Gorenstein case, so that the first part Theorem 2.5 should remain true by similar arguments. However major difficulties arise when X is only $\mathbb{Q}$-Gorenstein; here C need not be deformable.

Another difficulty is that we do not know whether we can take Y to be Kähler in (2). Of course this is crucial for performing the minimal model program. The only problematic case is when ϕ contracts a divisor to a curve. Then we need to make the right choice; it is easy to construct examples where Y is not Kähler when ϕ is not chosen appropriately. Certainly this question amounts to have a cone theorem in the Kähler case.

Here is another result, shown in (Demailly, Peternell and Schneider, 2000), concerning the existence of curves C with $K_X \cdot C < 0$ on a threefold with K_X not nef.

THEOREM 2.6. *Suppose that K_X is pseudo-effective, i.e. K_X has a singular metric whose positive curvature current. Assume that K_X is not nef. Then there exists a curve C with $K_X \cdot C < 0$.*

Note that if 2.1 is answered positively, then a threefold with K_X pseudo-effective will have non-negative Kodaira dimension. In that case the existence of C was already established in (Peternell, 1998).
Concerning the "abundance" part of 2.1 we have (Peternell, 2000)

THEOREM 2.7. *Let X be a minimal Kähler threefold. If X is not both simple non-Kummer, then mK_X is generated by global sections for suitable large m.*

Another result from (Demailly, Peternell and Schneider, 2000) is

THEOREM 2.8. *Let X be a smooth compact Kähler threefold with K_X pseudo-effective. Assume that K_X has a singular metric with "algebraic" singularities with positive curvature current, e.g. K_X is hermitian semi-positive. Then $\kappa(X) \geq 0$.*

This theorem uses in an essential way

THEOREM 2.9. *Suppose that $\kappa(X) = -\infty$ and that X is not both simple non-Kummer. Then X is uniruled.*

In the algebraic case this is an important result of Miyaoka, the non-algebraic case was estabished in (Campana and Peternell, 2000).

In a paper in preparation (Demailly and Peternell) finally the following is proved

THEOREM 2.10. *If X is minimal, then $\kappa(X) \geq 0$.*

In the proof a version of the Kawamata-Viehweg vanishing theorem on Kähler threefolds plays a crucial role. Of course the expectation is that mK_X should be spanned for suitable positive m. Here certain difficulties arise, which however might not be untractable. At least the Gorenstein case can be settled.

Algebraic geometers might only be interested in Kähler varieties which appear as deformations of projective varieties. Here is the relevant definition; we consider only varieties with terminal singularities.

DEFINITION 2.11. *Let X be a normal Kähler variety with only terminal singularities. X is nearly algebraic if there exists an algebraisation approximation of X. This is a proper surjective holomorphic map $\pi : \mathcal{X} \to \Delta$ from a normal complex space $\mathcal{X}$ where $\Delta \subset \mathbb{C}^m$ is the unit disc, where $X \simeq X_0$, where all complex analytic fibers $X_t = \pi^{-1}(t)$ are normal Kähler spaces with at most terminal singularities such that there is a sequence (t_j) in Δ converging to 0 so that all $X_j := X_{t_j}$ are projective.*

Of course, all X_t are smooth (after possibly shrinking Δ), in case X is smooth. The following well-known conjecture is attributed to Andreotti resp. Kodaira.

CONJECTURE 2.12. *Every compact Kähler manifold is nearly algebraic.*

Of course this should hold also in the singular (terminal) case. In the setting of 2.11 simple Kähler threefolds which are not bimeromorphic to a quotient of a torus are far from algebraic varieties:

THEOREM 2.13. *Let X be a nearly algebraic Kähler threefold with only terminal singularities. If X is simple, then $X \sim T/G$ with T a torus and G a finite group.*

The proof is found in (Demailly and Peternell). As a corollary of the previous results, we can state:

COROLLARY 2.14. *Let X be a nearly algebraic Kähler threefold with at most terminal singularities.*

- *If K_X is not nef, then there exists a curve C with $K_X \cdot C < 0$.*
- *If X is smooth and K_X is not nef, then there exists a contraction $\phi : X \to Y$ to a normal compact complex space such that $-K_X$ is ϕ-ample and such that $b_2(X) = b_2(Y) + 1$.*
- *If K_X is nef, then mK_X is spanned for a suitable positive integer m.*

3. Picard Numbers and Rationality

NOTATION 3.1. Let X be a compact Kähler manifold of dimension n. We let

$$\mathcal{K}(X) \subset H^{1,1}(X)$$

denote the Kähler cone of X and

$$\mathcal{K}^*(X) \subset H^{n-1,n-1}(X)$$

the interior of the dual Kähler cone (the dual cone is closed by definition). We define the Picard number $\rho(X)$ by

$$\rho(X) = \dim_{\mathbb{R}} c_1(\mathrm{Pic}(X) \otimes \mathbb{R})$$

where c_1 is considered as a mapping $c_1 : \mathrm{Pic}(X) \to H^2(X, \mathbb{R})$. In other words,

$$\rho(X) = \dim_{\mathbb{R}} (H^{1,1}(X) \cap H^2(X, \mathbb{Q})) \otimes \mathbb{R}.$$

Dually we define the *dual Picard number* of X:

$$\rho^*(X) = \dim_{\mathbb{R}} (H^{n-1,n-1}(X) \cap H^{2n-2}(X, \mathbb{Q})) \otimes \mathbb{R}.$$

Furthermore we let

$$\rho_d(X)$$

be the dimension of the subspace in $H^2(X, \mathbb{R})$ generated by the classes of irreducible hypersurfaces.

As usual (in the projective case) $\overline{NE}(X)$ denotes the closed cone generated by the classes of irreducible curves in $H^{2n-2}(X, \mathbb{R})$. We let $\rho_c(X)$ be the dimension of the subspace generated by $\overline{NE}(X)$.

REMARK 3.2. (1) If $\rho(X) = h^{1,1}(X)$, then X is projective. In fact, this assumption says that $H^{1,1}(X)$ generated by the Chern classes of line bundles, hence

$$\mathcal{K}(X) \cap H^2(X, \mathbb{Q}) \neq \emptyset,$$

and therefore X is projective by Hodge's theorem.

(2) If X is projective, then it is well known that

$$\rho(X) = \rho_d(X) = \rho_c(X) = \rho^*(X).$$

(3) In general we clearly have $\rho_d(X) \leq \rho(X)$ and $\rho_c(X) \leq \rho^*(X)$ but there are no more relations. E.g. there are threedimensional tori X with $a(X) = 0$ having $\rho(X) > 0$ but admitting no curves. So $\rho(X) > \rho_d(X) = \rho_c(X)$.
On the other hand, there are 3-dimensional tori X with $a(X) = 0 = \rho(X) = 0$ admitting an elliptic fiber bundle structure f, necessarily without multisections

(i.e. the projection map f is not projective; if f would have a multisection, then we could move this divisor and would obtain a covering family of divisors contradicting $a(X) = 0$ over a 2-dimensional torus T with $\rho(T) = 0$). This provides a 3-fold with $\rho(X) = 0$ and $\rho_c(X) = 1$. In particular $\rho(X) \leq \rho^*(X)$. Finally there are 3-dimensional tori X with $a(X) = 1$ whose algebraic reduction is a torus bundle without section over an elliptic curve such that all fibers F have $a(F) = 0$. This provides an example with $\rho_d(X) = 1$ and $\rho_c(X) = 0$.

We have the following dual version of (3.2)(1):

PROPOSITION 3.3. *Let X be a compact Kähler manifold. If $\rho^*(X) = h^{n-1,n-1}(X)$, then X is projective.*

PROOF. By assumption $H^{n-1,n-1}(X) \cap H^2(X,\mathbb{Q})$ generates the complex vector spaces $H^{n-1,n-1}(X)$. Let

$$\Phi : H^{n-1,n-1}(X) \cap H^2(X,\mathbb{Q}) \to \mathbb{Q}$$

be linear. Then Φ induces a unique extension

$$\Phi_0 : H^{n-1,n-1}(X) \to \mathbb{C}$$

and it also extends by 0 to a map

$$\Phi_1 : H^{2n-2}(X,\mathbb{Q}) \to \mathbb{Q}.$$

By duality, Φ therefore induces an element $\Lambda(\Phi) \in H^{1,1}(X) \cap H^2(X,\mathbb{Q})$. We obtain a linear isomorphism

$$\Lambda : H^{n-1,n-1}(X)^* \to H^{1,1}(X)$$

which creates a rationally defined generating subspace in $H^{1,1}(X)$ and therefore an application of (3.2) gives the projectivity of X. □

PROBLEM 3.4. Hodge's theorem, used in (3.2), says that a compact Kähler manifold X carrying a rational Kähler class, is projective. The dual problem asks whether a Kähler manifold with

$$\mathcal{K}^*(X) \cap H^{2n-2}(X,\mathbb{Q}) \neq \emptyset$$

is projective. For surfaces the answer is positive (Oguiso and Peternell, 2000), the threefold case is discussed in (Oguiso and Peternell, 2001).

If $\overline{NE}(X)$ contains an interior point, a much stronger condition, then (equivalently) $\rho^*(X) = h^{n-1,n-1}(X)$, so X is projective by (2.3).

4. Manifolds with vanishing Picard or dual Picard number

THEOREM 4.1. *Let X be a compact Kähler manifold with $\rho(X) = 0$. Then there exists a finite étale cover $\tilde{X} \longrightarrow X$ such that*

$$\tilde{X} \simeq A \times \Pi_i Y_i,$$

where A is a torus with $a(A) = 0$ and Y_i are irreducible symplectic manifolds with $a(Y_i) = 0$. If $\dim X = 3$ then $\tilde{X}$ is a torus.

PROOF. Since $\rho(X) = 0$, we have in particular $c_1(X) = 0$, so by (Beauville) there exists a finite étale cover $\tilde{X} \to X$ such that

$$\tilde{X} \simeq A \times \Pi_i Y_i,$$

where A is a torus and Y_i are irreducible Calabi-Yau or symplectic manifolds. Now all occuring varieties have vanishing algebraic dimension. Hence Y_i cannot be Calabi-Yau, these being projective.
In case $\dim X = 3$, we cannot have a symplectic factor Y, otherwise $X \simeq C \times Y$ with an elliptic curve C, contradicting $a(\tilde{X}) = 0$. □

Examples in 4.1 are provided by tori with $\rho = 0$ or K3 surfaces with $\rho = 0$. Turning to manifolds whose dual Picard number vanish, we have, assuming only that $\rho_c(X) = 0$:

THEOREM 4.2. *Let X be a smooth compact Kähler threefold with $\rho_c(X) = 0$. Then*

1. *If K_X is pseudo-effective, then X is a torus up to finite étale cover.*
2. *If K_X is not pseudo-effective (in particular $\kappa(X) = -\infty$), then X is simple. So if X is nearly algebraic, then K_X must be pseudo-effective, if $\rho_c(X) = 0$.*
3. *If $h^{1,1}(X) \leq 18$, then X is a torus up to finite étale cover.*

PROOF. Since $\rho_c(X) = 0$, X does not contain any curve. In particular $a(X) \leq 1$. By classification (Fujiki), cp. (Campana and Peternell), it also follows that $a(X) \neq 1$, so $a(X) = 0$.
(1) If K_X is pseudo-effective, then (Demailly, Peternell and Schneider, 2000, 5.4), see (2.6), shows that K_X is nef. Now (2.10) applies and X is Kummer. So $K_X \equiv 0$. From Riemann-Roch and $h^2(\mathcal{O}_X) > 0$ it follows $q(X) \geq 1$. Since $\rho_c(X) = 0$, either the Albanese map is an isomorphism or it maps to an elliptic curve. In this second case X is a torus after finite étale cover.
(2) Suppose K_X not pseudo-effective. Then either by direct classification (Fujiki, 1983) or by (Campana and Peternell, 2000), X must be simple.
(3) Suppose that X is not Kummer. By (1) and (2) we may assume X is simple. Then we conclude that $H^1(X, \mathcal{O}_X) = 0$ (consider the Albanese). Now, X being

Kähler but not projective, there exists a non-zero holomorphic 2-form ω on X. Let Z be the set of zeroes of ω. Then (set-theoretically)—recalling that X does not contain any curve—

$$Z = D_1 \cup \ldots \cup D_s \cup E$$

with irreducible hypersurfaces $D_i \subset X$ and a finite set E. Since D_i does not contain any curve, we observe $D_i \cap D_j = \emptyset$ for $i \neq j$ and moreover by Proposition 4.3, D_i is a torus (without curves) with numerically trivial normal bundle. Then

$$\omega \in H^0(X, \Omega^2_X(-\sum a_i D_i))$$

has only finitely many zeroes, for suitable choice of a_i. Let $L = \mathcal{O}_X(\sum a_i D_i)$. Since $N_{D_i} \equiv 0$ and $D_i \cdot D_j = 0$, we compute

$$0 \leq c_3(\Omega^2_X \otimes L^*) = c_3(\Omega^2_X) + c_2(\Omega^2_X) \cdot L = c_3(\Omega^2_X),$$

hence $c_3(X) \geq c_1(X) \cdot c_2(X)$. Since $\chi(X, \mathcal{O}_X) \geq 2$, we deduce from Riemann-Roch

$$\chi_{\mathrm{top}}(X) = c_3(X) \geq 48.$$

Since $b_1 = 0$, it follows $b_2 \geq 23$, hence by $h^{2,0} \leq 2$ we finally derive the contradiction $h^{1,1} \geq 19$. The inequality $h^{2,0} \leq 2$ is seen as follows: if $h^{2,0} \geq 3$, then by $a(X) = 0$, the bundle Ω^2_X is generically generated, hence K_X is pseudo-effective, contradiction. □

Of course one expects that the canonical bundle K_X of a Kähler threefold X with $\rho_c(X) = 0$ should be nef and therefore $K_X \equiv 0$ so that X should be a torus up to finite étale cover.

PROPOSITION 4.3. *Let X be a smooth compact Kähler threefold with $\rho_c(X) = 0$. Let $D \subset X$ be an irreducible divisor. Then D is a torus and $N_{D/X} \equiv 0$, actually N_D is torsion.*

PROOF. Since D does not contain any curve, classification implies that D is a torus, K3 or a K3 surface with all (-2)-curves blown down, i.e. with finitely many rational double points (these surfaces will be called singular K3).
Since X is not projective, X carries a non-zero holomorphic 2-form ω. Choose the unique nonnegative integer λ such that

$$0 \neq \omega | D \in H^0(D, \Omega^2_X | D \otimes N_D^{*\lambda})$$

(so λ is the vanishing order of ω along D).
(1) First assume D smooth. Then

$$0 \longrightarrow N_D^* \longrightarrow \Omega^1_X | D \longrightarrow \Omega^1_D \longrightarrow 0$$

yields an exact sequence

$$0 \longrightarrow N_D^* \otimes \Omega_D^1 \longrightarrow \Omega_X^2|D \longrightarrow K_D = \mathcal{O}_D, \tag{$*$}$$

so we obtain finally

$$0 \longrightarrow N_D^{*(\lambda+1)} \otimes \Omega_D^1 \longrightarrow \Omega_X^2|D \otimes N_D^{*\lambda} \xrightarrow{\kappa} N_D^{*\lambda}.$$

If $\kappa(\omega|D) \neq 0$, then* N_D^λ has a section without zeroes, so $N_D \equiv 0$ and actually is torsion. Suppose first D K3. Then by Proposition 4.5, we must have $\kappa(\omega|D) \neq 0$, so we conclude $N_D = \mathcal{O}_D$, hence D moves contrary to $a(X) = 0$. If D is a torus, we already may assume that $\omega|D$ is induced by a section of $\Omega_D^1 \otimes N_D^{*(\lambda+1)}$ which also shows that N_D is torsion.
(2) Now let D be singular K3. Then we still have the sequence (*) outside the finite singularity set of D. Moreover on all of D we have

$$0 \longrightarrow N_D^* \otimes (\Omega_D^1)^{**} \longrightarrow \Omega_X^2|D \longrightarrow (\bigwedge^2 \Omega_D^1)^{**} = K_D \longrightarrow 0.$$

Therefore the same arguments as above lead to the claim. □

In the next proposition $\mathcal{K}(S)$ denotes the Kähler cone in S, i.e. the cone in $H^{1,1}(S)$ generated by the Kähler classes. Moreover we let $\mathcal{K}^*(S)$ be the dual cone in $H^{n-1,n-1}(S)$. This cone is closed by definition.

PROPOSITION 4.4. *Let S be a K3 surface without curves. Then*

1. *Let T be a positive closed current on S. Then $[T] \in \overline{\mathcal{K}(S)}$.*
2. *$\mathcal{K}^*(S) = \overline{\mathcal{K}(S)}$.*
3. *If L is a line bundle on S such that $L \cdot \omega \geq 0$ for all Kähler forms ω, then $L \simeq \mathcal{O}_S$.*
4. *$H^0(S, \Omega_S^1 \otimes L) = 0$ for all line bundles L.*

PROOF. (1) Corollary 6.4 in (Demailly, 1992).
(2) One inclusion being trivial, we let $0 \neq x \in \mathcal{K}^*(S)$. Write $x = [T]$ with a positive $(1,1)$-current T, see (Oguiso and Peternell, 2001) for a proof using Hahn-Banach techniques. Then (1) implies the claim (2).
(3) By (2) L is nef, hence $L^2 = 0$ and Riemann-Roch yields the claim.
(4) Otherwise we have a locally free subsheaf $L^* \subset \Omega_S^1$ of rank 1. Since Ω_S^1 is ω-semi-stable for all ω, we have $L \cdot \omega \geq 0$ for all Kähler forms ω, hence L is nef, hence trivial by (3), which is absurd. □

The case of a singular K3 surface is taken care by

PROPOSITION 4.5. *Let S be singular K3, i.e. S arises by blowing down all $(-2)-$ curves in a K3 surface with $a = 0$. Then $H^0(S, (\Omega_S^1)^{**} \otimes L) = 0$ for all line bundles L on S.*

PROOF. Suppose $H^0(S,(\Omega^1_S)^{**}\otimes L)\neq 0$ for some L. Let $\pi:\hat{S}\longrightarrow S$ be the minimal desingularisation. Then also

$$H^0(\hat{S},\Omega^1_{\hat{S}}\otimes\hat{L})\neq 0,$$

where $\hat{L}=\pi^*(L)$. As in 4.3, we have $\hat{L}\cdot\hat{\omega}\geq 0$ for all Kähler forms $\hat{\omega}$ on $\hat{S}$. Thus $\hat{L}\in\overline{\mathcal{K}^*(\hat{S})}$. Hence $c_1(\hat{L})$ can be represented by a positive closed current $\hat{T}$. By Paun (1998), $[T]$, i.e. L, is nef, if $L\cdot C\geq 0$ for all curves $C\subset\hat{S}$. But this is clear since all C are contracted by π. Hence $\hat{L}$ is nef, therefore trivial as in 4.3. This is absurd. □

5. Threefolds with Picard number 1

We now will study threefolds with $\rho(X)=1$. To get precise results, we will assume also that the (algebraic) dual Picard number $\rho_c(X)$ is not too big, i.e. $\rho_c(X)\leq 1$. Sometimes it is even sufficient to make this last assumption without condition on ρ. The only general statement on threefolds just with $\rho(X)=1$ we make is

PROPOSITION 5.1. *Let X be a smooth compact Kähler threefold with $\rho(X)=1$.*

1. *Assume K_X not nef and that X is not simple with $\kappa(X)=-\infty$ (or that X is nearly algebraic with K_X not nef). Then either X is Fano with $\rho(X)=1$ or there exists a contraction $\varphi:X\longrightarrow Y$ with $\dim Y\geq 2$ such that $\rho(Y)=0$. Thus Y is Kummer with $a(Y)=0$ resp. a torus or K3 in case Y is a surface.*
2. *If K_X is nef, then $\kappa(X)\geq 0$. If moreover X is not both simple non-Kummer (with $\kappa(X)=0$), or if X is nearly algebraic, then K_X is semi-ample and we have a holomorphic Iitaka reduction $f:X\to B$ with $\rho(X)=\rho(B)=1$.*

PROOF. Everything is clear from sect. 1 except that we have to explain why the threefold Y is Kummer. Since $\rho(Y)=0$, we have $K_Y\equiv 0$. If Y is Gorenstein, then Riemann-Roch, see (Reid, 1987), gives

$$\chi(\mathcal{O}_Y)=0,$$

hence $q(Y)\geq 1$. Considering the Albanese, it follows from $\rho(Y)=0$ that $q(X)\geq 2$. If $q(Y)=3$ the Albanese map $\alpha:Y\to A$ must be finite and unramified, hence an isomorphism. If $q(Y)=2$, then we have an elliptic fibration over a 2-dimensional torus, so Y is not simple and (Campana and Peternell, 2000, 8.1) gives the claim. It remains to consider the non-Gorenstein case. Then by classification (Mori, 1982) the exceptional divisor E of φ is $\mathbb{P}_2$ with normal bundle $N_E=\mathcal{O}(-2)$. Since $K_Y\equiv 0$, it follows $2K_X\equiv E$. By Riemann-Roch $-K_X\cdot c_2(X)$ is divisible by 24, on the other hand we compute easily

$$-K_X\cdot c_2(X)=-\frac{1}{2}c_2(X)\cdot E=-\frac{3}{2},$$

that is a contradiction. □

PROPOSITION 5.2. *Let X be a smooth compact Kähler threefold with $\rho(X) = 1$ and $\rho_c(X) = 0$. Then either X is Kummer or X is simple, K_X is not pseudo-effective (in particular $\kappa(X) = -\infty$), $h^{1,1}(X) \geq 19$ and X has no divisors.*

PROOF. Everything follows from 4.2 and 4.3. The only thing to be careful on is the non-existence of divisors in the second case. In fact suppose X simple, K_X pseudo-effective and that D is an irreducible hypersurface on X. Then by 4.3, D is a torus with $N_D \equiv 0$, so D is nef. Since $\rho(X) = 1$ and since $K_X \not\equiv 0$, there is a rational number a such that $D \equiv aK_X$. K_X not being pseudo-effective and D being nef, we conclude $a < 0$. Since $K_X^3 = 0$, Riemann-Roch gives

$$\chi(X, -mK_X) = (2m-1)\chi(X, \mathcal{O}_X).$$

By an analytic version of the Kawamata-Viehweg vanishing theorem (Demailly and Peternell, 2001), we have $H^2(X, -mK_X) = 0$ for $m \geq 2$ (notice $K_X^2 \not\equiv 0$), so $h^0(-mK_X)$ grows at least linearly (since $\chi(X, \mathcal{O}_X)$ is clearly positive). So $a(X) \geq 1$, contradicting the simplicity of X. □

Of course, if K_X is not pseudo-effective, we expect to find a curve $K_X \cdot C < 0$ so that $\rho_c(X) \geq 1$.

In the rest of this section we assume $\rho_c(X) = 1$, so $\overline{NE}(X)$ is generated by a curve C. If $K_X \cdot C < 0$, we find a contraction $f : X \longrightarrow Y$ to a normal space Y contracting exactly the curves homologous to multiples of C, hence all curves. Moreover $-K_X$ is f-ample (2.5).

THEOREM 5.3. *Let X be a compact Kähler threefold with $\rho_c(X) = 1$. Suppose that $K_X \cdot C < 0$. Let $f : X \longrightarrow Y$ be the associated contraction. Then one of the following holds.*

1. $\dim Y = 0$ *and X is Fano with $\rho(X) = 1$.*
2. $\dim Y = 2$, *Y is a torus or K3 without curves, and f is a $\mathbb{P}_1$-bundle.*
3. $\dim Y = 3$ *and f contracts a divisor to a point; $\rho_c(Y) = 0$.*

If moreover $\rho(X) = 1$, then in (2) we have $\rho(Y) = 0$ and in (3) Y is a torus.

PROOF. (a) First we rule out $\dim Y = 1$. In fact in that case the general fiber is a rational surface and this implies that X is projective contradicting $\rho_c(X) = 1$.
(b) Let $\dim Y = 2$. Then f is a conic bundle. Therefore over any curve $D \subset Y$ we find many transversal curves contradicting $\rho_c(X) = 1$. So $\rho_c(Y) = 0$ and by classification the smooth surface Y is a torus or K3. Moreover f is a $\mathbb{P}_1$-bundle.
(c) Suppose the exceptional divisor E of f contracted to a curve B. Then $E \longrightarrow B$ is a $\mathbb{P}_1$-bundle and again we find transversal curves.
(d) Now suppose additionally that $\rho(X) = 1$. Then $\rho(Y) = 0$ and we conclude by (5.1) noticing that the proof of (5.1) implies that a Kummer Gorenstein threefold Y with $\rho(Y) = 0$ and without any curves must be a torus. □

We deal now with the case that $K_X \cdot C \geq 0$ By (2.5) this implies that K_X is nef unless possibly X is simple with $\kappa(X) = -\infty$. We will again ignore this at the moment untractable case. Then by (2.7) some multiple mK_X is spanned by global sections, unless X is simple non-Kummer with $\kappa(X) = 0$. We let $f : X \longrightarrow B$ be the *Iitaka fibration*, defined by the sections of mK_X. Before we formulate the theorem, let us consider (partly potential) examples.

EXAMPLES AND PROBLEMS 5.4. (1) Let A be a torus of dimension at least 3, $h : A \to A'$ an elliptic fiber bundle over the abelian variety A' such that $a(A) = a(A')$. Let $Z' \subset A'$ a surface, possibly singular, and let $Z = h^{-1}(Z')$. Assume that there is a smooth surface B and a finite map $g : B \to Z'$. Let $X = Z \times_{Z'} B$. Then X is smooth and we have an elliptic fiber bundle structure $f : X \to B$. If Z' is generic, then B is of general type. Then $\kappa(X) = 2$ and f is the Iitaka fibration.
Hopefully it is possible to arrange $\rho(B) = 1$ and that moreover $\rho(X) = 1$. Certainly it is reasonable to start with $\rho(A) = \rho(A') = 1$.

(2) We take a 3-dimensional torus X with $a(X) = \rho(X) = 1$ such that the fibers of the algebraic reduction have $a(F) = 1$. We can (equivalently) start with a 2-torus T with $a(T) = 1$ and then take an elliptic bundle $X \to T$ with $a(X) = a(T)$. Then clearly $\rho_c(X) = 1$. The connection between X and T is provided by the relative algebraic reduction.

(3) Let T be a 3-dimensional torus with $a(X) = \rho(X) = \rho_c(X) = 1$ such that the algebraic reduction $h : T \to C$ has algebraic fibers (i.e. abelian surfaces). Let $g : B \to C$ be a finite ramified cover. Then let $X = T \times_C B$. Then K_X is trivial on every curve. Under which circumstances is $\rho(X) = \rho_c(X) = 1$?

(4) Let B be a curve of genus at least 2 and F a K3 surface with $\rho(F) = 0$. Put $X = B \times F$. Then $\rho(X) = \rho_c(X) = 1$ and $a(X) = \kappa(X) = 1$. Furthermore K_X is ample on every curve.

In the next theorem we will use the following notations. If B is a normal compact surface (we will have only rational singularities), then we let $\tilde{q}(B)$ be the maximum of all irregularities $q(\tilde{B})$, where $\tilde{B} \to B$ is a finite cover, ramified at most over the singularities of B. Furthermore, given reflexive sheaves L_i over B, we let $L_1 \hat{\otimes} L_2 := (L_1 \otimes L_2)^{**}$.

THEOREM 5.5. *Let X be a smooth compact non-projective Kähler threefold which is not both simple non-Kummer. Assume that K_X is nef and that $\rho(X) = \rho_c(X) = 1$. Then using the above notations, we have the following.*

1. $\dim B \leq 2$,
2. *If* $\dim B = 0$, *then X is a torus,*
3. *If* $\dim B = 1$ *and if* $K_X \cdot C = 0$, *then all fibers of f are irreducible,* $a(X) = 1$ *and the general fiber of f is a torus T with* $a(T) \geq 1$.

4. *If* $\dim B = 2$, *then* f *is equidimensional. The set*

$$\Sigma = \{b \in B \mid \text{red}X_b \text{ is not elliptic }\}$$

is finite. If $\Delta_i \subset B$ *denotes the closure of the locus of the elliptic fibers of multiplicity* m_i, *and if we put*

$$\Delta = \sum_i (1 - \frac{1}{m_i})\Delta_i,$$

then $K_X \equiv f^*(K_B + \Delta)$. *The fibration* f *has constant* $j-$ *invariant. Moreover* (B, Δ) *is log terminal,* $\rho(B) = 1$, K_B *is* $\mathbb{Q}$*-Cartier and* K_B *and* Δ *are proportional. The canonical bundle* K_B *is ample or numerically trivial. Finally* B *has the following properties, where* $\pi : \hat{B} \to B$ *denotes the minimal desingularisation.*

   a) $q(X) = q(B)$ *unless* X *is of type (5.4(1)); and always* $\chi(X, \mathcal{O}_X) = 0$.
   b) $L := f_*(K_{X/B})^{**}$ *is* $\mathbb{Q}$*-Cartier and nef;* $f_*(K_X)$ *is reflexive with*

$$f_*(K_X) = L \hat{\otimes} K_B$$

   and $R^1 f_*(\mathcal{O}_X) \simeq I_A L^*$ *with a finite set* A, *in particular it is torsion free. If* $L \equiv 0$, *then* L *is torsion, i.e.* $mL = \mathcal{O}_B$ *for some positive integer* m.
   c) *If* B *contains a rational or elliptic curve, then* $q(B) = 0$.
   d) $\kappa(\hat{B}) \geq 0$ *and if* $\kappa(\hat{B}) = 0$, *then* B *can only be singular and birationally K3.*
   e) *If* $\kappa(\hat{B}) = 1$, *then* B *is singular and has a non-canonical singularity.*
   f) *If* $\kappa(\hat{B}) = 2$ *and if* B *is smooth, then possibly after étale base change,* X *is of type (5.4(1)).*

5. *Suppose* $K_X \cdot C > 0$. *Then either* $X = B \times F$ *where* F *is K3 with* $\rho(F) = 0$ *and* $C = B \times \{x\}$ *with some* $x \in F$ *or* F *is a torus and there exists a finite étale cover* $h : \tilde{X} \to X$ *of the form* $\tilde{X} \simeq \tilde{F} \times \tilde{B}$ *with* $\tilde{F}$ *a torus projecting onto* F *such that* $C = h(\tilde{B} \times \{x\})$.

PROOF. (1) $\dim B = 3$ is clearly not possible, since B would be a projective 3-fold, so $\rho_c(B) \geq 1$. On the other hand f contracts curves, so $\rho_c(X) \geq 2$.

(2) If $\dim B = 0$, then $mK_X = \mathcal{O}_X$ and the claim follows from Beauville's decomposition theorem.

(3) If some fiber of f would be reducible, then clearly $\rho(X) \geq 2$. If $a(X) = 2$, then consider an algebraic reduction $g : X \rightharpoonup S$ to a projective surface S. Now the elliptic fibration g is holomorphic near the general fiber F of f, at least if S is chosen appropriately. Let $\pi : \hat{X} \longrightarrow X$ be a sequence of blow-ups such that the induced map $\hat{g} : \hat{X} \to S$ is holomorphic. Then choose L ample on S and put $\hat{L} = \hat{g}^*(L)$. Then $G = (\pi_*(\hat{L}))^{**}$ is a line bundle on X and its restriction to F is ample on F.

This shows $\rho(X) \geq 2$, contradiction. So $a(X) = 1$ and f is the algebraic reduction contracting all curves of X. Using Fujiki's classification theorem (Fujiki, 1983), our assumption $K_X \cdot C = 0$ implies that X is not bimeromorphic to a quotient of a product, hence again by (Fujiki, 1983), the general fiber F of f has $a(F) \geq 1$.

(4) If $\dim B = 2$, then by $\rho_c(X) = 1$, we automatically have $K_X \cdot C = 0$ and all curves in X are contracted by f. From $\rho(X) = 1$, it is clear that f has to be equidimensional.
Next notice that there can be only finitely many fibers X_b which are not elliptic or multiples of an elliptic curve. Otherwise we would have an algebraic surface $S \subset X$ such that $f(S)$ is a curve and therefore we would obtain a curve D not contracted by f.
From (Nakayama, 1987, 0.4) it follows that $K_X \equiv f^*(K_B + \Delta)$ as $\mathbb{Q}$-divisors and that (S, Δ) has only log terminal singularities, in particular rational, singularities. Notice that (Nakayama, 1987) is applicable since f is a locally projective Kähler morphism.
It is also clear that $\rho(B) = 1$. Since B is $\mathbb{Q}$-factorial, K_X and Δ must be proportional. Concerning the j-invariant, we have a meromorphic map $j : B \setminus \Sigma \to \mathbb{P}_1$ which must be constant because of $\rho(B) = 1$.

(4.a) Let $\alpha : X \to A$ be the Albanese of X. Let F be a general fiber of f, an elliptic curve. If $\dim \alpha(F) = 0$, then α factors over f, hence $q(X) = q(B)$. So we may assume $\dim \alpha(F) = 1$. Let $Z = \alpha(A)$. Since $\rho(X) = 1$, we have $\dim Z \neq 1$.
We next rule out the case $\dim Z = 2$. So suppose $\dim Z = 2$. Then Z cannot be projective: since $\dim \alpha(F) = 1$, the map $X \to B \times Z$ is generically finite on to its image and then X would be projective. Moreover $a(Z) \neq 1$; otherwise we would have a map $X \to C$ to a curve C which would contradict $\rho(X) = 1$. But $a(Z) = 0$ is impossibly, too: if $a(Z) = 0$, then the inclusion $Z \subset A$ implies that Z itself must be a torus. But Z contains an elliptic curve and those tori have $a(Z) \geq 1$.
So $\dim Z = 3$, in particular $q(X) \geq 3$. Moreover $\alpha : X \to Z$ is finite. Now the images $\alpha(F)$ are elliptic curves in A and therefore we obtain an elliptic fiber bundle $h : A \to A'$ contracting all these elliptic curves. Then h induces an elliptic fiber bundle $Z \to Z'$ and a finite map $B \to Z'$ so that we are in case (5.4)(1).
Next we verify $\chi(X, \mathcal{O}_X) = 0$. Since $\rho(B) = 1$ and since mK_X is of the form $f^*(G)$ for suitable m, it is sufficient to show that

$$c_2(X) \cdot X_C = 0$$

for a general hyperplane section $C \subset B$. But

$$c_2(X) \cdot X_C = c_2(X_C) + N^*_{X_C} \cdot K_{X_C} = c_2(X_C)$$

and $c_2(X_C) = 0$, since X_C is an elliptic fibration over C with only multiple elliptic curves as singular fibers.

We finally prove that in case $L \equiv 0$, the bundle L is torsion. Take a general hyperplane section $C \subset B$ and consider the smooth elliptic surface $X_C \to C$. Since X_C has only multiple elliptic curves as singular fibers, L_C is torsion (e.g. III.18.3 in (Barth, Peters and Van de Ven, 1984)). Choose m positive such that mL is Cartier and that $mL_C = \mathcal{O}_C$. Then the exact sequence

$$H^0(B, mL) \to H^0(C, mL|C) \to H^1(B, mL - C) = 0$$

(by Kodaira vanishing) gives $H^0(mL) \neq 0$, hence $mL = \mathcal{O}_B$.

(4.b) L is $\mathbb{Q}$-Cartier since B is $\mathbb{Q}$-factorial. The nefness of L follows from (Fujita) and $\rho(B) = 1$. The reflexivity of $f_*(K_X)$ follows immediately by checking extendability of sections over codimension 2 sets using the equidimensionality of f. The formula for $f_*(K_X)$ being clear, it remains only to prove the torsion freeness of $R^1 f_*(\mathcal{O}_X)$, the rest being a consequence of relative duality. By our description of f, the torsion part of $R^1 f_*(\mathcal{O}_X)$ has finite support S. If however $S \neq \emptyset$, then $H^0(R^1 f_*(\mathcal{O}_X)) \neq 0$. Since $H^2(\mathcal{O}_B)$ injects canonically into $H^2(\mathcal{O}_X)$ (e.g. desingularise B, make a base change and consider 2-forms), the Leray spectral sequence implies $q(X) > q(B)$. So we are in the special case (5.4(1)), in which our claim is clear anyway!

(4.c) Suppose $q(B) > 0$. Since B has only rational singularities, we have a nontrivial Albanese map

$$\alpha : B \to A = \mathrm{Alb}(B) = \mathrm{Alb}(\hat{B}).$$

If $C \subset B$ is a rational curve, then $\dim \alpha(C) = 0$, hence we obtain a contradiction to $\rho(B) = 1$, since $\alpha(B)$ is projective. So let C be elliptic. Arguing as before, we see that $\alpha|B$ must be finite onto its image. Thus A contains an elliptic curve and then A has a torus fiber bundle structure $p : A \to E$ to an elliptic curve E such that $p(\alpha(C)) = E$. This again contradicts $\rho(B) = 1$.

(4.d) We first show the following.
CLAIM A. Suppose that $h^2(X, \mathcal{O}_X) > h^2(B, \mathcal{O}_B)$. Then $\tilde{q}(B) \neq 0$.
In fact, Serre duality applied to our assumption gives

$$h^1(K_X) > h^0(K_B). \qquad (*)$$

By the Leray spectral sequence,

$$H^1(K_X) \subset H^1(f_*(K_X)) \oplus H^0(R^1 f_*(K_X)).$$

Using (b) we have

$$H^1(f_*(K_X)) = H^1(L \hat{\otimes} K_B)).$$

Assume $L \not\equiv 0$. Then L is ample, and (Kawamata, Matsuda and Matsuki, 1987, 1-2-5) gives

$$H^1(L \hat{\otimes} K_B) = 0.$$

Noticing $R^1 f_*(K_X) \subset K_B$ by virtue of the relative duality

$$R^1 f_*(K_{X/B}) \simeq \mathcal{O}_B,$$

we obtain

$$H^1(K_X) \subset H^0(R^1 f_*(K_X)) \subset H^0(K_B).$$

But this contradicts (*) so that L must be numerically trivial and the above arguments also show $H^1(L \hat{\otimes} K_B) \neq 0$. If $L = \mathcal{O}_B$, then we conclude $H^1(K_B) \neq 0$, hence $q(B) \neq 0$ which implies our claim. So suppose L not trivial; also we may assume $q(B) = 0$. Choose n_0 such that $n_0 L$ is Cartier. Then $n_0 L$ is torsion: there exists m so that $mn_0 L = \mathcal{O}_B$ (either by $q(B) = 0$ or by (b)). By (Reid, 1987) e.g., there exists $h : \tilde{B} \to B$, unramified outside the singular points of B, such that $h^*(L)^{**} = \mathcal{O}_{\tilde{B}}$. Now we check that

$$H^1(L \hat{\otimes} K_B) \subset H^1(h^*(L)^{**} \otimes K_{\tilde{B}}) = H^1(K_{\tilde{B}}) \qquad (**)$$

which gives $\tilde{q}(B) \neq 0$. To prove (**), let $S \subset B$ denote the singular locus of B and let $\tilde{S} \subset \tilde{B}$ be the preimage. Then clearly (**) holds on outside $\tilde{S}$ resp. S. Now pick $\alpha \in H^1(L \hat{\otimes} K_B)$. Then $h^*(\alpha)$ induces a cohomology class in $H^1((h^*(L)^{**} \hat{\otimes} K_{\tilde{B}})$, which is non-zero by the previous remark and the injectivity part of Riemann's extension theorem. This establishes (**) and therefore Claim (A).

Claim A applies to all B with $H^2(B, \mathcal{O}_B) = 0$, since $H^2(X, \mathcal{O}_X) \neq 0$. So we have in that case $\tilde{q}(B) \neq 0$. Hence B cannot be a rational surface nor birationally an Enriques surface. But B cannot be birationally ruled over a curve C of positive genus either: since B has only rational singularities, π can only contract rational curves, hence B inherits the ruling map $\hat{p} : \hat{B} \to C$ contradicting $\rho(B) = 1$. A similar argument applies to hyperelliptic surface. So we are reduced to B being birationally a torus or K3.

(4.d.1) Suppose B is K3, so in particular $K_B = \mathcal{O}_B$. Then we distinguish the cases $L = \mathcal{O}_B$ and L ample. In the first case $h^0(L) \leq 1$, hence $h^0(K_{X/B}) \leq 1$ and therefore (by $K_B = 0$) $h^0(K_X) = h^3(\mathcal{O}_X) \leq 1$. Now $q(X) = q(B) = 0$ by (a), therefore

$$0 = \chi(X, \mathcal{O}_X) = 1 + h^2(\mathcal{O}_X) - h^3(\mathcal{O}_X)$$

produces a contradiction.
If L is ample, then (Kawamata, Matsuda and Matsuki, 1987, 1-2-5) gives

$$H^q(f_*(K_X)) = H^q(L \otimes K_B) = H^q(L) = 0$$

for $q = 1, 2$. So by Leray

$$H^2(\mathcal{O}_X) = H^0(R^1 f_* K_X) = H^0(K_B) = \mathbb{C}.$$

Since $q(X) = q(B) = 0$, we derive from (b) that

$$0 = \chi(\mathcal{O}_X) = 1 + 1 - h^0(K_X),$$

so $h^0(L) = 2$. Now Riemann-Roch reads

$$\chi(L) = \frac{L^2}{2} + \chi(\mathcal{O}_B) = \frac{L^2}{2} + 2.$$

By Kodaira vanishing, $h^0(L) = \chi(L)$, hence we obtain $L^2 = 0$. This contradicts the ampleness of L.

(4.d.2) Suppose B is birationally a torus. By (c) B does not contain any rational curve, hence B must be an abelian surface with $\rho(B) = 1$.
First assume $L = \mathcal{O}_B$. Then $h^0(K_X) = 1$; let $s \in H^0(K_X)$ be a non-zero section and $D = \sum \lambda_i D_i$ be the associated divisor. Then $f_*(s) \in H^0(L)$ has no zeroes, which means that D cannot contain any fiber of f. Thus all D_i must project onto B, so that f is projective and X is projective, too. So L cannot be trivial.
If $L \equiv 0$, then we notice that $f^* f_*(K_X) = f^*(L) \subset K_X$, so that $H^0(K_X \otimes f^*(L^*)) \neq 0$. Since $f_*(K_X \otimes f^*(L^*)) = f_*(K_X) \otimes L^* = \mathcal{O}_B$, we can apply the same arguments as before.
So we are reduced to the case L ample. Then Riemann-Roch and Kodaira vanishing give

$$h^0(L) = \chi(L) = \frac{L^2}{2}. \qquad (*)$$

The Leray spectral sequence gives

$$H^2(\mathcal{O}_X) = H^1(K_X) = H^0(R^1 f_*(K_X)) = \mathbb{C};$$

moreover we cannot be in the special case (5.4(1)), since then we would have $\kappa(X) = 0$ (B being a torus), so $q(X) = q(B) = 2$ by (a). From $\chi(X, \mathcal{O}_X) = 0$ we therefore derive $h^3(\mathcal{O}_X) = 0$, contradicting (*).

(4.e) Let $\kappa(\hat{B}) = 1$. In case B is smooth, then clearly $\rho(B) \geq 2$, ruling out that case. Suppose B has only canonical singularities. Since $\rho(B) = 1$ and B is projective, K_B is ample, negative or trivial. This is incompatible with $\kappa(K_B) = \kappa(\hat{B}) = 1$. Of course, this argument fails in the presence of non-canonical singularities.

(4.f) Now let B be smooth and of general type. Suppose first that $L = \mathcal{O}_B$ and that X is not of type (5.4(1)). Then $q(X) = q(B)$ implies that the canonical map

$$\tau : H^0(R^1 f_*(\mathcal{O}_X)) = H^0(I_A) \to H^2(\mathcal{O}_B)$$

is injective. Since $H^2(\mathcal{O}_B)$ injects into $H^2(\mathcal{O}_X)$, it follows that $\tau = 0$. So $A \neq \mathbf{0}$. Moreover

$$H^1(R^1 f_*(\mathcal{O}_X)) = H^1(I_A),$$

hence

$$h^2(\mathcal{O}_X) = h^2(\mathcal{O}_B) + h^1(\mathcal{O}_X) + l(A) - 1.$$

Finally

$$h^3(\mathcal{O}_X) = h^0(K_X) = h^0(K_B) = h^2(\mathcal{O}_B),$$

so, putting things together:

$$0 = \chi(X, \mathcal{O}_X) = l(A).$$

But $A \neq \emptyset$, as seen above, contradiction.
If $L \equiv 0$ but not trivial, then by (b), L is torsion, so performing an étale base change, L will be trivial, which just was seen to imply X to be in case (5.4(1)) (it was not necessary here to have $\rho(X) = 1$).
It remains to treat the case that L is ample. Then computing $h^q(\mathcal{O}_X)$ in the same way as before, but using Kodaira vanishing, we obtain

$$0 = \chi(X, \mathcal{O}_X) = \chi(\mathcal{O}_B) - \chi(L^*).$$

Hence $\chi(L^*) = \chi(\mathcal{O}_B)$; on the other hand by Riemann-Roch,

$$\chi(L^*) = \frac{L^2}{2} + L^* \cdot \frac{-K_B}{2} + \chi(\mathcal{O}_B),$$

so that in combination

$$L^2 + L \cdot K_B = 0$$

contradicting the ampleness of L and K_B.

(5) If $K_X \cdot C > 0$, we already know $\dim B = 1$. The case $a(X) = 2$ is ruled out as in (2). It is also clear that every curve in X projects onto B, so there are no curves in fibers. This means that every fiber is a torus, K3 or singular K3 with $a = 0$. Let F be the general fiber. From classification (Fujiki, 1983) it follows that X is bimeromorphic to $(\tilde{F} \times \tilde{B})/G$, where G is a finite group acting both on the K3 surface resp. torus $\tilde{F}$ and on the curve $\tilde{B}$ and on the product diagonally. We have $F = \tilde{F}/G$ and $B = \tilde{B}/G$. Now B is not rational, otherwise we would have a family of rational curves in X and X would be uniruled, hence $\kappa(X) = -\infty$. Therefore any bimeromorphic map

$$\phi : (\tilde{F} \times \tilde{B})/G \rightharpoonup X$$

must be holomorphic. Since both $\tilde{F}/G$ and F are minimal, ϕ is actually an isomorphism and $\tilde{F}/G = F$. Now the covering $\tilde{F} \to F$ cannot be ramified since F has no curves. If F is simply connected, $\tilde{F} = F$ and consequently $X = F \times B$. If $\tilde{F}$ is a torus, then F itself is a torus. □

We expect that the cases (4.d) and (4.e) in Theorem 5.5 do not exist.

References

Beauville, A. (1983) Variétés kählériennes dont la premiere classe de Chern est nulle, *J. Diff. Geom.* **18**, 755–782.

Barth, W., Peters, C., and Van de Ven, A. (1984) *Compact complex surfaces*, Erg. d. Math. 4, Springer.

Campana, F., and Peternell, Th. (1997) Towards a Mori theory on compact Kähler threefolds, *I. Math. Nachr.* **187**, 29–59.

Campana, F., and Peternell, Th. (2000) Holomorphic 2-forms on complex 3-folds, *J. Alg. Geom.* **9**, 223–264.

Demailly, J.P. (1992) Regularization of positive closed currents and intersection theory, *J. Alg. Geom.* **1**, 361–409.

Demailly, J.P., and Peternell, Th. (2001) A Kawamata-Viehweg vanishing theorem on compact Kähler manifolds, preprint in preparation.

Demailly, J.P., Peternell, Th., and Schneider, M. (1994) Compact complex manifolds with numerically effective tangent bundles, *J. Alg. Geom.* **3**, 295–345.

Demailly, J.P., Peternell, Th., and Schneider, M. (2000) Pseudo-effective line bundles on projective varieties and Kähler manifolds, preprint, to appear in *Intl. J. Math.*

Fujita, T. (1978) Kähler fiber spaces over curves, *J. Math. Soc. Japan* **30**, 779–794.

Fujiki, A. (1983) On the structure of compact complex manifolds in class *C*, *Adv. Stud. Pure Math.* **1**, 231–302.

Kawamata, Y., Matsuda, K., and Matsuki, K. (1987) Introduction to the minimal model program, *Adv. Stud. Pure Math.* **10**, 283–360.

Kollař, J., et al. (1992) *Flips and abundance for algebraic 3-folds*, Astérisque 211, Soc. Math. France.

Mori, S. (1982) Threefolds whose canonical bundle are not numerically effective, *Ann. Math.* **116**, 133–176.

Mori, S. (1988) Flip theorem and existence of minimal models of 3-folds, *J. Amer. Math. Soc.* **1**, 177–353.

Miyaoka, Y., and Peternell, Th. (1997) *Geometry of higher dimensional algebraic varieties*, DMV Seminar vol. 26, Birkhäuser.

Nakayama, N. (1987) On Weierstrass models, in *Algebraic Geometry and Commutative Algebra 2, volume in honour of Nagata*, Kinokuniya, Tokyo, pp. 405–431.

Oguiso, K., and Peternell, Th. (2000) Projectivity via the dual Kähler cone - Huybrechts' criterion, *Asian J. Math.* **4**, 213–220.

Oguiso, K., and Peternell, Th. (2001) The dual cone of compact Kähler threefolds, preprint in preparation.

Paun, M. (1998) Sur l'effectivité numérique des images inverses de fibrés en droites, *Math. Ann.* **310**, 411–421.

Peternell, Th. (1998) Towards a Mori theory on compact Kähler 3-folds, II, *Math. Ann.* **311**, 729–764.

Peternell, Th. (2000) Towards a Mori theory on compact Kähler 3-folds, III, preprint, to appear in *Bull. Soc. Math. France.*

Reid, M. (1987) Young person's guide to canonical singularities, *Proc. Symp. Pure Math.* **46**, 345–414.

ABELIAN VARIETIES OVER THE FIELD OF THE 20TH ROOTS OF UNITY THAT HAVE GOOD REDUCTION EVERYWHERE

R. SCHOOF
University of Rome "Tor Vergata", Italy

Abstract. The elliptic curve E given by $Y^2+(i+1)XY+iY=X^3+iX^2$ acquires good reduction everywhere over the cyclotomic field $\mathbb{Q}(\zeta_{20})$. We show, under assumption of GRH, that every abelian variety over $\mathbb{Q}(\zeta_{20})$ with good reduction everywhere is isogenous to E^g for some $g \geq 0$.

Key words: abelian varieties, group schemes, good reduction.

Mathematics Subject Classification (2000): 11G10, 14L05, 14L15.

1. Introduction

For $f = 1,3,4,5,7,8,9$ and 12 there do not exist any abelian varieties over the cyclotomic field $\mathbb{Q}(\zeta_f)$ that have good reduction modulo every prime. Under assumption of the Generalized Riemann Hypothesis (GRH) the same can be proved for $\mathbb{Q}(\zeta_{11})$ and $\mathbb{Q}(\zeta_{15})$. These are the main results of (Schoof, 2001b). For all other conductors f there do exist abelian varieties over $\mathbb{Q}(\zeta_f)$ with good reduction everywhere.

The techniques of (Schoof, 2001b) still give substantial information about abelian varieties with good reduction everywhere over cyclotomic fields that are not in this list. Ordering the fields with respect to their root discriminants, the first field not in the list is $\mathbb{Q}(\zeta_{20})$. Over this field the elliptic curve E given by the equation

$$Y^2+(i+1)XY+iY=X^3+iX^2.$$

has good reduction everywhere. This can be seen as follows. The discriminant of the equation is equal to $-(1+2i)^3$. Therefore E has good reduction outside the prime $1+2i$ of the ring $\mathbb{Z}[i]$. Since the reduction at $1+2i$ is of Kodaira type III, the curve E acquires good reduction everywhere over any Galois extension of $\mathbb{Q}(i)$ for which the ramification indices of the primes over $1+2i$ are divisible by 4. In particular, E acquires good reduction everywhere over $\mathbb{Q}(\zeta_{20})$.

C. Ciliberto et al. (eds.),
***Applications of Algebraic Geometry to Coding Theory, Physics and Computation*, 291–296.**

The main result of this paper is the following.

THEOREM 1.1. *(GRH) Every abelian variety over $\mathbb{Q}(\zeta_{20})$ that has good reduction everywhere is isogenous to a power of E.*

The theorem is proved under assumption of the Generalized Riemann Hypothesis (GRH) for zeta functions of number fields. The main ingredients are the results of (Schoof, 2001b). See (Schoof, 2001a) for a similar result, proved without assuming GRH, for abelian varieties over $\mathbb{Q}(\sqrt{6})$.

The proof of the theorem proceeds by analyzing finite flat commutative group schemes of 2-power order over the ring $\mathbb{Z}[\zeta_{20}]$. We deduce that the 2-divisible group associated to any abelian variety A over $\mathbb{Q}(\zeta_{20})$ with good reduction everywhere is isogenous to the 2-divisible group associated to E^g for some $g \geq 0$. Faltings' isogeny theorem implies then that the abelian varieties A and E^g are isogenous over $\mathbb{Q}(\zeta_{20})$.

2. 2-group schemes over $\mathbb{Z}[\zeta_{20}]$

Let F denote the number field $\mathbb{Q}(\zeta_{20})$ and let $O_F = \mathbb{Z}(\zeta_{20})$. In this section we study finite flat commutative group schemes of 2-power order over the ring O_F or 2-*group schemes* for short. Finite flat group schemes of order 2 are examples of 2-group schemes. Since (2) is the square of the prime ideal generated by $1+i$ in O_F, it follows from the discussion on the first page of (Oort and Tate) there are three of these. Apart from the group schemes $\mathbb{Z}/2\mathbb{Z}$ and μ_2, there is an order 2 group scheme that is local-local at the prime 2. It can be described as follows. The elliptic curve E of section 1 has j-invariant 1728 and endomorphism ring isomorphic to $\mathbb{Z}[i]$. The kernel $E[f]$ of the endomorphism $f = 1+i$ in the ring $\mathrm{End}(E)$ is a finite flat group scheme of order 2. Its Hopf algebra is isomorphic to $O_F[T]/(T^2-(1+i)T)$ with group law $t+t'-(i-1)tt'$.

Group schemes of order 2 are *simple*. Under assumption of GRH, the converse is true over the ring O_F.

THEOREM 2.1. *(GRH) The only simple 2-group schemes over $\mathbb{Z}[\zeta_{20}]$ are $\mathbb{Z}/2\mathbb{Z}$, μ_2 and $E[f]$.*

PROOF. It suffices to show that all simple 2-group schemes have order 2 by the above discussion. We apply Proposition 2.2 of (Schoof, 2001b) to the prime $p=2$ and check "condition (A)". The root discriminant of the field L that appears in that proposition satisfies $\delta_L < 8\cdot 5^{3/4} = 26.749\ldots$. Since this number is larger than the asymptotic value $4\pi e^{\gamma} \approx 22.3\ldots$ of Odlyzko's unconditional discriminant bound, we use his GRH bounds (Martinet, pp. 178–179). These imply that $[L:\mathbb{Q}] < 600$. By the assumptions of Proposition 2.2 in (Schoof, 2001b), we have the following inclusions of fields

$$\mathbb{Q} \underset{16}{\subset} F(i) \underset{2\times 2\times 2}{\subset} K \underset{\leq 4}{\subset} L.$$

Here $K = \mathbb{Q}(\zeta_{40}, \sqrt{\varepsilon_1}, \sqrt{\varepsilon_2}, \sqrt{\varepsilon_3})$ where the ε_i are a basis for the unit group of O_F modulo torsion. Let $\Gamma = \mathrm{Gal}(L/\mathbb{Q})$. The field $F(i) = \mathbb{Q}(\zeta_{40})$ is the largest abelian extension of $\mathbb{Q}$ inside L. By Theorem 11.1 in (Washington) the class number of $\mathbb{Q}(\zeta_{40})$ is 1. In addition, the unit $1 - \zeta_{40}$ generates the multiplicative group of the residue field $\mathbb{F}_{16}$ of the unique prime over 2. Therefore, by class field theory, the field $\mathbb{Q}(\zeta_{40})$ admits no abelian odd degree extension inside L. This implies that Γ'/Γ'' is a 2-group. By Proposition 3.2 in (Schoof, 2001b) we then see that Γ'' and hence $\mathrm{Gal}(L/F)$ are 2-groups.

It follows from Proposition 2.2 in (Schoof, 2001b) that all simple 2-group schemes over O_F have order 2 as required. □

The next theorem describes various extensions of the three group schemes of order 2 by one another.

THEOREM 2.2. *Over the ring O_F we have that*

(i) Any extension of constant 2-group schemes is constant; any extension of diagonalizable 2-group schemes is diagonalizable;
(ii) $\mathrm{Ext}^1_{O_F}(\mu_2, \mathbb{Z}/2\mathbb{Z}) = 0$;
(iii) the groups $\mathrm{Ext}^1_{O_F}(E[f], \mathbb{Z}/2\mathbb{Z})$ *and* $\mathrm{Ext}^1_{O_F}(\mu_2, E[f])$ *are zero;*
(iv) the group $\mathrm{Ext}^1_{O_F}(E[f], E[f])$ *has order 2; its non-trivial element is represented by the group scheme $E[2]$ of 2-torsion points of the elliptic curve E.*

PROOF. By Proposition 2.6 in (Schoof, 2001b), parts *(i)* and *(ii)* follow from the facts that the class number of O_F is 1 and that there lies only one prime over 2 in O_F. To prove *(iii)*, consider an extension

$$0 \longrightarrow \mathbb{Z}/2\mathbb{Z} \longrightarrow G \longrightarrow E[f] \longrightarrow 0$$

over O_F. Since $E[f]$ is local, the extension is split over the completion $\widehat{O}_F$ at the prime $\pi = 1 + i$. Therefore G is killed by 2. The quadratic character that gives the Galois action is everywhere unramified and hence trivial. It follows that G is locally and generically trivial. Since there lies only one prime over 2, the Mayer-Vietoris sequence (Corollary 2.4 in (Schoof, 2001b)) then implies that G is split over O_F. This proves *(iii)*.

To prove *(iv)* consider an extension

$$0 \longrightarrow E[f] \longrightarrow G \longrightarrow E[f] \longrightarrow 0$$

over O_F. We use local results of Cornelius Greither (1992). For $n \geq 0$ let U_n denote the multiplicative group $\{\varepsilon \in \widehat{O}_F^* : \varepsilon \equiv 1 \pmod{\pi^n}\}$. By Greither's theorem, $\mathrm{Ext}^1_{\widehat{O}_F}(E[f], E[f]) \cong U_3/U_2^2 = U_3/(U_3 \cap (\widehat{O}_F^*)^2)$. Moreover, it follows from the arguments in (Greither) that the points of the extension G corresponding to a unit $\varepsilon \in U_3$, generate the field $\mathbb{Q}_2(\zeta_{20}, \sqrt{\varepsilon})$. This implies that G is determined by its Galois module.

Since $\pm\varepsilon \equiv 1 \pmod{\pi^2}$, it follows that the number field generated by the points of G has conductor at most π^2 over F. The ray class field of F of conductor π^2 has degree 2. This follows from the fact that the global units 1 and $(1-\zeta_{20}^a)^{15}$ generate a subgroup of $(1+(\pi))/(1+(\pi^2)) \cong \mathbb{F}_{16}$ of index 2. Therefore the Galois action on the points of G is either trivial or is given by the unique character of conductor (2). If it is trivial, the action is also locally trivial and hence G is trivial over $\widehat{O}_F$. By the equivalence of categories of Proposition 2.3 in (Schoof, 2001b), the extension G is then trivial. If the Galois action is not trivial, then locally it is also non-trivial. This follows from the fact that the class number of O_F is 1. This fixes the structure of G over $\widehat{O}_F$ and hence, by Proposition 2.3 in (Schoof, 2001b), there is only one choice for G. The exact sequence

$$0 \longrightarrow E[f] \longrightarrow E[2] \longrightarrow E[f] \longrightarrow 0$$

is non-split since the 2-torsion points of E generate the quadratic extension $F(\sqrt{\eta})$ of F. Here η denotes the unit $(1+\sqrt{5})/2 \in O_F^*$. Therefore $E[2]$ provides the non-trivial class in $\mathrm{Ext}^1_{O_F}(E[f], E[f])$. □

3. 2-divisible groups over $\mathbb{Z}[\zeta_{20}]$

In this section we prove Theorem 1.1. For any abelian variety A over $F = \mathbb{Q}(\zeta_{20})$ with good reduction everywhere, the group scheme $A[2^n]$ is a 2-group scheme over $O_F = \mathbb{Z}[\zeta_{20}]$.

PROPOSITION 3.1. *Let A be an abelian variety over F with good reduction everywhere. Then the 2-group scheme $A[2^n]$ admits for each $n \geq 1$, a filtration with subquotients isomorphic to the group scheme $E[f]$.*

PROOF. First consider the 2-torsion subgroup scheme $A[2]$ over O_F. It admits a filtration with simple subquotients. By Theorem 2.1 the simple subquotients of this filtration are isomorphic to $\mathbb{Z}/2\mathbb{Z}$, μ_2 or $E[f]$. By Theorem 2.2 *(ii)* and *(iii)* we can modify the filtration and obtain closed flat subgroup schemes G_1 and G_2 of $A[2]$ for which

$$0 \hookrightarrow G_1 \hookrightarrow G_2 \hookrightarrow A[2], \qquad (*)$$

and where G_1 is filtered with group schemes isomorphic to μ_2, the quotient G_1/G_2 is filtered with group schemes isomorphic to $E[f]$ and $A[2]/G_2$ is filtered with group schemes isomorphic to $\mathbb{Z}/2\mathbb{Z}$. Let 2^a and 2^b denote the orders of G_1 and $A[2]/G_2$ respectively.

Next consider the group scheme $A[2^n]$. We filter it with its closed subgroup schemes $A[2^i]$ for $i = 1, \ldots, m$. All subquotients in this filtration are isomorphic to $A[2]$ and we filter these as in (*). By Theorem 2.2, *(ii)* and *(iii)*, we can modify this filtration and obtain closed flat subgroup schemes H_1 and H_2 of $A[2^n]$ for which

$$0 \hookrightarrow H_1 \hookrightarrow H_2 \hookrightarrow A[2^n], \qquad (*)$$

and where H_1 is filtered with group schemes isomorphic to μ_2 and has order 2na, where the quotient H_1/H_2 is filtered with group schemes isomorphic to $E[f]$ and where $A[2]/H_2$ is filtered with group schemes isomorphic to $\mathbb{Z}/2\mathbb{Z}$ and has order 2^{nb}.

By Theorem 2.2, *(i)*, the group scheme $A[2^n]/H_2$ is constant. It is a closed subgroup scheme of the abelian variety A/H_2. It follows from Weil's Riemann Hypothesis that the order 2^{bn} is bounded as $n \to \infty$. This is only possible when $b = 0$. Applying the same argument to the Cartier dual $H_1^\vee$ of H_1 and the abelian variety $A^{\text{dual}}/H_1^\vee$ we see that $a = 0$ as well. It follows that $H_2 = A[2^n]$ and $H_1 = 0$ respectively. □

Theorem 1.1 is now proved by an application of Proposition 3.2 below. The endomorphism ring $\text{End}(\mathcal{E})$ of the 2-divisible group $\mathcal{E}$ of the elliptic curve E is isomorphic to the discrete valuation ring $\mathbb{Z}_2[i]$. Let f denote the prime element $1+i \in \text{End}(\mathcal{E})$. The kernel of the morphsim $f : \mathcal{E} \longrightarrow \mathcal{E}$ is denoted by $\mathcal{E}[f]$. It is the group scheme $E[f]$ of order 2 that appears in section 2. By Proposition 3.1 the subgroup schemes $A[2^n]$ of an abelian variety A with good reduction everywhere over $\mathbb{Q}(\zeta_{20})$ is filtered by group schemes isomorphic to $\mathcal{E}[f]$. By Theorem 2.2 *(iv)* and the fact that the points of $E[2]$ are not defined over F, the condition of Proposition 3.2 below is satisfied. It follows that the 2-divisible group of A is isogenous to $\mathcal{E}^g$ for some $g \geq 0$. Faltings' Theorem implies then that A and E^g are isogenous over F as required, see (Faltings).

PROPOSITION 3.2. *Let O be a Noetherian domain of characteristic 0 and let $\mathcal{G}$ be a p-divisible group over O. Suppose that $R = \text{End}(\mathcal{G})$ is a discrete valuation ring with prime element f. Let $k = R/fR$. If the connecting homomorphism*

$$\text{Hom}_O(\mathcal{G}[f], \mathcal{G}[f]) \longrightarrow \text{Ext}^1_O(\mathcal{G}[f], \mathcal{G}[f])$$

associated to the exact sequence $0 \to \mathcal{G}[f] \to \mathcal{G}[f^2] \to \mathcal{G}[f] \to 0$ is an isomorphism of 1-dimensional k-vector spaces, then every p-divisible group over O that can be filtered with group schemes isomorphic to $\mathcal{G}[f]$, is isogenous to a power of $\mathcal{G}$.

Here a p-divisible group $\mathcal{H}$ is said to be filtered by $\mathcal{G}[f]$ if the group scheme $\mathcal{H}[p]$ of p-torsion points admits a filtration by closed flat subgroup schemes with successive subquotients isomorphic to $\mathcal{G}[f]$. The condition implies that all group schemes $\mathcal{H}[2^n]$ admit such filtrations. See (Schoof, 2001c) for a proof of Proposition 3.2.

References

Faltings, G. (1983) Endlichkeitssätze für abelsche Varietäten über Zahlkörpern, *Invent. Math.* **73**, 349–366.

Fontaine, J.-M. (1985) Il n'y a pas de variété abélienne sur $\mathbb{Z}$, *Invent. Math.* **81**, 515–538.

Greither, C. (1992) Extensions of finite group schemes, and Hopf Galois theory over a complete discrete valuation ring, *Math. Zeitschrift* **210**, 37–67.

Martinet, J. (1981) Petits discriminants des corps de nombres, in J.V. Armitage, *Journées Arithmetiques 1980*, Lecture Notes Series 56, Cambridge University Press, Cambridge, pp. 151–193.

Schoof, R. (2001a) Abelian varieties over $\mathbb{Q}(\sqrt{6})$ with good reduction everywhere, in K. Miyake (ed.), *Class Field Theory — Its Centenary and Prospect*, Advanced Studies in Pure Mathematics, Tokyo.

Schoof, R. (2001b) Abelian varieties over cyclotomic fields with good reduction everywhere, to appear in *Math. Ann.*

Schoof, R. (2001c) Semi-stable abelian varieties over $\mathbb{Q}$, preprint.

Oort, F., and Tate, J.T. (1970) Group schemes of prime order, *Ann. Scient. École Norm. Sup.* **3**, 1–21.

Washington, L.C. (1982) *Introduction to cyclotomic fields*, Graduate Texts in Math. 83, Springer-Verlag, New York.

USING MONODROMY TO DECOMPOSE SOLUTION SETS OF POLYNOMIAL SYSTEMS INTO IRREDUCIBLE COMPONENTS

A. J. SOMMESE*
University of Notre Dame, Indiana, U.S.A

J. VERSCHELDE†
University of Illinois at Chicago, U.S.A.

C. W. WAMPLER‡
General Motors Research Laboratories, Warren, Michigan, U.S.A.

Abstract. To decompose solution sets of polynomial systems into irreducible components, homotopy continuation methods generate the action of a natural monodromy group which partially classifies generic points onto their respective irreducible components. As illustrated by the performance on several test examples, this new method achieves a great increase in speed and accuracy, as well as improved numerical conditioning of the multivariate interpolation problem.

Key words: components of solutions, embedding, generic points, homotopy continuation, irreducible components, monodromy group, numerical algebraic geometry, polynomial system, primary decomposition.

Mathematics Subject Classification (2000): *Primary* 65H10; *Secondary* 13P05, 14Q99, 68W30.

1. Introduction

To describe positive dimensional solution sets of polynomial systems, we work with *generic points* (Sommese and Wampler) of irreducible components. These points are produced by slicing the solution set with general linear subspaces of the ambient Euclidean space. In particular, slicing the union Z_i of the i-dimensional

* The first author thanks the Duncan Chair of the University of Notre Dame for its support.
Address: Department of Mathematics, Notre Dame, IN 46556, U.S.A., e-mail: sommese@nd.edu, URL: http://www.nd.edu/~sommese.

† Department of Mathematics, 851 South Morgan (M/C 249), Chicago, IL 60607-7045, U.S.A., e-mail: jan@math.uic.edu, jan.verschelde@na-net.ornl.gov, URL: http://www.math.uic.edu/~jan.

‡ General Motors Research Laboratories, Mail Code 480-106-359, 30500 Mound Road, Warren, MI 48090-9055, U.S.A., e-mail: Charles.W.Wampler@gm.com.

C. Ciliberto et al. (eds.),
Applications of Algebraic Geometry to Coding Theory, Physics and Computation, 297–315.

components of the solution set in $\mathbb{C}^n$ with a general linear space of dimension $n-i$ will result in a set W_i of smooth points of the components. The cardinality of this set equals the degree of Z_i. A decomposition of a positive dimensional solution set into irreducible components is realized by a partition of the whole set W_i of generic points into subsets of points that lie on the same irreducible component.

In (Sommese, Verschelde, and Wampler, 2001a and 2001b), this decomposition was achieved by incrementally building up interpolating polynomials. In this paper we propose to apply the actions of a natural monodromy group to find a partition of all generic points that is compatible with the partition into irreducible components, i.e., a subset generated by this new decomposition consists of points on an irreducible component, but not necessarily all of them. This reduces, in many cases significantly, the work needed to compute the filtering polynomials. Given a grouping G of generic points from the breakup of W_i induced by the monodromy action, we construct a polynomial p, that would be a filtering polynomial, if the grouping did coincide with the points of W_i lying on an irreducible component of the solution set. Evaluation of p on additional test points obtained by homotopy paths starting in G checks whether the grouping G is the grouping associated to an irreducible component. If so, then p is a filtering polynomial for the irreducible component. If not, then p can be used as a starting point for constructing a filtering polynomial of the irreducible component containing G.

An important added advantage is that a set G of generic points on which the monodromy action is transitive, is well suited for use of generalized divided differences (Isaacson and Keller) to write down a well-conditioned multivariate polynomial p that interpolates the points. This will be covered in a sequel.

The method applies to all components, but because we only have an efficient implementation of path tracking for paths on components having multiplicity one, in this paper, we apply the new technique only to such components.

This paper is organized in four parts. In section 2 we present the fundamentals on monodromy actions, and we outline the algorithm to decompose solution sets with the monodromy group. In section 3 we discuss the extra processing needed if the breakup achieved, using monodromy, does not equal the breakup corresponding to the decomposition of Z_i into its irreducible components. Section 4 addresses the condition of the decomposition problem on the special problem of factoring multivariate polynomials. In section 5, we apply our new approach to systems from the literature.

2. Monodromy Group Actions

2.1. FUNDAMENTALS ON MONODROMY ACTIONS

Let Z denote the reduction of the algebraic set defined by a system of polynomials f on $\mathbb{C}^N$, that is, Z is the underlying set of points of the solution set of $f = \mathbf{0}$. Z is an algebraic set. Let us consider Z_i, the union of the i-dimensional irreducible components of Z. Let S_i denote the algebraic subset of Z_i consisting of all points of Z_i that lie on at least two distinct components of Z. We consider the space U of all affine linear subspaces $\mathbb{C}^{N-i} \subset \mathbb{C}^N$. Most $L \in U$ meet Z_i in a set of $\deg Z_i$ distinct points contained in $Z_i \setminus S_i$. The distinctness of the points implies, in particular, that the points are smooth points of the set Z_i. Let $\mathcal{D}$ denote the set of points of U for which this is not true. Choose an $L \in U \setminus \mathcal{D}$. Taking a piecewise smooth map of the unit circle $\gamma : S^1 \to U \setminus \mathcal{D}$ with $\gamma(0) = \gamma(1) = L$, we can trace the homotopy paths of the points $L \cap Z_i$ as we traverse around the unit circle. The mapping from the starting points at $\gamma(0)$ to the corresponding endpoints at $\gamma(1)$ is a bijection of the set $L \cap Z_i$. Since the elements of $L \cap Z_i$ stay on the same irreducible component under the continuation along γ, it follows that if a point $\mathbf{a}$ of $L \cap Z_i$ is taken to a point $\mathbf{b}$ of $L \cap Z_i$, then $\mathbf{a}$, $\mathbf{b}$ are on the same irreducible component. The key observation of this paper is that it is easy — and numerically stable — to generate many loops γ, and the associated bijections can be used to decompose $L \cap Z_i$ into "monodromy groupings." These groupings are compatible with the grouping according to irreducible components, which is the objective of our previous papers (Sommese and Wampler; Sommese, Verschelde, and Wampler, 2001a and 2001b).

It is an elementary fact that the converse is also true; that is, two points of $L \cap Z_i$ are on the same irreducible component of Z_i only if they are connected by a monodromy action. Hence, a complete irreducible decomposition is determined by a sufficient set of monodromy loops. We do not know of any efficient algorithm that is guaranteed to find such a set of loops, but we can test whether a given monodromy grouping is sufficient in this sense. If not, one can generate another round of loops, or one can proceed to an approach like that in (Sommese, Verschelde, and Wampler, 2001a and 2001b), which may be expensive, but which is guaranteed to terminate.

2.2. AN ILLUSTRATIVE EXAMPLE

For example, let us take as system one equation $f(x,y) = xy - 1$ and slice it with $L = x + y - t = 0$, where t is some parameter. Then, for all $t \neq \pm 2$, the system

$$\left\{ \begin{array}{l} xy - 1 = 0 \\ x + y - t = 0 \end{array} \right. \tag{1}$$

has two solutions $(x_1(t), y_1(t))$ and $(x_2(t), y_2(t))$:

$$\left(x_1(t) = \frac{t}{2} + \frac{1}{2}\sqrt{t^2-4}, y_1(t) = \frac{t}{2} - \frac{1}{2}\sqrt{t^2-4}\right) \tag{2}$$

and

$$\left(x_2(t) = \frac{t}{2} - \frac{1}{2}\sqrt{t^2-4}, y_2(t) = \frac{t}{2} + \frac{1}{2}\sqrt{t^2-4}\right). \tag{3}$$

For $t = \pm 2$ the two solutions coincide.

To illustrate a monodromy action, we choose a parameterized closed loop $t = 2(1+e^{i\theta})$, which starts and ends at $t = 4$ as θ takes real values going from 0 to 2π. This loop does not contain either of the double roots, and so it qualifies as a monodromy loop. Substitution of the parameterization for t into Eq. (2) gives, after some simplification,

$$x_1(t) = (1+e^{i\theta}) + e^{i\theta/2}\sqrt{2+e^{i\theta}}. \tag{4}$$

For continuity, the square root in this expression is taken to lie always in the right half of the complex plane, while $e^{i\theta/2}$ goes from 1 to -1 as θ goes from 0 to 2π. The other expressions in Eqs. (2, 3) follow a similar pattern, and as a consequence we see that after one trip around the loop, the first solution $(x_1(\theta), y_1(\theta))$ has moved to the second solution $(x_2(\theta), y_2(\theta))$, and vice versa. This shows that the two solutions lie on the same connected component.

2.3. THE DECOMPOSITION ALGORITHM

Suppose that f is a polynomial system that has an i-dimensional solution set $Z_i = \cup_{j\in I_i} Z_{ij} \subset \mathbb{C}^N$, where the Z_{ij} are the irreducible components of dimension i. We follow the convention of letting L stand for both a system of i linear equations of rank i on $\mathbb{C}^N$ and for the $(N-i)$-dimensional linear subspace of solutions of L on $\mathbb{C}^N$. Let us pick two such subspaces, L_0 and L_1, at random, and suppose that we have the set of solution points $W_i := Z_i \cap L_0$. (These are the "witness points" for dimension i as described in (Sommese, Verschelde, and Wampler, 2001a).) With probability one, $\deg(Z_i) = \#(Z_i \cap L_0) = \#(Z_i \cap L_1)$. For general subspaces L_j and L_k, we define the homotopy

$$H_{jk\lambda}(\mathbf{x}(t),t) = \lambda(1-t)\begin{pmatrix} f \\ L_k \end{pmatrix} + t\begin{pmatrix} f \\ L_j \end{pmatrix} = \mathbf{0}, \quad \lambda \in \mathbb{C}, t \in [0,1]. \tag{5}$$

For generic λ, the solution set to $H_{jk\lambda}(\mathbf{x}(t),t) = \mathbf{0}$ on Z_i consists of exactly $\deg(Z_i)$ paths $\mathbf{x}(t)$ starting at points of $Z_i \cap L_j$ and ending at points of $Z_i \cap L_k$ as t goes from 1 to 0. This gives a bijection from $Z_i \cap L_j$ to $Z_i \cap L_k$. A bijection mapping of W onto itself can be constructed by concatenating two or more such homotopy mappings, such as $H_{01\lambda_1}(\mathbf{x}(t),t) = \mathbf{0}$ and $H_{10\lambda_2}(\mathbf{x}(t),t) = \mathbf{0}$. (If $\lambda_1 \neq \lambda_2$, the forward and return paths are not identical, and the bijection is not necessarily the identity mapping.) In general, we can concatenate any number of such homotopies, returning at the end to the same $Z_i \cap L_j$ from which we start, to generate additional bijections.

There are many ways one could choose to set up bijections using homotopies. We choose to proceed as follows. We start with $W_i := Z_i \cap L_0$ given. For simplicity, we refer to W_i as W. At stage k, we pick a new L_k and λ_{0k} at random and compute the homotopy paths for $H_{0k\lambda_{0k}}(\mathbf{x}(t),t) = \mathbf{0}$, getting the ordered list of solutions $X_k := Z_i \cap L_k$ and the mapping ${}_k h_0 : W \to X_k$. (Here and below, the left and right subscripts of ${}_k h_0$ indicate, respectively, the output and input spaces.) Then, choosing new random constants λ_{kj}, we compute k homotopy paths $H_{kj\lambda_{kj}}(\mathbf{x}(t),t) = \mathbf{0}$, for $j = 0,\ldots,k-1$, thereby obtaining mappings ${}_j h_k : X_k \to X_j$. Altogether, we have k new bijections being maps from W to itself, namely, $T_{0,k,0} := {}_0 h_k \circ {}_k h_0$ and $T_{0,k,j,0} := {}_0 h_j \circ {}_j h_k \circ {}_k h_0$, for $j = 1,\ldots,k-1$. The L_k for $k \neq 0$ do not have to be distinct, but we find that the algorithm is more robust if we choose each one independently.

We accumulate the monodromy groupings from the individual bijections as follows. Begin with a partition in which each point in W is assigned to its own subset. For a bijection generated as above, check if any point is mapped to a point from a different subset. If so, a new partition is formed by joining the two subsets into one. The partition is updated by each subsequent bijection in a similar manner.

A basic version of an implementation of this idea is given below in Algorithm MONODROMYGROUPING. The subroutine HOMOTOPYMAP uses the homotopy (5) to determine mappings ${}_k h_j$ as described above and procedure PARTITION forms a new partition from a previous one according to any connections between subsets implied by the associated bijections. It may be that a newly computed bijection does not find any new connections so that the partitioning is unchanged. We call this a "stable" iteration. If the partitioning achieves a complete irreducible decomposition, then all further iterations must be stable, but a stable iteration does not imply that the irreducible decomposition is in hand: it may just be that the random constants chosen in the homotopies are not fortuitous. Therefore, a termination condition S, an integer, is specified, such that MONODROMYGROUPING terminates when S consecutive stable iterations are computed. The larger S is, the longer the algorithm will persist in the face of fruitless iterations. The algorithm must terminate in at most $(\#(W)-1)S$ iterations, since at least one connection must be found every S iterations and after $(\#(W)-1)$ connections are made, the algorithm terminates with all points in a single group. In practice, the algorithm generally terminates much sooner.

ALGORITHM 2.1. $[P] = $ MONODROMYGROUPING(f, L_0, W, S)

Input: Polynomial system f on $\mathbb{C}^N$;
Affine linear space $L_0 \subset \mathbb{C}^N$ of dimension $N-i$;
Generic points W of f on L_0; Termination condition S.

Output: P is partition of W.

```
s := 0;                                     [counter for stable iterations]
P := { {w} | w ∈ W };                       [initial partition is finest one]
k := 0;                                     [counter for forward paths]

loop
  k := k + 1;                               [increment counter]
  L_k := RandomLinear(N − i);               [random (N − i)-space]
  _k h_0 := HomotopyMap(f, W, L_0, L_k);    [paths from L_0 to L_k]
  j := 0;                                   [counter for return paths]
  while j < k do
    _j h_k := HomotopyMap(f, W, L_k, L_j);  [return paths]
    if j = 0                                [form bijection T from maps]
     then T := _0 h_k ∘ _k h_0;             [path 0 to k and back]
     else T := _0 h_j ∘ _j h_k ∘ _k h_0;    [path 0 to k and back via j]
    end if;
   P' := Partition(P, T);                   [merge subsets connected by paths]
   if #P' = #P                              [compare with previous partition]
    then s := s + 1;                        [another stable iteration]
    else s := 0;                            [reset stable iteration counter]
   end if;
   P := P';                                 [update partition]
   exit when ((s = S) or (#P = 1));         [termination condition]
   j := j + 1;                              [increment counter]
  end while;
end loop.
```

An additional output of MonodromyGrouping could be the list of samples on the new random slices L_k used in (5). This list can be used intermediately to determine the linear span of the solution components. In case all components are linear, no further computation of filtering polynomials is necessary.

To classify the generic points on solution sets of several different dimensions, we apply MonodromyGrouping at each dimension. Before starting at each dimension, we filter out points on higher dimensional components using the homotopy membership test in (Sommese, Verschelde, and Wampler, 2001b).

3. Further Processing and Validation

A basic operation in our approach to computing an irreducible decomposition is the determination of a filtering polynomial that vanishes on an irreducible component and whose degree is equal to the degree of the component. In (Sommese,

Verschelde, and Wampler, 2001a and 2001b), this was accomplished by sampling a component via homotopy paths extending from a single solution point and successively testing for higher and higher degrees for the filtering polynomial. The monodromy grouping described in the previous section gives us two advantages in finding a filtering polynomial. First, the number of points in a monodromy group is a lower bound on the degree of the irreducible component. Second, we may sample the component via homotopy paths extending from all of the points in the group. We will show in a sequel that by organizing these sample points into a grid, we can use divided differences to more efficiently compute the interpolating polynomial.

Assume that MONODROMYGROUPING gives a partition P of W into disjoint subsets $G_1,\ldots,G_m$. We can construct a polynomial p_j for each G_j, which would be a filtering polynomial, if $G_j \subset W$ is the set of $\deg Z_{ij}$ generic points of an irreducible component Z_{ij} of Z_i. By further sampling, we can check which, if any, of the p_j are not filtering polynomials. Since further processing is needed only for these groupings, we can assume by renaming that we have

1. a set W of generic points of some set of i-dimensional irreducible components of $f^{-1}(\mathbf{0})$;
2. a partition P of W into disjoint subsets $G_1,\ldots,G_m$; and
3. polynomials q_j of degree $\#(G_j)$ vanishing on G_j and a subset of samples from the irreducible component Z_{ij} that contains G_j and satisfies $\deg Z_{ij} > \#(G_j)$.

Observe that if $m = 2$, we are done since the only possibility is that the irreducible component Z_{i1} containing G_1 must also contain G_2, and has degree $\#(G_1)+\#(G_2)$. So we can assume $m \geq 3$. Order the groupings by size

$$\#(G_1) \geq \#(G_2) \geq \ldots \geq \#(G_m).$$

Find the filtering polynomial p_1 for the irreducible component Z_{1j} containing G_1. This can be done by the technique of (Sommese, Verschelde, and Wampler, 2001a). (In a sequel we will show how to take advantage of having already produced q_1.) Using the filtering polynomial p_1, we can check which G_j lie on Z_{i1}. This lets us remove at least two of the G_j from consideration. We now repeat this procedure each time decreasing m by at least two.

4. Numerical Conditioning: A Sensitivity Experiment

In this section, we have created some special systems to test the numerical behavior of our algorithms. We examine how the monodromy grouping algorithm behaves on polynomials in the neighborhood of a polynomial that factors.

Suppose we have a polynomial whose coefficients are very near to one that factors into several components. This could correspond to two opposing scenarios: in one, the given polynomial could be exact, and we hope to find that it does not factor, while in the other, the polynomial is a numerical approximation to the nearby factorizable polynomial, and we hope to find the approximate factorization. We wish to examine the behavior of the monodromy grouping algorithm under such conditions.

To conduct our experiment, we generated three dense quartics in two variables with random coefficients on the complex unit circle and multiplied them together to form a factorizable polynomial of degree 12. This polynomial is then perturbed, adding a random complex number of modulus ε to each coefficient. With mathematical exactness, none of the polynomials factors for any positive ε. For $\varepsilon = 10^{-i}$, $i = 0, 1, \ldots, 14$, a collection of 15 test polynomials is obtained. The test suite consisted of six such collections. The algorithm MONODROMYGROUPING was applied to all polynomials in the test suite, with termination condition $S = 10$. All calculations here were done with standard machine arithmetic (16 decimal places with double precision floating-point numbers). The results are summarized in Table III.

TABLE III. Six experiments on a perturbed product of three quartics, for various values ε of the magnitude of the error. The column header "c" lists the number of components, while "n" is the number of iterations used in computing this factorization.

ε	c	n	c	n	c	n	c	n	c	n	c	n
1.0E+00	1	4	1	3	1	3	1	3	1	3	1	5
1.0E–01	1	6	1	3	1	3	1	7	1	4	1	4
1.0E–02	1	7	1	3	1	3	1	9	1	14	1	5
1.0E–03	1	11	1	16	2	14	1	5	1	14	1	7
1.0E–04	2	19	2	13	1	11	3	12	2	13	3	15
1.0E–05	3	12	2	13	3	13	3	15	3	14	2	19
1.0E–06	3	15	2	21	3	13	3	12	3	14	3	13
1.0E–07	3	17	2	24	3	14	3	13	3	14	3	15
1.0E–08	3	16	3	14	2	13	3	15	3	15	3	13
1.0E–09	3	17	3	17	3	15	3	12	3	13	2	16
1.0E–10	3	14	3	13	3	16	3	14	3	15	3	16
1.0E–11	3	14	3	17	3	12	3	12	3	12	3	19
1.0E–12	3	13	3	16	3	13	3	14	2	25	3	14
1.0E–13	3	14	3	13	3	14	3	13	3	12	3	13
1.0E–14	3	13	3	18	3	14	3	14	3	12	3	12

For the case of perturbations with $\varepsilon = 1$, we see from the top row of Table III, that in all six cases MONODROMYGROUPING found that all twelve roots belong to a single component; that is, the polynomial does not factor. In contrast, for $\varepsilon = 10^{-14}$, the polynomial was always predicted to break up into three factors. As ε

gets smaller, the test polynomial resembles more and more a product, monodromy actions are harder to find and the algorithm terminates without connecting the points into a single component.

This behavior of the algorithm is expected: for sufficiently small perturbations the test polynomial is numerically indistinguishable from the nearby factorizable polynomial. For different perturbations on different coefficients it is hard to quantify this effect precisely — we cannot really point out a clear threshold for ε. When the perturbations are numerical artifacts, such as roundoff, it is useful to discover a nearby factored form.

5. Applications

The algorithms in this paper have been implemented as a separate module of PHCpack (Verschelde). All computations where done on a dual processor Pentium III 800 Mhz Linux machine.

5.1. THE CYCLIC *N*-ROOTS PROBLEM

The cyclic n-roots problems is one of the most notorious benchmark systems for polynomial system solvers, brought to the computer algebra community in (Davenport, 1987). The systems come from an application involving Fourier transforms, see (Björck, 1985 and 1989; Björck and Fröberg, 1991). Fröberg conjectured (reported in (Möller)) that, if n has a quadratic divisor, then there are infinitely many solutions and that, in case the number of solutions is finite, this number is $\frac{(2n-2)!}{(n-1)!^2}$. (Haagerup) showed that for n prime, the number of solutions is always finite and confirmed the conjectured number of solutions.

In this section we confirm earlier results obtained in (Björck and Fröberg, 1994) for $n = 8$ and in (Faugère) for $n = 9$. Our former methods of (Sommese, Verschelde, and Wampler, 2001a and 2001b) were limited to the reduced version (credited to Canny, see (Emiris)) of this problem. See (Emiris and Canny) for polyhedral root counts. For $n = 10$ and $n = 11$, all solutions are isolated and thus "easier" to solve numerically (see the companion web site to (Verschelde) with test examples for the solution to these problems). Note that Björck found all distinct isolated 184,756 cyclic 11-roots (reported as unpublished result in (Haagerup)). Here we report on how homotopy methods bridged the gap for this problem between $n = 7$ and $n = 10$. Faster computers are needed for $n = 12$.

5.1.1. *The Cyclic 8-roots Problem*

Starting from the given list of generic points, the decomposition of the one dimensional component of the cyclic 8-roots system (Björck and Fröberg, 1994) of degree 144 was achieved in 6m 24s 930ms user CPU time. This took 21 iterations of the algorithm. The drop in cardinalities of the partition went as follows:

$$\begin{aligned} 144 \to 102 \to 70 \to 63 \to 33 \to 30 \to 29 \to 18 \\ \to 18 \to 18 \to 18 \to 16 \to \cdots \end{aligned} \tag{6}$$

where the last ten iterations were stable. So the monodromy breakup predicted 16 components: eight quadrics and eight curves of degree 16. This breakup was subsequently confirmed with Newton interpolation. The numerical results of the certification by interpolating filtering polynomials are presented in Table IV. The quadrics were computed with standard arithmetic, while 32 decimal places were used for the 16-th degree polynomials. It took 41m 54s 780ms user CPU time to complete the certification process.

TABLE IV. Numerical results of the certification of cyclic 8-roots. The columns contain the degree d, the maximal error (eps) on the samples in the grid, the minimal distance between the samples, the largest value of the interpolating filter evaluated at all samples (grid res) and at the test points (test res) used to compute the linear span of the component.

d	eps	distance	grid res	test res
2	6.877E–16	3.680E+00	1.665E–16	–14,–14,–14,–14
2	1.113E–15	3.961E+00	4.663E–15	–14,–14,–14,–15
2	4.909E–16	6.719E+00	4.330E–15	–16,–16,–15,–14
2	5.532E–16	3.639E+00	5.801E–15	–14,–13,–14,–14
2	2.211E–15	5.456E+00	1.665E–15	–15,–15,–15,–15
2	1.717E–15	7.517E+00	5.551E–15	–15,–14,–14,–13
2	4.116E–16	4.129E+00	1.941E–16	–13,–14,–13,–14
2	7.944E–16	7.286E+00	2.442E–15	–15,–15,–15,–15
16	1.387E–27	1.718E+00	9.700E–21	–16,–24,–24,–24,–31,–16
16	2.584E–28	1.026E+00	2.200E–21	–17,–23,–15,–22,–23,–17
16	1.547E–28	1.026E+00	1.130E–21	–18,–18,–24,–21,–19,–15
16	1.199E–25	1.732E+00	5.000E–20	–28,–28,–23,–25,–28,–23
16	6.074E–27	1.734E+00	1.600E–20	–27,–26,–25,–29,–26,–24
16	1.201E–27	1.695E+00	2.700E–20	–29,–30,–17,–30,–23,–23
16	2.290E–28	1.053E+00	5.100E–21	–12,–20,–20,–21,–17,–16
16	3.399E–27	1.053E+00	1.100E–21	–25,–17,–19,–11,–15,–16

The test points are samples used to determine the linear span of the component. In case of the quadrics, we got four test points and with the 16 degree polynomials six test points were used. Since only the magnitude of the residual is important, we list –16 instead of 3.254E–16.

5.1.2. *The Cyclic 9-roots Problem*

The cyclic 9-roots problem has a two dimensional solution component of degree 18. We found this to break up into six components each of degree three. The cardinalities in the partition reduce as follows:

$$18 \to 18 \to 14 \to 11 \to 6 \to \cdots \tag{7}$$

where the last ten iterations were stable.

TABLE V. Numerical results of the certification of cyclic 9-roots, first done with standard floating-point arithmetic and in the second half of the table redone with 32 decimal places. The columns contain the degree d, the maximal error (eps) on the samples in the grid, the minimal distance between the samples, the largest value of the interpolating filter evaluated at all samples (grid res) and at five test points (test res).

d	eps	distance	grid res	test res
3	1.618E–11	2.310E+00	2.383E–11	–12,–10,–11,–11,–11
3	2.484E–11	2.335E+00	9.526E–11	–10,–11,–10,–11,–11
3	4.631E–11	1.837E+00	7.218E–11	–12,–12,–12,–11,–11
3	4.561E–11	1.818E+00	2.360E–09	–10,–9,–9,–9,–9
3	6.438E–11	2.597E+00	3.986E–10	–11,–12,–11,–12,–11
3	2.193E–11	1.515E+00	4.687E–11	–11,–8,–9,–9,–8
3	1.470E–26	2.103E+00	2.712E–26	–26,–26,–26,–26,–26
3	1.063E–26	2.812E+00	1.642E–24	–24,–25,–24,–25,–24
3	9.400E–27	1.972E+00	7.565E–27	–28,–29,–27,–27,–27
3	4.283E–27	2.363E+00	5.765E–26	–26,–26,–25,–25,–26
3	1.238E–26	2.158E+00	7.215E–25	–26,–26,–26,–26,–25
3	1.493E–26	2.243E+00	5.202E–26	–25,–27,–27,–25,–26

Achieving this breakup takes only 2m 32s 400ms. The approach of (Sommese, Verschelde, and Wampler, 2001a and 2001b) finishes in about the same time. Table V displays the numerical results of the symbolic certification, once with standard floating-point machine arithmetic and once with multi-precision floating numbers of 32 decimal places long. This validation required 59s 250ms and 14m 56s 570ms for the respective machine and multi-precision numbers. The results in the first half of Table V are somehow "lucky": in many computed instances machine arithmetic did not lead to small residuals. At the expense of a slowdown with a factor 15 we always get reliable results with 32 decimal places.

Polyhedral homotopies are required to exploit the sparse structure of the cyclic n-roots problem. For $n = 9$, the program available with the paper (Li, T.Y., and

Li, X.) has been used to compute the mixed volume of the polynomial system, sliced and embedded according to the techniques of (Sommese and Verschelde). With the program of (Li, T.Y., and Li, X.), it took 13m 4s 540ms to compute the mixed volume of the embedded system. Tracing all 20,376 paths to solve a random coefficient start system required with PHC 4h 4m 29s 730ms, and solving the embedded system to reach the 18 generic witness points took an additional 4h 54m 53s 550ms. Thus, we see that the computation of the decomposition by monodromy, even when using multiple precision in the validation, is small compared to the overhead of computing the witness points.

TABLE VI. Results of the monodromy breakup algorithm on the systems of all adjacent minors of a general $2 \times (n+1)$-matrix. The sum of the degrees $d = 2^n$, c is the number of components, "it" the number of iterations, and lastly the User CPU time.

n	d	c	it	User CPU time
3	8	3	15	3s 260ms
4	16	5	16	15s 670ms
5	32	8	17	43s 340ms
6	64	13	20	2m 19s 140ms
7	128	21	27	8m 47s 940ms
8	256	34	22	20m 20s 420ms
9	512	55	20	45m 44s 50ms
10	1024	89	35	3h 6m 48s 750ms
11	2048	144	24	6h 6m 27s 890ms

5.2. ADJACENT MINORS OF A GENERAL $2 \times (N+1)$-MATRIX

In (Diaconis, Eisenbud, and Sturmfels), it was shown that the number of components of the ideal of all adjacent 2×2-minors of a general $2 \times (n+1)$-matrix is radical, of degree 2^n and that it breaks up into F_n components, F_n being the nth Fibonacci number. We found this system an interesting benchmark. See (Hosten and Shapiro) for methods dedicated to binomial ideals. Table VI illustrates the performance of MONODROMYGROUPING on this class of systems. Compared to our earlier methods in (Sommese, Verschelde, and Wampler, 2001a and 2001b), three more cases could be solved.

Results on the symbolic certification for the case $n = 10$ are presented here. There are 20 components of degree less than or equal to five which were treated with standard machine arithmetic, see Table VII. Note that in Table VII, there is no interpolating polynomial constructed for $d = 1$, since in that case the linear span completely describes the component. To interpolate the other 69 higher degree components 64 decimal places were used, see Table VIII and Table IX. This certification took 23h 56m 28s 850ms, and is thus a lot more expensive than the prediction of the breakup which took 3h 6m 48s 750ms.

TABLE VII. Numerical results of the certification of the system of adjacent minors, for $n = 10$, for components of degree $d \leq 5$. The columns contain the maximal error (eps) on the samples in the grid, the minimal distance between the samples, the largest value of the interpolating filter evaluated at all samples (grid res) and at the test points (test res) used to compute the linear span of the component.

d	eps	distance	grid res	test res
4	1.343E–14	9.002E–01	7.730E–14	–13,–16,–14,–12,–15
3	3.068E–14	7.500E–01	6.772E–13	–12,–10,–9,–10,-12
4	4.646E–14	7.567E–01	5.144E–13	–11,–9,–12,–13,–12
3	1.856E–12	5.789E–01	1.778E–13	–14,–15,–11,–12,–14
4	1.496E–14	6.836E–01	8.475E–11	–11,–10,–9,–6,–7
4	5.168E–14	1.378E+00	1.776E–13	–10,–9,–14,–13,–11
4	2.457E–14	5.508E–01	8.704E–14	–15,–12,–12,–15,–14
4	6.305E–15	3.400E–01	3.708E–14	–14,–15,–10,–12,–10
5	8.814E–14	5.100E–01	1.819E–10	–12,–13,–12,–13,–13,–13,–12
4	7.439E–15	1.122E+00	7.849E–14	–13,–14,–15,–15,–15
3	1.940E–15	6.817E–01	1.035E–13	–11,–12,–12,–13,–12
5	5.953E–12	1.145E+00	2.612E–11	–3,–9,–10,–6,–2,–11,–9
3	1.103E–13	1.028E+00	6.972E–14	–11,–11,–12,–14,–12
1	–	–	–	–
3	1.446E–14	1.327E+00	1.730E–12	–11,–10,–9,–11,–10
4	4.056E–14	1.357E+00	1.674E–13	–14,–10,–14,–12,–11
4	1.095E–14	5.225E–01	3.908E–14	–11,–12,–9,–12,–11
5	3.672E–14	1.780E+00	5.586E–13	–5,–8,–3,–8,–13,–9,–11
5	6.821E–12	9.090E–01	2.274E–11	–9,–13,–8,–9,–13,–6,–10
4	1.828E–14	7.402E–01	8.527E–14	–12,–14,–14,–13,–14

5.3. A MOVING STEWART-GOUGH PLATFORM

Stewart-Gough platforms are mechanical devices consisting of a rigid base and a rigid endplate, joined via six legs using ball joint connections. In motion simulators and other robotic applications, the lengths of the legs are actuated under computer control to move the endplate with respect to the base. Generally, once the leg lengths are fixed, the entire structure becomes rigid, but the same set of leg lengths may be compatible with multiple endplate locations. The problem of determining all possible endplate locations given the leg lengths has a long history.

TABLE VIII. Numerical results of the certification of the system of adjacent minors, for $n = 10$, for components of degree $d > 5$, part A. The columns contain the maximal error (eps) on the samples in the grid, the minimal distance between the samples, the largest value of the interpolating filter evaluated at all samples (grid res) and at the test points (test res) used to compute the linear span of the component.

d	eps	distance	grid res	test res
12	8.361E–57	1.007E–01	1.085E–42	–45,–44,–44,–41,–44,–43,–44,–43,–39
20	9.967E–57	2.416E–01	4.340E–45	–56,–58,–41,–49,–45,–48,–45,–37,–53
8	9.344E–57	4.516E–01	5.000E–64	–61,–57,–63,–63,–61,–53,–63
12	4.368E–55	1.071E+00	6.300E–54	–51,–52,–50,–50,–49,–52,–51,–50,–49
8	9.911E–57	4.453E–01	7.600E–64	–59,–55,–53,–54,–57,–60,–61
8	6.377E–57	3.961E–01	4.346E–54	–57,–55,–54,–55,–57,–60,–55
15	8.365E–57	5.533E–01	9.200E–55	–45,–41,–26,–55,–60,–57,–54,–54,–56
15	9.385E–57	5.574E–01	4.016E–54	–47,–50,–46,–42,–51,–45,–46,–49,–46
24	9.712E–57	1.490E–01	4.700E–44	–57,–38,–6,–31,–20,–22,–51,–48,–42
21	4.148E–59	2.994E–01	3.900E–47	–22,–37,–45,–46,–39,–43,–45,–49,–33,–51,–41
16	9.370E–57	6.769E–01	3.901E–49	–53,–51,–56,–36,–36,–53,–36
8	6.826E–57	5.912E–01	1.110E–56	–50,–50,–50,–51,–52,–52,–52
16	1.089E–58	2.743E–01	5.270E–46	–42,–54,–37,–50,–56,–52,–53,–49,–38,–58,–42
20	9.991E–57	2.057E–01	1.920E–47	–45,–42,–59,–55,–43,–58,–46,–46,–49
12	9.229E–59	7.635E–01	1.623E–44	–37,–38,–37,–38,–37,–39,–37
16	8.607E–57	2.514E–01	4.800E–51	–62,–54,–49,–65,–44,–67,–65,–46,–57
16	3.796E–59	1.796E–01	2.620E–49	–58,–53,–54,–53,–49,–29,–53,–45,–39,,–22,–57
24	4.618E–57	4.110E–01	6.660E–36	–30,–31,–29,–36,–34,–26,–32,–27,–28
24	2.141E–59	2.436E–01	2.894E–42	–51,–57,–56,–51,–58,–60,–43,–52,–58,–55,–58
8	9.755E–57	7.382E–01	7.000E–58	–59,–54,–58,–57,–56,–55,–55
9	9.659E–57	5.337E–01	1.020E–58	–59,–54,–49,–57,–57,–52,–53
16	5.846E–59	7.444E–01	2.500E–48	–22,–45,–38,–28,–40,–35,–50,–38,–44
25	9.358E–57	1.996E–01	1.160E–47	–47,–9,–9,–25,–56,–39,–26,–54,–56,–56,–26
8	9.990E–57	4.346E–01	1.600E–57	–59,–61,–56,–61,–57,–54,–62
12	9.564E–57	2.609E–01	4.654E–48	–47,–51,–50,–50,–49,–48,–50
11	1.059E–57	7.889E–01	7.100E–49	–50,–49,–39,–54,–47,–53,–52,–48,–51,–54,–46,–54,–40
24	9.732E–57	3.537E–01	3.800E–46	–54,–19,–52,–42,–54,–38,–22,–20,–50
9	9.872E–57	2.681E–01	1.500E–57	–58,–60,–55,–59,–52,–58,–60
12	9.526E–57	3.511E–01	1.690E–57	–60,–56,–59,–55,–60,–49,–60
12	9.421E–57	6.753E–01	1.938E–53	–60,–56,–54,–49,–47,–60,–61,–55,–59
12	9.432E–57	3.142E–01	2.200E–57	–48,–45,–32,–60,–53,–47,–61
8	9.179E–57	6.102E–01	1.300E–61	–62,–57,–57,–60,–50,–60,–59
15	2.898E–59	4.977E–01	3.100E–51	–48,–44,–47,–43,–50,–49,–43,–45,–49
24	9.766E–57	2.674E–01	5.000E–41	–47,–39,–23,–36,–44,–47,–51,–39,–29
15	8.335E–57	2.226E–01	9.200E–52	–56,–54,–58,–59,–54,–53,–49,–55,–50

TABLE IX. Numerical results of the certification of the system of adjacent minors, for $n = 10$, for components of degree $d > 5$, part B. The columns contain the maximal error (eps) on the samples in the grid, the minimal distance between the samples, the largest value of the interpolating filter evaluated at all samples (grid res) and at the test points (test res) used to compute the linear span of the component.

d	eps	distance	grid res	test res
12	9.840E–57	5.607E–01	1.111E–56	–55,–39,–41,–42,–49,–53,–41
24	2.029E–56	4.163E–01	6.310E–17	–24,–29,–21,–23,–20,–27,–27,–30,–18
12	9.009E–57	4.460E–01	7.500E–54	–40,–51,–49,–50,–55,–53,–44
27	9.903E–57	2.949E–01	1.690E–33	–30,–24,–30,–19,–47,–34,–41,–34,–40
9	3.759E–59	9.179E–01	7.600E–60	–54,–51,–58,–50,–54,–59,–55,–50,–49,–50,–48
8	9.172E–60	3.264E–01	3.955E–58	–53,–48,–52,–52,–55,–53,–50
16	9.685E–57	3.339E–01	1.200E–52	–56,–57,–55,–54,–44,–46,–51,–53,–53
7	1.553E–58	1.340E+00	1.990E–54	–56,–53,–58,–57,–53,–57,–58,–55,–56
12	9.030E–57	9.606E–01	1.000E–54	–52,–42,–54,–60,–45,–56,–59,–52,–53
15	9.849E–57	1.732E–01	3.973E–55	–40,–59,–52,–59,–59,–55,–54,–43,–33
9	1.678E–58	7.574E–01	2.700E–59	–53,–47,–53,–55,–53,–56,–52,–50,–51,–44,–53
12	9.657E–57	2.657E–01	6.000E–57	–38,–58,–55,–53,–62,–49,–50
12	9.948E–57	6.428E–01	1.717E–53	–49,–58,–50,–51,–46,–51,–54
9	9.934E–57	4.570E–01	3.500E–58	–56,–56,–56,–57,–57,–57,–53
12	9.886E–57	1.096E+00	1.803E–53	–53,–42,–52,–31,–56,–27,–54
8	9.920E–57	4.307E–01	2.700E–61	–61,–54,–59,–60,–59,–59,–58
21	4.976E–57	2.592E–01	1.486E–49	–45,–46,–43,–43,–45,–46,–41,–49,–42,–49,–49
9	8.021E–57	8.084E–01	2.200E–57	–42,–56,–52,–57,–52,–57,–55
12	1.824E–57	7.706E–01	9.940E–50	–44,–49,–47,–43,–44,–46,–48,–53,–47
7	2.164E–57	6.063E–01	3.500E–59	–54,–51,–50,–47,–54,–57,–58,–52,–50
7	5.919E–59	3.709E–01	1.180E–57	–55,–55,–53,–51,–49,–55,–55,–51,–55
24	4.255E–58	1.023E–01	6.830E–26	–32,–36,–25,–29,–32,–31,–36,–28,–32,–39,–27
9	1.114E–58	5.179E–01	1.087E–47	–43,–40,–43,–41,–43,–42,–43
12	8.536E–59	4.588E–01	5.020E–47	–50,–42,–37,–47,–41,–39,–43,–45,–42
12	7.665E–57	5.159E–01	3.828E–57	–56,–49,–58,–54,–48,–50,–53
8	8.207E–57	1.138E–01	4.100E–59	–62,–58,–58,–62,–63,–62,–55
12	9.968E–57	3.490E–01	2.000E–50	–54,–50,–46,–46,–50,–47,–44
9	2.355E–57	8.551E–02	2.000E–56	–49,–53,–51,–56,–50,–52,–54
24	9.932E–57	1.234E–01	3.166E–13	–48,–46,–47,–47,–40,–41,–41,–48,–41
12	9.167E–57	4.959E–01	1.139E–50	–48,–50,–44,–48,–46,–33,–44
8	1.498E–58	7.819E–01	6.425E–53	–57,–55,–56,–54,–57,–55,–55
8	9.153E–57	8.786E–01	1.190E–56	–53,–55,–54,–59,–48,–54,–55
15	4.202E–58	4.008E–01	1.971E–37	–23,–21,–23,–23,–20,–21,–23,–23,–22
20	9.680E–57	3.669E–01	1.000E–43	–55,–30,–39,–38,–40,–40,–47,–43,–34

For generic choices of the mechanical parameters, the problem has forty isolated solutions, a fact first established by continuation (Raghavan) and later proven analytically (Wampler; Husty). One of the more recent results (Dietmaier) involved the demonstration (obtained by methods of numerical homotopy continuation) that platforms exist that have forty real solutions.

For special choices of the parameters, a Stewart-Gough platform may have solution curves or other higher dimensional solution components, instead of only isolated solutions. We have tested a special case called "Griffis-Duffy type" by (Husty and Karger). This mechanism has, besides 12 lines (which correspond to degenerate assemblies), a single solution curve of degree 28. Our monodromy method confirmed this result by finding connections between all of the generic points on that solution curve. Computing the 40 generic points with PHC took 1m 12s 480ms cpu time. The drop in cardinalities in the partitions generated by MONODROMYGROUPING went as follows

$$40 \rightarrow 28 \rightarrow 27 \rightarrow 22 \rightarrow 17 \rightarrow 16 \rightarrow 15 \rightarrow 13 \rightarrow \cdots \tag{8}$$

where the last 10 iterations were stable. It took only 33s 430ms to achieve this result. At first glance, the result of degree 28 for the curve may appear to be in conflict with Husty and Karger, who claim that the curve is of degree 20. The conflict is resolved by noting that the curve is degree 28 in the full space of rotation and translation (represented in Study coordinates), but its degree falls to 20 when the curve is projected onto its rotational component only.

Since the lines correspond to degenerate assemblies (validating a line requires only three samples anyway), no further validation with interpolation is needed for this problem. This is fortunate, because validation for high degree components can be expensive due to the number of monomials that appear and the numerical sensitivity of high degree equations, forcing the use of multi-precision arithmetic. If one were to compute an interpolating polynomial for the case at hand (even though this is not necessary), one would find that a general polynomial of degree 28 in two variables has 435 monomials. Different methods to construct the interpolating polynomial require between 435 (direct approach with linear system) and 812 (Newton interpolation) samples. While these numbers are modest for homotopies on modern machines, the use of software driven multi-precision arithmetic imposed by the relatively high degree will constitute a serious speed bump.

This is illustrated in Table X, which displays the results of the validation for the curve of degree 28, executed with 64 decimal places as working precision. The Newton form of the interpolating polynomial constructed with generalized divided differences required 812 samples. Creating the grid of 812 samples took 39m 53s 960ms, which is of the same magnitude as 37m 47s 920ms, which is the time it took to evaluate all 812 samples in that high degree interpolating polynomial. While other operations (finding linear span of the component, construction Newton form, and evaluating at extra test samples) were relatively small, the total time to get the results in Table X was 1h 19m 13s 110ms.

TABLE X. Numerical results of the certification of the system of a moving Stewart-Gough platform, done with 64 decimal places. The columns contain the maximal error (eps) on the samples in the grid, the minimal distance between the samples, the largest value of the interpolating filter evaluated at all samples (grid res) and at the test points (test res) used to compute the linear span of the component.

d	eps	distance	grid res	test res
28	1.316E-59	2.395E-01	3.800E-37	−49,−44,−48,−56,−40,−41,−20,−40,−57

6. Conclusions

Using the monodromy we presented an algorithm to predict the breakup of a positive dimensional component into irreducible ones. This predicted breakup is then subsequently validated by computing interpolating polynomials.

Compared to our previous approaches described in (Sommese, Verschelde, and Wampler, 2001a and 2001b), we point out several advantages. First of all, in almost all cases, standard floating-point machine arithmetic suffices to execute the algorithm MONODROMYGROUPING. This has put more difficult applications within our reach. Related to this issue is the experience that the running times for this breakup remain of the same order of magnitude regardless of the geometry of the breakup. For example, whether a curve of degree forty breaks up in two, or in twenty pieces does not cause major fluctuations in the needed running time. By contrast, the symbolic validation of two curves of degree 20 is much more expensive than the interpolation of twenty quadrics. Thirdly, with the predicted breakup we can set up a structured grid of samples and apply generalized divided differences (Isaacson and Keller) to construct the polynomial equations that cut out the components with Newton interpolation. We observed an improved conditioning of the interpolation problem and will elaborate on this in a sequel.

References

Björck, G. (1985) Functions of modulus one on Z_pb whose Fourier transforms have constant modulus, in *Proceedings of the Alfred Haar Memorial Conference, Budapest*, volume 49 of Colloquia Mathematica Societatis János Bolyai, pp. 193–197.

Björck, G. (1989) Functions of modulus one on Z_n whose Fourier transforms have constant modulus, and "cyclic n-roots", in J.S. Byrnes and J.F. Byrnes (eds.), *Recent Advances in Fourier Analysis and its Applications*, volume 315 of *NATO Adv. Sci. Inst. Ser. C: Math. Phys. Sci.*, Kluwer, pp. 131–140.

Björck, G., and Fröberg, R. (1991) A faster way to count the solutions of inhomogeneous systems of algebraic equations, with applications to cyclic n-roots, *J. Symbolic Computation* **12** (3), 329–336.

Björck, G., and Fröberg, R. (1994) Methods to "divide out" certain solutions from systems of algebraic equations, applied to find all cyclic 8-roots, in M. Gyllenberg and L.E. Persson (eds.), *Analysis, Algebra and Computers in Math. research*, volume 564 of Lecture Notes in Applied Mathematics, Marcel Dekker, pp. 57–70.

Davenport, J. (1987) Looking at a set of equations, Technical report 87-06, Bath Computer Science.

Diaconis, P., Eisenbud, D., and Sturmfels, B. (1998) Lattice Walks and Primary Decomposition, in B.E. Sagan and R.P. Stanley (eds.), *Mathematical Essays in Honor of Gian-Carlo Rota*, volume 161 of Progress in Mathematics, Birkhäuser, pp. 173–193.

Dietmaier, P. (1998) The Stewart-Gough platform of general geometry can have 40 real postures, in J. Lenarcic and M.L. Husty (eds.), *Advances in Robot Kinematics: Analysis and Control*, Kluwer Academic Publishers, Dordrecht, pp. 1–10.

Emiris, I.Z. (1994) Sparse Elimination and Applications in Kinematics, PhD thesis, Computer Science Division, Dept. of Electrical Engineering and Computer Science, University of California, Berkeley.

Emiris, I.Z., and Canny, J.F. (1995) Efficient incremental algorithms for the sparse resultant and the mixed volume, *J. Symbolic Computation* **20** (2), 117–149. Software available at http://www.inria.fr/saga/emiris.

Faugère, J.C. (1999) A new efficient algorithm for computing Gröbner bases (F_4), *Journal of Pure and Applied Algebra* **139** (1-3), 61–88. Proceedings of MEGA'98, 22–27 June 1998, Saint-Malo, France.

Haagerup, U. (1996) Orthogonal maximal abelian $*$-algebras of the $n \times n$ matrices and cyclic n-roots. in *Operator Algebras and Quantum Field Theory*, International Press, Cambridge, MA, pp. 296–322.

Hosten, S., and Shapiro, J. (2000) Primary Decomposition of Lattice Basis Ideals, *Journal of Symbolic Computation* **29** (4&5), 625–639.

Husty, M.L. (1996) An algorithm for solving the direct kinematics of general Stewart-Gough Platforms, *Mech. Mach. Theory* **31** (4), 365–380.

Husty, M.L., and Karger, A. (2000) Self-motions of Griffis-Duffy type parallel manipulators, *Proc. 2000 IEEE Int. Conf. Robotics and Automation*, CDROM, 24–28 April 2000, San Francisco, CA.

Isaacson, E., and Keller, H.B. (1994) *Analysis of Numerical Methods*, Dover Publications.

Li, T.Y., and Li, X. (2001) Finding mixed cells in the mixed volume computation, *Found. Comput. Math.* **1** (2), 161–181. Software available at http://www.math.msu.edu/~li.

Möller, H.M. (1998) Gröbner bases and numerical analysis, in B. Buchberger and F. Winkler (eds.), *Gröbner Bases and Applications*, volume 251 of London Mathematical Lecture Note Series, Cambridge University Press, pp. 159–178.

Raghavan, M. (1993) The Stewart platform of general geometry has 40 configurations, *ASME J. Mech. Design* **115**, 277–282.

Sommese, A.J., and Verschelde, J. (2000) Numerical homotopies to compute generic points on positive dimensional algebraic sets, *Journal of Complexity* **16** (3), 572–602.

Sommese, A.J., Verschelde, J., and Wampler, C.W. (2001a) Numerical decomposition of the solution sets of polynomial systems into irreducible components, *SIAM J. Numer. Anal.* **38** (6), 2022–2046.

Sommese, A.J., Verschelde, J., and Wampler, C.W. (2001b) Numerical irreducible decomposition using projections from points on the components, accepted by *Contemporary Mathematics*, available at http://www.nd.edu/~sommese and http://www.math.uic.edu/~jan.

Sommese, A.J., and Wampler, C.W. (1995) Numerical algebraic geometry, in J. Renegar, M. Shub and S. Smale (eds.), *The Mathematics of Numerical Analysis*, Proceedings of the AMS-SIAM Summer Seminar in Applied Mathematics, July 17–August 11, 1995, Park City, Utah, volume 32 of Lectures in Applied Mathematics, pp. 749–763.

Verschelde, J. (1999) Algorithm 795: PHCpack: A general-purpose solver for polynomial systems by homotopy continuation, *ACM Transactions on Mathematical Software* **25** (2), 251–276. Software available at http://www.math.uic.edu/~jan.

Wampler, C.W. (1996) Forward displacement analysis of general six-in-parallel SPS (Stewart) platform manipulators using soma coordinates, *Mech. Mach. Theory* **31** (3), 331–337.

DIFFEOMORPHISMS AND FAMILIES OF FOURIER-MUKAI TRANSFORMS IN MIRROR SYMMETRY

B. SZENDRŐI
University of Warwick, United Kingdom, and *Alfred Renyi Institute of Mathematics, Hungarian Academy of Sciences*

Abstract. Assuming the standard framework of mirror symmetry, a conjecture is formulated describing how the diffeomorphism group of a Calabi-Yau manifold Y should act by families of Fourier-Mukai transforms over the complex moduli space of the mirror X. The conjecture generalizes a proposal of Kontsevich relating monodromy transformations and self-equivalences. Supporting evidence is given in the case of elliptic curves, lattice-polarized K3 surfaces and Calabi-Yau threefolds. A relation to the global Torelli problem is discussed.

Key words: mirror symmetry, Fourier-Mukai transform, K3 surface, Calabi-Yau threefold.

Mathematics Subject Classification (2000): 14J32, 81T30.

Introduction

Derived categories of coherent sheaves entered the mirror symmetry scene with the paper of Kontsevich (1995). The mirror relationship as envisaged by Kontsevich is an equivalence of A_∞-categories built out of two manifolds X and Y, both equipped with (complexified) symplectic forms and complex Calabi-Yau structures. One of these categories is $D^b(X_t)$, the bounded derived category of coherent sheaves on the complex manifold X_t, or a twisted version thereof. The other category is the derived Fukaya category $D^b\mathrm{Fuk}(Y,\omega_0)$ of the symplectic manifold (Y,ω_0), a category constructed from Lagrangian submanifolds and local systems on them, with morphism spaces given by Floer homology.

The origin of the equivalence of the two categories was not specified by Kontsevich. The picture was filled in by a proposal of (Strominger, Yau and Zaslow), based on arguments coming from non-perturbative string theory. According to (Strominger, Yau and Zaslow), mirror manifolds should be fibered into middle-dimensional real special Lagrangian tori over a common base space; moreover, the two fibrations should in a suitable sense be dual to each other. As it was

C. Ciliberto et al. (eds.),
Applications of Algebraic Geometry to Coding Theory, Physics and Computation, 317–337.

subsequently realized, this should supply the equivalence of categories proposed by Kontsevich by an analytic form of the Fourier-Mukai transform, converting Lagrangian submanifolds equipped with a local system on one manifold into holomorphic bundles or more generally sheaves on the mirror; see e.g. (Leung, Yau and Zaslow).

The next question that arises is the origin of the torus fibrations; recently, some proposals have been put forward by (Gross and Wilson) and (Kontsevich and Soibelman) to the effect that these should arise from certain degenerations of the complex structure on the Calabi-Yau manifolds. These degenerations, to so-called maximally degenerate boundary points or complex cusps in the Calabi-Yau moduli space, have been known to play a major role in mirror symmetry since the heroic age (Candelas, de la Ossa, Green and Parkes); the ideas of (Gross and Wilson) and (Kontsevich and Soibelman) show their significance in the new order of things.

Many of the details of the torus fibrations, the definitions of the categories and their equivalence, and the relation to more traditional notions such as curve-counting generation functions and variations of Hodge structures are missing or at best conjectural. However, this should not necessarily prevent one from investigating some further issues such as symmetries on the two sides of the mirror map. The study of these issues was also initiated by Kontsevich. He pointed out that the categorical equivalence implies a correspondence between symplectomorphisms on a Calabi-Yau manifold and self-equivalences of the derived category of coherent sheaves of its mirror, and gave some explicit examples of this relationship. In particular, he related symplectomorphisms arising from certain monodromy transformations to certain relatively simple self-equivalences on the mirror side in the case of the quintic, and proposed some more general constructions of self-equivalences. These ideas were generalized by (Seidel and Thomas) and in a slightly different direction by (Horja, 1999 and 2001). (Aspinwall) and (Björn, Curio, Hernández Ruipérez and Yau) studied the relation between monodromies and self-equivalences in certain concrete cases.

A common feature of these works is that they present a static picture: the symplectic form and complex structure are frozen, and symmetries compared. The main contribution of the present work is that by investigating these notions in families, some additional insight can be gained.

The language of kernels (derived correspondences) is used to define a group that I term the categorical mapping group of a family of complex manifolds. The main conjecture of the paper states that for a pair (X,Y) of mirror Calabi-Yau manifolds, there should be a homomorphism from the diffeomorphism group of Y to the categorical mapping group of the total space of the complex moduli space of X. A weaker version of the conjecture concerns cohomology actions on the two sides of the mirror map.

I investigate the proposed conjecture for families of Calabi-Yau manifolds of dimension at most three. After a brief look at the elliptic curve case, where the relevant group action was already known to Mukai[1], I consider the case of K3 surfaces. Derived categories of sheaves on K3s have only been studied in the projective case, to which I restrict; this can be done elegantly in families by considering lattice-polarized K3 surfaces, a notion due to Nikulin. The diffeomorphism group has to be restricted also; in effect, the lattice polarization partitions H^2 (generically) into algebraic and transcendental parts a priori, and the allowable diffeomorphisms are supposed to respect this choice. One further issue that arises is an analogue of a theorem of Donaldson on the complex side, regarding orientation on the cohomology of K3 surfaces. Unfortunately, I cannot prove the required statement at present; it is formulated as Conjecture 5.4. Assuming this, Theorem 5.5 confirms that for any allowable diffeomorphism, there is a family of categorical equivalences with the correct cohomology action. As an illustration I investigate this correspondence in some detail for toric families.

In the threefold case, I only discuss examples of elements of the categorical mapping group, mostly coming from birational contractions on the threefolds. A typical case of Horja's work emerges in a novel fashion, showing that indeed, the present version of the conjecture is needed to get a full picture. To conclude, I point out a relation to (counterexamples to) the global Torelli problem.

The purpose of this paper is to discuss the proposed framework. Detailed proofs of the assertions, together with some further examples and applications, will be given elsewhere.

Acknowledgments. Conversations and correspondence with Tom Bridgeland, Andrei Căldăraru, Mark Gross, Paul Seidel and Richard Thomas during the course of this work were very helpful indeed. I thank the MPI for hospitality during a brief stay in September 2000, where some of these ideas took shape. I especially thank Yuri Manin for the warm welcome and stimulating conversations; he also provided the starting point for these ideas with questions in (Manin).

1. Basic definitions

For the purposes of the paper, a *Calabi-Yau manifold* is complex manifold X_0 with holonomy exactly $SU(n)$, in particular trivial canonical bundle. I assume everywhere below that $n = \dim_{\mathbb{C}}(X_0) \leq 3$. The subscript means that a particular

[1] The reader will excuse me for a short historical digression. Mukai remarks in the introduction of (Mukai, 1981) that the result that the derived category of an elliptic curve carries an action of $SL(2,\mathbb{Z})$ modulo the shift 'seems to be significant'. This must be one of the first explicit hints to mirror symmetry in the mathematics literature. Another early hint is discussed in Section 8 of (Morrison): a picture of Mori about the cone of curves of an abelian surface closely resembles a picture of Mumford about the compactification of a Hilbert modular surface; these cones become duals under mirror symmetry. It is a curious fact that Mori drew his picture during the fall of 1979 which is exactly the time when (Mukai, 1981) was submitted.

complex structure is chosen on the differentiable manifold X; sometimes I write X_0 as a pair (X, I_0) to make the complex structure explicit in notation. $D^b(X_0)$ or $D^b(X, I_0)$ denote the bounded derived category of coherent sheaves on X_0. The *Mukai map*

$$v\colon D^b(X_0) \to H^*(X_0, \mathbb{C})$$

is defined for $U \in D^b(X_0)$ by

$$v(U) = \mathrm{ch}(U) \cup \sqrt{\mathrm{td}(X_0)}.$$

The *Mukai pairing* on cohomology

$$H^*(X, \mathbb{Z}) \times H^*(X, \mathbb{Z}) \to \mathbb{Z} + i\mathbb{Z}$$

is defined by

$$(\alpha_0 + \ldots + \alpha_{2n}) \cdot (\beta_0 + \ldots + \beta_{2n}) = (-1)^{n-1} \sum_{j=0}^{2n} i^j \int_Y \alpha_j \cup \beta_{2n-j}$$

where $\alpha_i, \beta_i \in H^i(X, \mathbb{Z})$; this pairing extends linearly to rational and complex cohomology. Let $\tilde{L} = H^*(X, \mathbb{Z})$ be the $\mathbb{Z}$-module equipped with the Mukai pairing; a *marked family* is a smooth family $\pi\colon X \to S$ with topological fibre X and with a fixed isomorphism ϕ of the local system $\mathrm{R}^*\pi_*\mathbb{Z}$ with the constant local system $\tilde{L}$ on S respecting pairings.

A *kernel* (derived correspondence) between smooth projective varieties X_i, $i = 1, 2$, is by definition an object $U \in D^b(X_1 \times X_2)$. There is a composition product on kernels given for $U \in D^b(X_1 \times X_2)$ and $V \in D^b(X_2 \times X_3)$ by the standard formula

$$U \circ V = \mathbf{R}p_{13*}\left(\mathbf{L}p_{12}^*(U) \overset{\mathbf{L}}{\otimes} \mathbf{L}p_{23}^*(V)\right) \in D^b(X_1 \times X_3);$$

here $p_{ij}\colon X_1 \times X_2 \times X_3 \to X_i \times X_j$ are the projection maps. A kernel $U \in D^b(X_1 \times X_2)$ is *invertible*, if there is a kernel $V \in D^b(X_2 \times X_1)$ such that the products $U \circ V$ and $V \circ U$ are isomorphic in $D^b(X_i \times X_i)$ to $\mathcal{O}_{\Delta_{X_i}}$, the (complexes consisting of) the structure sheaves of the diagonals.

A kernel $U \in D^b(X_1 \times X_2)$ defines a functor

$$\Psi^U\colon D^b(X_2) \to D^b(X_1)$$

by

$$\Psi^U(-) = \mathbf{R}p_{1*}(U \overset{\mathbf{L}}{\otimes} p_2^*(-)),$$

If U is invertible then Ψ^U is a *Fourier-Mukai functor*, an equivalence of triangulated categories. An invertible kernel also induces an isomorphism

$$\psi^U\colon H^*(X_2, \mathbb{Q}) \to H^*(X_1, \mathbb{Q})$$

on cohomology, defined by

$$\psi^U(-) = p_{1*}(v(U) \cup p_2^*(-)).$$

Ψ^U and ψ^U are compatible via the Mukai map v. For Calabi-Yau manifolds of dimension at most four, ψ^U is an isometry with respect to the Mukai pairing.

Finally if X_0 is a complex manifold and $\alpha \in H^2(X_0, \mathcal{O}^*_{X_0})$, then there is a notion of an α-*twisted sheaf* on X_0. This can either be thought of as a sheaf living on the gerbe over X_0 defined by α, or more explicitly as a collection of sheaves on an open cover where the gluing conditions are twisted by a Cech representative of α.

2. Mirror symmetry and diffeomorphisms

The mirror symmetry story begins with the data (Y, I_0, ω_0), where Y is a differentiable manifold, I_0 is a Calabi-Yau complex structure on Y, and ω_0 is a Kähler form in this complex structure giving rise to a unique Ricci-flat metric[2]. The mirror of (Y, I_0, ω_0) is conjecturally found using the following procedure.

PROCEDURE 2.1.
STEP 1. Choose a degeneration $\mathcal{Y} \to \Delta^*$ of Y_0 over a punctured multi-disc Δ^* with mid-point $P \in \Delta$, which is a complex cusp (large complex structure limit point), an intersection of boundary divisors of the complex moduli space with prescribed Hodge theoretic behaviour (Morrison).

STEP 2. The choice of complex cusp P induces a torus fibration on Y_0 via a degeneration process (Gross and Wilson; Kontsevich and Soibelman). Arrange this fibration so that its fibres are (special) Lagrangian with respect to the complex structure I_0 and the symplectic form ω_0.

STEP 3. Dualizing this fibration (Strominger, Yau and Zaslow) should yield, after an appropriate compactification, a compact differentiable manifold X, together with an isomorphism

$$\mathrm{mir}_P \colon H^*(Y, \mathbb{Q}) \to H^*(X, \mathbb{Q}),$$

which relates algebraic cohomology to transcendental cohomology, and is compatible with filtrations coming from the Leray spectral sequence of the torus fibration on the transcendental part and degree on the algebraic part.

STEP 4. The symplectic form ω_0 and complex structure I_0 on Y induce a Calabi-Yau structure (X, I_0') and a symplectic form ω_0' on the manifold X.

STEP 5. According to the homological mirror symmetry conjecture of (Kontsevich), cf. also (Manin), the statement that (Y, I_0, ω_0) and (X, I_0', ω_0') are mirror

[2] The B-field is discussed in Remark 2.4 below.

symmetric is expressed by a equivalence of categories determined by the torus fibration,

$$\mathrm{Mir}_P\colon D^b\mathrm{Fuk}(Y,\omega_0) \longrightarrow D^b(X_0),$$

where $D^b\mathrm{Fuk}(Y,\omega_0)$ is the derived Fukaya category, which is (conjecturally) constructed from Lagrangian submanifolds and flat bundles on them. This equivalence is compatible with the cohomology isomorphism mir_P of Step 3.

My main interest in this paper lies in an understanding of the action of the diffeomorphism group $\mathrm{Diff}^+(Y)$ of Y on the above data. A diffeomorphism $\gamma \in \mathrm{Diff}^+(Y)$ maps a triple (Y, I_0, ω_0) to a new triple $(Y, \gamma^* I_0, \gamma^*\omega_0)$ and in particular it defines a symplectomorphism $(Y, \gamma^*\omega_0) \cong (Y, \omega_0)$. Lifting γ to a graded symplectomorphism (Seidel) induces an equivalence

$$\tilde{\gamma}\colon D^b\mathrm{Fuk}(Y,\gamma^*\omega_0) \xrightarrow{\sim} D^b\mathrm{Fuk}(Y,\omega_0) \bmod [1]$$

i.e. well-defined up to translation; this comes from the fact that the lifting of γ to a graded symplectomorphism is only well-defined up to shift.

Perform Procedure 2.1 on the Calabi-Yau triples (Y, I_0, ω_0) and $(Y, \gamma^* I_0, \gamma^*\omega_0)$ with respect to the same complex cusp P in order to obtain mirrors (X, I_0', ω_0') and (X, I_0'', ω_0''). Then there is a diagram of categorical equivalences up to translation, where the bottom arrow is defined by the others:

$$\begin{array}{ccc} D^b\mathrm{Fuk}(Y,\gamma^*\omega_0) & \xrightarrow{\tilde{\gamma}} & D^b\mathrm{Fuk}(Y,\omega_0) \\ \downarrow{\scriptstyle \mathrm{Mir}_P} & & \downarrow{\scriptstyle \mathrm{Mir}_P} \\ D^b(X, I_0'') & \xrightarrow{\Psi_0} & D^b(X, I_0') \end{array}$$

Further, the categorical equivalence Ψ_0 induces a cohomology isomorphism ψ_0 which must be compatible via the isomorphism mir_P with the action of γ^* on $H^*(Y, \mathbb{Q})$.

To push this further, assume that (I_t, ω_t) is a family of complex structures and compatible symplectic forms on Y deforming (I_0, ω_0). The diffeomorphism γ induces a family of equivalences

$$\tilde{\gamma}\colon D^b\mathrm{Fuk}(Y,\gamma^*\omega_t) \xrightarrow{\sim} D^b\mathrm{Fuk}(Y,\omega_t) \bmod [1].$$

The mirrors of (Y, I_t, ω_t) and $(Y, \gamma^* I_t, \gamma^*\omega_t)$ with respect to the cusp P are deformations (X, I_t', ω_t') and (X, I_t'', ω_t'') of the original (X, I_0', ω_t') and (X, I_0'', ω_t''). Consequently, there are induced equivalences

$$\Psi_t\colon D^b(X, I_t'') \xrightarrow{\sim} D^b(X, I_t') \bmod [1].$$

SLOGAN 2.2. *Diffeomorphisms of the manifold Y induce families of equivalences of derived categories up to translation over the complex moduli space of the mirror manifold X.*

In the next Section, this slogan will be translated into a precise conjecture. The rest of the paper is providing evidence and examples. Before that however, there are two important points to clear up.

REMARK 2.3. In order to be able to compare cohomology actions, it is necessary to work with marked moduli spaces of Calabi-Yau varieties in the procedure described above. This implies that the moduli spaces arising will typically be non-connected; this issue already arises for K3 surfaces. The procedure outlined above tacitly assumed that for $\gamma \in \mathrm{Diff}^+(Y)$, the complex structures I_0 and $\gamma^* I_0$ will be in the same connected component of the (marked) moduli space of complex structures on Y, and thus the same complex cusp P may be used to form the mirror. In general, several components of the marked moduli space will contain complex structures related by diffeomorphisms. The way out is to pick a full set of components of the complex structure moduli space of Y containing structures related by diffeomorphisms, and fix a complex cusp P_i in each one of them in a compatible way so that the topological mirror X and the cohomology map mir_P are the same for all cusps. Then the procedure described above is accurate. I tacitly assume this extension of the setup everywhere below.

REMARK 2.4. The above discussion ignores one crucial piece of data present in mirror symmetry, namely the B-field. Here I want to argue that, possibly under an extra assumption in the case of threefolds, I can consistently restrict to the case of vanishing B-field on the holomorphic side of mirror symmetry. The generalization of the present ideas to the case of arbitrary B-fields is left for future work.

Physical mirror symmetry deals with quadruples (Y, I_0, ω_0, B_0); here $B_0 \in H^2(Y, \mathbb{R}/\mathbb{Z})$ is the B-field on the Calabi-Yau manifold (Y, I_0). The mirror of (Y, I_0, ω_0, B_0) is a quadruple $(X, I_0', \omega_0', B_0')$. According to a physics proposal originally formulated in the context of K-theory, for nonzero B-fields Kontsevich' homological mirror symmetry conjecture should take a form of an equivalence of categories

$$\mathrm{Mir}_P \colon D^b\mathrm{Fuk}(Y, \omega_0, B_0) \longrightarrow D^b(X, I_0', B_0');$$

see for example (Kapustin and Orlov). The category $D^b\mathrm{Fuk}(Y, \omega_0, B_0)$ on the left hand side should be a derived Fukaya-type category where the bundles on Lagrangian submanifolds are equipped with connections whose curvature is given by the restriction of B. For the current picture this is not a serious issue as diffeomorphisms should still give equivalences between these generalized derived Fukaya categories.

More importantly, the category on the right hand side should be a version of the derived category of coherent sheaves on (X, I_0'). More precisely, consider the natural map

$$\delta \colon H^2(X_0, \mathbb{R}/\mathbb{Z}) \to H^2(X_0, \mathcal{O}^*_{X_0}).$$

Suppose first that the class B has torsion image in $H^2(X_0, \mathcal{O}^*_{X_0})$. Then $D^b(X_0, B_0')$ should be the derived category of $\delta(B_0')$-twisted coherent sheaves $D^b(X_0, \delta(B_0'))$.

If $\delta(B)$ is non-torsion, then $D^b(X_0, B'_0)$ should be a suitable subcategory of a category of sheaves over a gerbe; for more discussion see Remark 2.6 in (Kapustin and Orlov).

Note that for elliptic curves, $H^2(X, \mathcal{O}_X^*) = 0$ from the exponential sequence, so the derived category is not affected. However in higher dimensions, there are nontrivial twists.

Consider the case of K3 surfaces. The induced action of $\mathrm{Diff}^+(Y)$ on the space parameterizing quadruples (X, I', ω', B') is compatible via the cohomology isomorphism mir_P with the action on the rational cohomology $H^*(Y, \mathbb{Q})$. Moreover, by Proposition 5.2, the cohomology isomorphism mir_P is defined over $\mathbb{Z}$. So the induced action of $\mathrm{Diff}^+(Y)$ maps a zero B'-field to a zero B'-field.

In the case of Calabi-Yau threefolds, the isomorphism mir_P is not expected to be defined over $\mathbb{Z}$ but it is still defined over $\mathbb{Q}$ so it preserves torsion B'-fields. On the other hand, in this case the torsion subgroup of $H^2(X, \mathcal{O}_X^*)$ is isomorphic to the torsion subgroup of $H^3(X, \mathbb{Z})$. Assuming $H^3(X, \mathbb{Z})_{\mathrm{tors}} = 0$, there are no possible torsion twists at all. As there is no known example of a smooth Calabi-Yau threefold with torsion in H^3, this restriction does not appear to be very serious.

I conclude therefore that for elliptic curves, K3 surfaces and Calabi-Yau threefolds under the assumption $H^3(X, \mathbb{Z})_{\mathrm{tors}} = 0$, it is legitimate to set $B' = 0$, as was effectively done in the above discussion.

3. The categorical mapping group

To translate Slogan 2.2 into a conjecture, I formalize the notion of a family of equivalences for complex structures on X. Assume that $\pi\colon \mathcal{X} \to \mathcal{D}$ is a family of complex projective varieties over a smooth complex base $\mathcal{D}$. Consider triples (ϕ, α, U) where

- S is an open subset of $\mathcal{D}$, the complement of a countable number of closed analytic submanifolds,
- $\phi\colon S \to \mathcal{D}$ is an analytic injection (not necessarily the identity), giving rise to the fibre product diagram

$$\begin{array}{ccc} \mathcal{X}_\phi & \longrightarrow & \mathcal{X}_S \\ \downarrow & & \downarrow \pi \\ \mathcal{X}_S & \xrightarrow{\phi\circ\pi} & S \end{array}$$

  where $\pi\colon \mathcal{X}_S \to S$ is the restriction of the original family to S;
- $\alpha \in H^2(S, \mathcal{O}_S^*)$ is a twisting class on the base;
- U is an object of the bounded derived category of quasi-coherent $p^*\alpha$-twisted sheaves on $\mathcal{X}_\phi$, whose derived restriction to the fibres of p is isomorphic to a bounded complex of coherent sheaves on $X_s \times X_{\phi(s)}$; here $p\colon \mathcal{X}_\phi \to S$ is the natural map.

Let $M(\mathcal{X},\mathcal{D})$ be the set of triples under obvious identifications, namely extension of ϕ, and tensoring by a line bundle pulled back from the base S. $M(\mathcal{X},\mathcal{D})$ can be given a monoid structure by a standard procedure generalizing the composition structure on kernels; the maps ϕ simply compose under the multiplication rule. The multiplication has a two-sided unit $(\mathrm{id}_S, 0, \mathcal{O}_{\Delta_X})$ where $\Delta_X \subset \mathcal{X} \times_S \mathcal{X}$ is the relative diagonal.

Suppose further that $\mathcal{X} \to \mathcal{D}$ is a marked family with topological fibre X, and let $\tilde{L}_{\mathbb{Q}} = H^*(X,\mathbb{Q})$ be the rational cohomology of the fibres.

DEFINITION 3.1. *The* categorical mapping group $G(\mathcal{X},\mathcal{D})$ *of the marked family* $\pi\colon \mathcal{X} \to \mathcal{D}$ *is the group of invertible elements* $g=(\phi,\alpha,U)$ *of the monoid* $M(\mathcal{X},\mathcal{D})$ *satisfying the following cohomological condition*[3]*: there should exist an isometry* $\psi_g \in \mathrm{Aut}(\tilde{L}_{\mathbb{Q}})$ *and a commutative diagram of local systems on* S

$$\begin{array}{ccc} R^*(\phi\circ\pi)_*\mathbb{Q}_{\mathcal{X}} & \longrightarrow & R^*\pi_*\mathbb{Q}_{\mathcal{X}} \\ \downarrow & & \downarrow \\ \tilde{L}_{\mathbb{Q}} & \xrightarrow{\psi_g} & \tilde{L}_{\mathbb{Q}}. \end{array}$$

Here the vertical arrows come from the markings, and the top horizontal arrow is the isomorphism of local systems induced by the invertible relative kernel U.

The image $G^{coh}(\mathcal{X},\mathcal{D})$ *of the resulting homomorphism*

$$\tilde{\psi}\colon G(\mathcal{X},\mathcal{D}) \to \mathrm{Aut}(\tilde{L}_{\mathbb{Q}}).$$

is the cohomological mapping group *of the family* $\mathcal{X} \to \mathcal{D}$.

For an element $g=(\phi,\alpha,U) \in G(\mathcal{X},\mathcal{D})$, derived restriction of U to the fibre over $s \in S$ gives a well-defined untwisted object U_s in $D^b(X_s \times X_{\phi(s)})$. An easy base change argument shows that as (ϕ,α,U) is invertible, U_s is also invertible, and hence there is a Fourier-Mukai transform

$$\Psi^{U_s}\colon D^b(X_{\phi(s)}) \to D^b(X_s)$$

for every $s \in S$, with a well-defined cohomology action ψ_g independent of $s \in S$.

The following main conjecture summarizes the discussion of the preceding sections.

CONJECTURE 3.2. *Let* $(\mathcal{Y} \to T, P)$ *be a family of Calabi-Yau manifolds together with a choice of complex cusp* $P \in \partial T$. *Let* X *be the topological mirror of* Y *with respect to* P*; assume that* $H^3(X,\mathbb{Z})_{\mathrm{tors}} = 0$. *Let* $\mathcal{X} \to \mathcal{M}_X$ *be the marked Calabi-Yau moduli space (Teichmüller space) of* X. *Then there exists a homomorphism* ξ *fitting into a commutative diagram*

$$\begin{array}{ccc} \mathrm{Diff}^+(Y) & \xrightarrow{\xi} & G(\mathcal{X},\mathcal{M}_X)/(\text{translations}) \\ \beta\downarrow & & \downarrow\psi \\ \mathrm{Aut}(H^*(Y,\mathbb{Q})) & \longrightarrow & \mathrm{Aut}(H^*(X,\mathbb{Q}))/(\pm 1), \end{array}$$

[3] Note that the cohomology condition is vacuous if $\mathcal{D}$ is connected.

where

- *translations mean relative translations over connected components on the base $\mathcal{M}_X$; moreover,*
- *β is the natural action,*
- *ψ is the map associating to an element g of the categorical mapping group its cohomology action ψ_g and*
- *the bottom horizontal arrow is the canonical map induced by the isomorphism* mir_P *of Procedure 2.1.*

The diagram gives rise to an isomorphism of groups

$$\bar{\xi}\colon \langle \mathrm{Diff}^{\mathrm{coh}}(Y), (-1)\rangle/(\pm 1) \longrightarrow G^{\mathrm{coh,filtr}}(\mathcal{X}, \mathcal{M}_X)/(\pm 1)$$

where

- $\mathrm{Diff}^{\mathrm{coh}}(Y)$ *is the image of* $\mathrm{Diff}^+(Y)$ *in* $\mathrm{Aut}(H^*(Y,\mathbb{Q}))$, *and*
- $G^{\mathrm{coh,filtr}}(\mathcal{X}, \mathcal{M}_X)$ *is the subgroup of* $G^{\mathrm{coh}}(\mathcal{X}, \mathcal{M}_X)$ *consisting of elements whose associated cohomology action preserves the Leray filtration on the transcendental cohomology of* X.

REMARK 3.3. As discussed before, a diffeomorphism only acts on the derived Fukaya category up to translation. Hence in going from diffeomorphisms to their cohomology action, I have to take account of the cohomology action of translation [1] on the derived Fukaya category which is simply multiplication by (-1). This is reflected on the left hand side of the cohomological form of the conjecture. On the right hand side, the cohomology actions of relevant families of Fourier-Mukai transforms must preserve the Leray filtration on the transcendental part of the cohomology of X; this is mirror to the statement that diffeomorphisms preserve the degree filtration on the algebraic cohomology of Y.

4. Elliptic curves

The diffeomorphism group $\mathrm{Diff}^+(T^2)$ of the two-torus acts via the standard action of $SL(2,\mathbb{Z})$ on $H^1(T^2,\mathbb{Z})$ and trivially on $H^2(T^2,\mathbb{R})$. Thus, recalling the discussion of Section 2, $\mathrm{Diff}^+(T^2)$ should be realized by self-equivalences in the mirror family.

The elliptic curve is self-mirror. Choose cohomology classes dual to a collapsing circle fibre and a section of a circle fibration corresponding to the decomposition $T^2 = (\mathbb{R}/\mathbb{Z})^2$ to fix an isomorphism

$$\mathrm{mir}_P\colon H^*(T^2,\mathbb{Z}) \to H^*(T^2,\mathbb{Z}) \tag{1}$$

interchanging even and odd cohomology. Let

$$\mathcal{X} = \mathbb{C} \times \mathcal{H}/[(z,\tau) \sim (z+n+m\tau, \tau)], \quad m,n \in \mathbb{Z}$$

be the universal family of marked elliptic curves together with the map

$$\pi\colon \mathcal{X} \to \mathcal{H}$$

to the marked moduli space, the upper half plane $\mathcal{H}$.

THEOREM 4.1. *There exists a group homomorphism*

$$\xi\colon \mathrm{Diff}^+(T^2) \to G(\mathcal{X},\mathcal{H})/(\text{translations})$$

compatible with the isomorphism mir_P *and descending to an isomorphism*

$$\bar{\xi}\colon PSL(2,\mathbb{Z}) \tilde{\to} G^{\mathrm{coh,filtr}}(\mathcal{X},\mathcal{H})/(\pm 1).$$

SKETCH PROOF. One only has to relativize the discussion of Section 3d in (Seidel and Thomas); the $PSL(2,\mathbb{Z})$-action on the derived category of an elliptic curve was of course already known to (Mukai, 1981). There are no nontrivial twists over the base $\mathcal{H}$ and for all elements $(\phi, U) \in G(\mathcal{X},\mathcal{H})$ in the image of ξ, the map ϕ is the identity. □

5. K3 surfaces

5.1. LATTICE POLARIZED MIRROR SYMMETRY

Let M be an even non-degenerate sublattice of signature $(1,t)$ of the K3 lattice $L = H^2(X,\mathbb{Z})$. A *marked ample M-polarized K3 surface* is a marked $K3$ surface (Y_t,ψ), which satisfies $\psi^{-1}(M) \subset \mathrm{Pic}(Y)$ and moreover $\psi^{-1}(C)$ contains an ample class for C a chosen chamber of $M_{\mathbb{R}}$; for the precise definition see (Dolgachev). For a pair (Y_t,ψ), Hodge decomposition on the second cohomology gives the period point in

$$\mathcal{D}_M = \left\{ z \in \mathbb{P}(M^{\perp}_{\mathbb{C}}) : z^2 = 0, z \cdot \bar{z} > 0 \right\} \setminus \bigcup_{\delta \in \Delta(M^{\perp})} \langle \delta \rangle^{\perp}$$

where $\Delta(M^{\perp})$ is the set of vectors in $M^{\perp}$ of length -2. The period domain $\mathcal{D}_M$ has two connected components.

PROPOSITION 5.1 (Dolgachev, Corollary 3.2). *The period map realizes* $\mathcal{D}_M$ *as the marked moduli space of marked ample M-polarized K3 surfaces; in particular, there exists a universal marked family* $\pi\colon \mathcal{Y} \to \mathcal{D}_M$.

The mirror construction works best for a sublattice M of L with $M^\perp$ containing hyperbolic planes; I assume this from now on. The choice of a pair of vectors $\{f,f'\} \subset M^\perp$ spanning a hyperbolic plane amounts to choosing a complex cusp[4] $P \in \partial\mathcal{D}_M$. The choice of $\{f,f'\}$ in $M^\perp$ also induces a splitting $M^\perp = \check{M} \perp \langle f,f' \rangle$. This defines a new sublattice $\check{M}$ of L of signature $(1,20-t)$ and gives rise to a family $\mathcal{X} \to \mathcal{D}_{\check{M}}$. This is the mirror family of $\mathcal{Y} \to \mathcal{D}_M$. The mirror construction, under the assumption on M, is an involution: starting with the pair $\{f,f'\}$ thought of as defining a complex cusp in the boundary of $\mathcal{D}_{\check{M}}$, one recovers the original M-polarized family $\mathcal{Y} \to \mathcal{D}_M$. Monodromy and filtration considerations show

PROPOSITION 5.2. *The cohomology isometry*

$$\mathrm{mir}_P \colon H^*(Y,\mathbb{Q}) \to H^*(X,\mathbb{Q}),$$

where Y, X are thought of as fibres in M-, respectively $\check{M}$-polarized families, is given by

$$\mathrm{mir}_P(h_0) = f',\quad \mathrm{mir}_P(h_4) = -f,\quad \mathrm{mir}_P(f) = -h_4,\quad \mathrm{mir}_P(f') = h_0,$$
$$\mathrm{mir}_P(m) = m \text{ for } m \in \langle h_i, f_i \rangle^\perp.$$

In particular mir_P *is defined over* $\mathbb{Z}$.

In the framework of Procedure 2.1, I want to start with a general ample M-polarized K3 (Y, I_0, ω_0) and I want to use the degeneration to the complex cusp $P \in \partial\mathcal{D}_M$ defined by $\{f,f'\}$ in the M-polarized family $\mathcal{Y} \to \mathcal{D}_M$. The cohomology class of the the Kähler form has to be orthogonal to the period, so ω_0 is restricted to live in the subspace $M_\mathbb{R} \subset H^2(Y,\mathbb{R})$. This condition has to be preserved throughout the whole procedure, hence the set of diffeomorphisms has to be restricted as well. Call $\gamma \in \mathrm{Diff}^+(Y)$ *M-allowable*, if $\gamma^* \in O(L)$ satisfies $\gamma^*(M) = M$. Let $\mathrm{Diff}^+_M(Y)$ be the group of M-allowable diffeomorphisms of Y and $\mathrm{Diff}^{\mathrm{coh}}_M(Y)$ the image of $\mathrm{Diff}^+_M(Y)$ in the isometry group of $\tilde{L}$. Note that monodromy transformations around various rational boundary components of the marked moduli space $\mathcal{D}_M$ are certainly M-allowable.

5.2. ACTION ON COHOMOLOGY

Before I proceed further, let me recall

THEOREM 5.3 (Donaldson). *Let $\gamma \in \mathrm{Diff}^+(Y)$ be an orientation-preserving diffeomorphism of a K3 surface. Then the induced action on cohomology preserves the orientation of positive definite three-planes in $H^2(Y,\mathbb{R})$.*

[4] The choice in fact specifies a complex cusp in both components of $\mathcal{D}_M$; cf. Remark 2.3.

Comparing this theorem and the cohomology isomorphism mir_P given above leads to an important observation. Fix once and for all an orientation for the positive part of $H^0(X) \oplus H^4(X)$. Then for an algebraic K3 surface X_0 this, together with the real and imaginary parts of the period of X_0 and an ample class, gives an orientation to positive definite four-planes in the inner product space $H^*(X_0, \mathbb{R})$, that I call for want of a better expression the *canonical orientation*. If the diffeomorphism group maps, at least on the level of cohomology, surjectively onto the cohomological mapping group, then Donaldson's theorem must have an analogue for Fourier-Mukai functors. I formulate this as[5]

CONJECTURE 5.4. *Let $\Psi\colon D^b(Z_0) \to D^b(X_0)$ be a Fourier-Mukai equivalence between derived categories of smooth projective K3 surfaces. Then the induced cohomology action $\psi\colon H^*(Z_0, \mathbb{Z}) \to H^*(X_0, \mathbb{Z})$ preserves the canonical orientation of positive definite four-planes in cohomology.*

5.3. COHOMOLOGICAL REALIZATION OF THE MAIN CONJECTURE

All the pieces are together; the following theorem confirms that Conjecture 3.2 can be realized on a cohomological level for mirror families of polarized K3 surfaces.

THEOREM 5.5. *Assume Conjecture 5.4. Let P be a complex cusp in the boundary $\partial\mathcal{D}_M$ of the marked moduli space of the family of ample M-polarized K3 surfaces. Let $\mathcal{X} \to \mathcal{D}_{\check{M}}$ be the mirror family. Then there exists an isomorphism*

$$\bar{\xi}\colon \left\langle \mathrm{Diff}_M^{\mathrm{coh}}(Y), (-1) \right\rangle \Big/ (\pm 1) \xrightarrow{\sim} G^{\mathrm{coh,filtr}}(\mathcal{X}, \mathcal{D}_{\check{M}})/(\pm 1)$$

compatible with the isometry mir_P.

SKETCH PROOF. Let

$$\sigma \in \mathrm{mir}_P \left\langle \mathrm{Diff}_M^{\mathrm{coh}}(Y), (-1) \right\rangle \mathrm{mir}_P^{-1} \subset O(\tilde{L})$$

be an isometry of the Mukai lattice $\tilde{L}$. The task is to find an element $(\phi, \alpha, U) \in G(\mathcal{X}, \mathcal{D}_{\check{M}})$ with cohomology action σ.

Proposition 5.2 shows that σ fixes $\check{M}^{\perp} = M \perp \langle f, f' \rangle$. Therefore the map σ^{-1} induces an automorphism

$$\phi\colon \mathcal{D}_{\check{M}} \to \mathcal{D}_{\check{M}}$$

mapping $[z] \in \mathcal{D}_{\check{M}}$ to $[\sigma_{\mathbb{C}}^{-1}(z)]$. The point about this map is that for $s \in \mathcal{D}_{\check{M}}$, σ induces a Hodge isometry from $H^*(X_{\phi(s)}, \mathbb{Z})$ to $H^*(X_s, \mathbb{Z})$ that by Donaldson's Theorem 5.3 preserves the canonical orientation.

[5] A weaker form of this conjecture was proposed independently in a letter of E. Markman to R. Thomas, in January 1999.

Work of Mukai (1987) and global Torelli identifies $X_{\phi(s)}$ with a moduli space of sheaves on X_s and this can be done in a relative way over an open subset $S \subset \mathcal{D}_{\check{M}}$ in the family $\mathcal{X} \to \mathcal{D}_{\check{M}}$. Some further arguments show that there exists a (suitably twisted) universal sheaf U which defines an element of $G(\mathcal{X}, \mathcal{D}_{\check{M}})$ with cohomology action σ. □

5.4. TORIC FAMILIES

I want to give some examples of the correspondence between diffeomorphisms and families of Fourier-Mukai transformations in toric families. Let $\Delta \subset \mathbb{Z}^3 \otimes \mathbb{R}$ be a three-dimensional reflexive lattice polyhedron (Batyrev). It gives rise to a toric threefold $\mathbb{P}_\Delta$, which has a crepant toric resolution $\tilde{\mathbb{P}}_\Delta \to \mathbb{P}_\Delta$.

Let $\mathcal{X}_\Delta \to \mathcal{F}_\Delta$ be the family of anticanonical hypersurfaces $X_t \subset \tilde{\mathbb{P}}_\Delta$. For $t \in \mathcal{F}_\Delta$, X_t is a smooth K3 surface and there is a restriction map $\operatorname{Pic}(\tilde{\mathbb{P}}_\Delta) \to \operatorname{Pic}(X_t)$ with image $M_\Delta \subset \operatorname{Pic}(X_t)$, the space of toric divisors on X_t. The equality $M_\Delta = \operatorname{Pic}(X_t)$ for general t holds only under an extra condition, namely if for all one-dimensional faces Γ of Δ, neither Γ nor its dual $\Gamma^* \subset \Delta^*$ contain extra lattice points in their interior. If this condition fails, and it frequently does, then the Picard lattice of X_t is not spanned by toric divisors.

Associated to the lattice M_Δ is a family of ample M_Δ-polarized surfaces $\mathcal{X} \to \mathcal{D}_{M_\Delta}$; $\mathcal{X}_\Delta \to \mathcal{F}_\Delta$ is a subfamily of this family. However, if the above condition fails, then the general ample M_Δ-polarized K3 is not toric and $\mathcal{X}_\Delta \to \mathcal{F}_\Delta$ is a proper subfamily of $\mathcal{X} \to \mathcal{D}_{M_\Delta}$.

Let Δ^* be the dual polyhedron of Δ, and let M_{Δ^*} be the lattice of toric divisors of the Batyrev mirror family $\mathcal{Y}_{\Delta^*} \to \mathcal{F}_{\Delta^*}$. According to a conjecture of (Dolgachev), there should exist a decomposition $M_\Delta^\perp = \langle f, f' \rangle \perp \check{M}_\Delta$ as before, together with a primitive embedding

$$i \colon M_{\Delta^*} \hookrightarrow \check{M}_\Delta.$$

This conjecture is supported by numerical evidence and can be checked in several concrete cases (Dolgachev, Section 8); I assume that Δ is chosen so that the embedding exists. The embedding i realizes inside the period domain

$$\mathcal{D}_{M_\Delta} \subset \mathbb{P}(M_\Delta^\perp)_\mathbb{C} = \mathbb{P}(\check{M}_\Delta \perp \langle f, f' \rangle)_\mathbb{C}$$

the toric deformation space $\mathcal{F}_\Delta$ as an explicit subdomain

$$\mathcal{F}_\Delta = \mathcal{D}_{M_\Delta} \cap \mathbb{P}(M_{\Delta^*} \perp \langle f, f' \rangle)_\mathbb{C},$$

cf. (Kobayashi, Section 4.3).

The decomposition $M_\Delta^\perp = \langle f, f' \rangle \perp \check{M}_\Delta$ gives the family $\mathcal{Y} \to \mathcal{D}_{\check{M}}$ of marked ample $\check{M}_\Delta$-polarized surfaces as the mirror family of $\mathcal{X} \to \mathcal{D}_{M_\Delta}$.

Turning to the specific issue of diffeomorphisms, Theorem 5.5 realizes, on the level of cohomology, $\check{M}_\Delta$-allowable diffeomorphisms on the family $\mathcal{X} \to \mathcal{D}_{M_\Delta}$ by Fourier-Mukai functors[6]. The first examples are $\check{M}_\Delta$-allowable diffeomorphisms arising as monodromy transformations around boundary components meeting at the large complex structure limit point in the boundary of $\mathcal{D}_{\check{M}_\Delta}$.

LEMMA 5.6 (Dolgachev, 6.2). *The monodromy transformations in boundary components meeting at the complex cusp $P \subset \partial\mathcal{D}_{\check{M}_\Delta}$ generate a group isomorphic to M_Δ. The transformation corresponding to $d \in M_\Delta$ acts on the K3 lattice $L = H^2(Y,\mathbb{Z})$ by*

$$\begin{aligned} T_d(f) &= f \\ T_d(f') &= f' + d - \tfrac{d^2}{2} f \\ T_d(m) &= m - (d \cdot m) f \text{ for } m \in \langle f, f' \rangle^\perp. \end{aligned}$$

This action fixes the lattice $\check{M}_\Delta \perp \langle h_0, h_4 \rangle$ pointwise, so a glance at the definition the map ϕ in the sketch proof of Theorem 5.5 shows that ϕ acts trivially on $\mathcal{D}_{M_\Delta}$. Indeed, by common wisdom, the monodromy transformation T_d is mirrored by tensoring by the line bundle $\mathcal{O}_{X_t}(d)$ for $d \in M_\Delta \subset \text{Pic}(X_t)$ and $t \in \mathcal{D}_{M_\Delta}$. In the relative framework, this is simply tensoring by a relative line bundle (twisted over the base if necessary).

However, as one moves away from these boundary components, diffeomorphisms with more complicated cohomology action crop up. In particular, there are diffeomorphisms $\gamma \in \text{Diff}^+_{\check{M}_\Delta}(Y)$ whose associated cohomology action $\gamma^* \subset O(L)$ fixes every element of $\check{M}_\Delta$ but does not fix the whole of M_Δ pointwise. Hence these diffeomorphisms have mirrors with a nontrivial map $\phi\colon \mathcal{D}_{M_\Delta} \to \mathcal{D}_{M_\Delta}$. This nontrivial map has a fixed locus which includes the toric locus $\mathcal{F}_\Delta \subset \mathcal{D}_{M_\Delta}$. The interpretation is that these diffeomorphisms are mirrored by families of Fourier-Mukai transforms acting as self-equivalences over the toric locus $\mathcal{F}_\Delta$, but acting by equivalences between derived categories of different K3s away from the toric locus. Compare Example 6.7.

6. Threefolds

6.1. EXAMPLES OF FAMILIES OF SELF-EQUIVALENCES

EXAMPLE 6.1. Let $\mathcal{X} \to \mathcal{M}_X$ be a marked family of Calabi-Yau threefolds. As in the previous section, some immediate examples of elements of $G(\mathcal{X}, \mathcal{M}_X)$ come from line bundles. Let $d \in H^2(X,\mathbb{Z})$; in the threefold case d is automatically the first Chern class of a line bundle $\mathcal{O}_{X_s}(d) \in \text{Pic}(X_s)$ on the fibres X_s for $s \in \mathcal{M}_X$.

[6] A word of warning: the notation used here is related to that used in Theorem 5.5 by $M = \check{M}_\Delta$ and consequently $\check{M} = M_\Delta$.

There is a relative line bundle $\mathcal{O}_{\mathcal{X}}(d)$, possibly twisted over the base, which when thought of as a sheaf on the diagonal $\Delta_{\mathcal{X}} \subset \mathcal{X} \times_{\mathcal{M}_{\mathcal{X}}} \mathcal{X}$ gives an invertible relative kernel and so an element of $G(\mathcal{X}, \mathcal{M}_{\mathcal{X}})$ with ϕ the identity on $\mathcal{M}_{\mathcal{X}}$. The fibrewise Fourier-Mukai transform is tensoring by $\mathcal{O}_{X_s}(d)$ in the derived category of X_s. These elements of $G(\mathcal{X}, \mathcal{M}_{\mathcal{X}})$ should again mirror monodromy transformations around components of the boundary of $\mathcal{M}_{\mathcal{X}}$ meeting at the large structure limit point $P' \in \partial\mathcal{M}_{\mathcal{X}}$.

EXAMPLE 6.2. A more interesting case where families of Fourier-Mukai self-equivalences arise is that of families of twist functors coming from contractions. Consider a Calabi-Yau threefold X_0 together with a flopping (or Type I) contraction $X_0 \to \bar{X}_0$, a birational contraction with exceptional locus of dimension one. There is a diagram

$$\begin{array}{ccccc} \mathcal{C} & \to & \mathcal{X} & \to & \bar{\mathcal{X}} \\ & \searrow & \downarrow & \swarrow & \\ & & S & & \end{array}$$

where $S \subset \mathcal{M}_{\mathcal{X}}$ is a dense open subset in a connected component of the marked Calabi-Yau moduli space of X_0, and $\mathcal{C}$ is flat over S intersecting every fibre in a chain of rational curves. The structure sheaf $\mathcal{O}_{\mathcal{C}}$ restricts to the fibre X_s as $\mathcal{O}_{C_s}$, a spherical sheaf in the sense of (Seidel and Thomas). This sheaf gives rise to a twist functor, a Fourier-Mukai self-equivalence of $D^b(X_s)$. There is an invertible relative kernel $\mathcal{U}$, giving rise to an element $(\phi, 0, \mathcal{U})$ of $G(\mathcal{X}, \mathcal{M}_{\mathcal{X}})$ with ϕ the identical embedding of S in $\mathcal{M}_{\mathcal{X}}$ and restricting to the twisting kernel on $X_s \times X_s$ for $s \in S$.

EXAMPLE 6.3. Suppose that a Calabi-Yau threefold X_0 has a Type II contraction, a birational morphism $X_0 \to \bar{X}_0$ contracting an irreducible divisor E_0 to a point. E_0 is possibly non-normal but in any case it is a del Pezzo surface i.e. ω_{E_0} is ample. In this case, there is a diagram

$$\begin{array}{ccccc} \mathcal{E} & \to & \mathcal{X} & \to & \bar{\mathcal{X}} \\ & \searrow & \downarrow & \swarrow & \\ & & S & & \end{array}$$

where $S \subset \mathcal{M}_{\mathcal{X}}$ is dense in a connected component of the marked moduli space and the exceptional locus $\mathcal{E}$ is a flat family of surfaces over S intersecting X_t in a del Pezzo E_s.

The structure sheaf $\mathcal{O}_{E_s}$ of a del Pezzo surface is exceptional in the sense that there are no higher Ext's, so by (Seidel and Thomas, Proposition 3.15), its pushforward to X_s is spherical. It defines a twist functor, a self-equivalence of the derived category of X_s. The construction works again over S; there is an invertible relative kernel defined by the family of surfaces $\mathcal{E}$ over S which gives an element of $G(\mathcal{X}, \mathcal{M}_{\mathcal{X}})$ with ϕ still being the identity.

The elements of $G(\mathcal{X}, \mathcal{M}_{\mathcal{X}})$ given in Examples 6.2-6.3 should mirror monodromies around various other components of the boundary of moduli as discussed by Horja (1999) and more recently by (Aspinwall). In particular, some of these transformations should mirror Dehn twists in Lagrangian three-spheres in the mirror, which are known to arise in degenerations to threefolds with double points; this was one of the starting points for (Seidel and Thomas).

6.2. AN EXAMPLE WITH A NONTRIVIAL ACTION ON MODULI

In this section, I give an example of an element of the categorical mapping group with a nontrivial map ϕ on the base of moduli. The example comes from birational contractions on Calabi-Yau threefolds of Type III.

EXAMPLE 6.4. Let $f_0\colon X_0 \to \bar{X}_0$ be a birational contraction contracting a $\mathbb{P}^1$-bundle E_0 to a curve C_0 of genus at least two. On a dense open subset S of the marked moduli space of X, the contraction extends to a diagram

$$\begin{array}{ccc} f\colon \mathcal{X} & \to & \bar{\mathcal{X}} \\ \downarrow & \swarrow & \\ S. & & \end{array}$$

The main new feature is that f_s remains divisorial only for s in a proper submanifold $S^{\mathrm{div}} \subset S$. At a general point $s \in S$, the contraction $f_s\colon X_s \to \bar{X}_s$ is of flopping type, i.e. it contracts a codimension two locus; I assume for simplicity that it contracts disjoint $\mathbb{P}^1$'s to nodes.

By (Kollár and Mori, Theorem 11.10), there is a flop in the family

$$\begin{array}{ccc} \mathcal{X} & \dashrightarrow & \mathcal{X}^+ \\ & \searrow \quad \swarrow & \\ & S & \end{array}$$

For $s \in S \setminus S^{\mathrm{div}}$ where f_s is of flopping type, $X_s \dashrightarrow X_s^+$ is a classical flop. For $s \in S^{\mathrm{div}}$, the map $X_s \dashrightarrow X_s^+$ is the identity morphism.

The crucial point is that $\mathcal{X}^+ \to S$ is the pullback of the original family $\mathcal{X} \to S$ under a nontrivial map $\phi\colon S \to S$. The fixed locus of ϕ is exactly S^{div}. There is an embedding

$$\mathcal{X} \times_{\bar{\mathcal{X}}} \mathcal{X}^+ \hookrightarrow \mathcal{X} \times_S \mathcal{X}^+$$

and by definition, the latter space is isomorphic to $\mathcal{X} \times_\phi \mathcal{X}$. Thus the structure sheaf of $\mathcal{X} \times_{\bar{\mathcal{X}}} \mathcal{X}^+$ gives an element $U \in D^b(\mathcal{X} \times_\phi \mathcal{X})$.

THEOREM 6.5. *The triple* $(\phi, 0, U)$ *gives an element of* $G(\mathcal{X}, \mathcal{M}_{\mathcal{X}})$.

SKETCH PROOF. An easy argument shows that to prove that $(\phi, 0, U)$ is invertible, it is enough to show that the restriction is invertible on fibres. For $s \in S \setminus S^{\mathrm{div}}$,

U restricts to $X_s \times X_s^+$ as the structure sheaf of $X_s \times_{\bar{X}_s} X_s^+$. This is the well-known kernel of (Bondal and Orlov, Theorem 3.6), giving the isomorphism of derived categories under a classical flop.

For points $s \in S^{\text{div}}$, the restriction of U to $X_s \times X_s$ is the kernel $\mathcal{O}_{X_s \times_{\bar{X}_s} X_s}$. The method of (Bridgeland) can be used to show that this sheaf is a universal sheaf for a certain moduli problem on X_s with fine moduli space X_s, and hence is an invertible kernel on $X_s \times X_s$. □

REMARK 6.6. The existence of the above kernel arising from Type III contractions was conjectured by Horja, although he used a different expression, cf. (4.35) in (Horja, 1999). In (Horja, 2001) he proves independently and by a different method that it is invertible. However, he does not discuss the relation to flops.

I illustrate the above discussion by a well-studied toric family.

EXAMPLE 6.7. Start with a general degree eight hypersurface

$$\bar{X}_0 \subset \mathbb{P}^4[1^2, 2^3]$$

in weighted projective four-space. A simple blow-up gives a crepant resolution $f_0 \colon X_0 \to \bar{X}_0$, which is a Type III contraction contracting a $\mathbb{P}^1$-bundle E_0 in X_0 to a genus 3 curve C_0 in $\bar{X}_0$. All toric deformations are resolutions of degree eight hypersurfaces in the weighted projective space; however, $h^{2,1}(X_0) = 86$ whereas the octic has only 83 deformations, so some deformations of X_0 are missing.

To find the other deformations, embed $\mathbb{P}^4[1^2, 2^3]$ into $\mathbb{P}^5$ using $\mathcal{O}(2)$ as a hypersurface given by the equation $\{z_1 z_2 = z_3^2\}$. The variety $\bar{X}_0$ becomes a complete intersection

$$\bar{X}_0 \cong \left\{ \begin{array}{ll} z_1 z_2 & = z_3^2 \\ f_4(z_i) & = 0 \end{array} \right\} \subset \mathbb{P}^5,$$

of a general quartic and a quadric of rank three in $\mathbb{P}^5$. The curve of singularities is along $\{z_1 = z_2 = z_3 = 0\}$. Here more deformations of X_0 are visible: I can deform either the quadric or the quartic. Deformations of f_4 contribute 83 moduli, which are exactly the toric deformations; there are 3 further deformation directions deforming the quadric to a quadric of rank four[7]. The generic such deformation

$$\bar{X}_t = \left\{ \begin{array}{ll} z_1 z_2 & = z_3^2 - t^2 z_4^2 \\ f_4(z_i) & = 0 \end{array} \right\} \subset \mathbb{P}^5$$

has four nodes at $\{z_1 = z_2 = z_3 = z_4 = 0\}$ and a pair of small resolutions $X_t^{\pm}$, which are deformations of X_0:

[7] Deformations of $\bar{X}_0$ to non-singular complete intersections do not lift to deformations of the crepant resolution X_0.

$$X_t^{\pm} = \left\{ \begin{array}{ll} z_1 u_1 & = (z_3 \pm t z_4) u_2 \\ (z_3 \mp t z_4) u_1 & = z_2 u_2 \\ f_4(z_i) & = 0 \end{array} \right\} \subset \mathbb{P}^1 \times \mathbb{P}^5.$$

For $t \neq 0$, the birational map

$$\psi_t : X_t^+ \dashrightarrow X_t^-$$

is a simple flop, flopping four copies of $\mathbb{P}^1$ over the four nodes of $\bar{X}_t$. By the main result of (Szendrői), for general t the varieties $X_t^{\pm}$ are not isomorphic; what is more important from the point of view of the present discussion is that the flopped family explicitly appears as the original family under the base change $\phi : t \mapsto -t$.

In the connected component $\mathcal{X} \to \mathcal{M}_X$ of the Calabi-Yau moduli space of X_0, there is a sublocus $\mathcal{M}_X^{\mathrm{div}} \subset \mathcal{M}_X$ where the contraction f_0 deforms as a divisorial contraction; this is also the fixed locus of the base change map ϕ. On the other hand, as discussed above, $\mathcal{M}_X^{\mathrm{div}}$ also equals $\mathcal{M}_X^{\mathrm{toric}}$, the toric deformation space of X_0. The element $(\phi, 0, U)$ of $G(\mathcal{X}, \mathcal{M}_X)$ guaranteed by Theorem 6.5 acts as a self-equivalence on the toric deformations of X_0, but acts by Fourier-Mukai transforms between different (generically non-isomorphic) deformations away from the toric locus. This is the same behaviour as that found at the end of Section 5.4.

6.3. RELATION TO THE GLOBAL TORELLI PROBLEM

The global Torelli theorem for K3 surfaces is an important tool in the proof of the cohomological form of Conjecture 3.2 for K3 surfaces. For Calabi-Yau threefolds, the situation appears to be exactly the opposite. As an invertible kernel induces a Hodge isometry on H^3, the following holds:

PROPOSITION 6.8. *Suppose that $\mathcal{X} \to \mathcal{M}_X$ is a component of the marked moduli space of a Calabi-Yau threefold. Assume that there is an element $(\phi, \alpha, U) \in G(\mathcal{X}, \mathcal{M}_X)$ such that for general $s \in \mathcal{M}_X$, the varieties X_s and $X_{\phi(s)}$ are not isomorphic. (In particular, ϕ is not the identity.) Then the global Torelli theorem fails for the family: the polarized (rational) Hodge structure on middle cohomology does not determine the general variety in the family up to isomorphism.*

The counterexample to global Torelli given in (Szendrői) indeed comes from the nontrivial element of $G(\mathcal{X}, \mathcal{M}_X)$ of Theorem 6.5. Some conjectural examples of Calabi-Yau threefold families with a family of Fourier-Mukai transforms with nontrivial ϕ have been constructed by (Căldăraru, Chapter 6). These examples are of a very different flavour; they are elliptic fibrations without a section, the Fourier-Mukai transform coming from changing to a different elliptic fibration with the same Jacobian. The framework presented here suggests that counterexamples to global Torelli arising from families of Fourier-Mukai transforms should be rather more common than hitherto suspected. However, one may speculate that the isomorphism of the transcendental part of Hodge structures (in this case simply H^3) may imply the isomorphism of derived categories; this would be at the end of the day in complete agreement with the K3 case. At least for cohomology with integral coefficients, there appears to be no known counterexample to this speculation.

References

Aspinwall, P.S. (2001) Some navigation rules for D-Brane monodromy, hep-th/0102198.

Batyrev, V.V. (1994) Dual polyhedra and mirror symmetry for Calabi-Yau hypersurfaces in toric varieties, *J. Alg. Geom.* **3**, 493–535.

Björn, A., Curio, G., Hernández Ruipérez, D., and Yau, S.-T. (2001) Fourier-Mukai transform and mirror symmetry for D-branes on elliptic Calabi-Yau, preprint, math.AG/0012196.

Bondal, A., and Orlov, D. (1995) Semi-orthogonal decomposition for algebraic varieties, preprint, math.AG/9506012.

Bridgeland, T. (2000) Flops and derived categories, preprint, math.AG/0009053.

Căldăraru, A. (2000) Derived categories of twisted sheaves on Calabi-Yau threefolds, Cornell Thesis, available from http://www.math.umass.edu/~andreic.

Candelas, P., de la Ossa, X., Green, P.S., and Parkes, L. (1991) A pair of Calabi-Yau manifolds as an exactly soluble superconformal theory, *Nuclear Phys. B* **359**, 21–74.

Dolgachev, I. (1996) Mirror symmetry for lattice polarized $K3$ surfaces, Algebraic geometry, 4., *J. Math. Sci.* **81**, 2599–2630.

Donaldson, S.K. (1990) Polynomial invariants for smooth four-manifolds, *Topology* **29**, 257–315.

Gross, M. (1998) Special Lagrangian fibrations I: Topology, in M.-H. Saito, Y. Shimizu and K. Ueno (eds.), *Integrable systems and algebraic geometry (Kobe/Kyoto, 1997)*, World Sci. Publishing, River Edge, NJ, pp. 156–193.

Gross, M., and Wilson, P.M.H. (2000) Large complex structure limits of K3 Surfaces, preprint, math.DG/0008018.

Horja, P. (1999) Hypergeometric functions and mirror symmetry in toric varieties, preprint, math.AG/9912109.

Horja, P. (2001) Derived category automorphisms from mirror symmetry, preprint February 2001.

Kapustin, A., and Orlov, D. (2001) Vertex algebras, mirror symmetry and D-branes: the case of complex tori, preprint, hep-th/0010293.

Kobayashi, M. (1995) Duality of weights, mirror symmetry and Arnold's strange duality, preprint, math.AG/9502004.

Kollár, J., and Mori, S. (1992) Classification of three-dimensional flips, *J. Amer. Math. Soc.* **5**, 533–703.

Kontsevich, M. (1995) Homological algebra of mirror symmetry, in S.D. Chatterji (ed.), *Proceedings of the International Congress of Mathematicians, Zurich, 1994*, Birkhauser, Basel, pp. 120–139.

Kontsevich, M., and Soibelman, Y. (2001) Homological mirror symmetry and torus fibrations, preprint, math.SG/0011041.

Leung, N.C., Yau, S.-T., and Zaslow, E. (2000) From special Lagrangian to Hermitian-Yang-Mills via Fourier-Mukai transform, preprint, math.DG/0005118.

Manin, Yu. (2000) Moduli, motives, mirrors, ECM 2000 talk, available at math.AG/0005144.

Morrison, D.R. (1993) Compactifications of moduli spaces inspired by mirror symmetry, in *Journées de Géométrie Algébrique d'Orsay (Orsay, 1992)*, Astérisque 218, pp. 243–271.

Mukai, S. (1981) Duality between $D(X)$ and $D(\hat{X})$ with its application to Picard sheaves, *Nagoya Math. J.* **81**, 153–175.

Mukai, S. (1987) On the moduli space of bundles on $K3$ surfaces. I, in *Vector bundles on algebraic varieties (Bombay, 1984)*, Tata Inst., Bombay, pp. 341–413.

Seidel, P. (2000) Graded Lagrangian submanifolds, *Bull. Soc. Math. France* **128**, 103–149.

Seidel, P., and Thomas, R. (2000) Braid group actions on derived categories of coherent sheaves, preprint, math.AG/0001043.

Strominger, A., Yau, S.-T., and Zaslow, E. (1996) Mirror symmetry is T-duality, *Nuclear Phys. B* **479**, 243–259.

Szendrői, B. (2000) Calabi-Yau threefolds with a curve of singularities and counterexamples to the Torelli problem, *Int. J. Math.* **11** (2000), 449–459.